2.5D Printing

Wiley-IS&T Series in Imaging Science and Technology

Reproduction of Colour (6th Edition)
R.W.G. Hunt

Colorimetry: Fundamentals and Applications
Noburu Ohta and Alan R. Robertson

Color Constancy
Marc Ebner

Color Gamut Mapping
Ján Morovič

Panoramic Imaging: Sensor-Line Cameras and Laser Range-Finders
Fay Huang, Reinhard Klette and Karsten Scheibe

Digital Color Management (2nd Edition)
Edward J. Giorgianni and Thomas E. Madden

The JPEG 2000 Suite
Peter Schelkens, Athanassios Skodras and Touradj Ebrahimi (Eds)

Color Management: Understanding and Using ICC Profiles
Phil Green (Ed.)

Fourier Methods in Imaging
Roger L. Easton, Jr

Measuring Colour (4th Edition)
R.W.G. Hunt and M.R. Pointer

The Art and Science of HDR Imaging
John McCann and Alessandro Rizzi

Computational Colour Science Using MATLAB (2nd Edition)
Stephen Westland, Caterina Ripamonti and Vien Cheung

Color in Computer Vision: Fundamentals and Applications
Theo Gevers, Arjan Gijsenij, Joost van de Weijer and Jan-Mark Geusebroek

Color Appearance Models (3rd Edition)
Mark D. Fairchild

2.5D Printing: Bridging the Gap between 2D and 3D Applications
Carinna Parraman and Maria V. Ortiz Segovia

Published in Association with the Society for
Imaging Science and Technology

IS&T
imaging.org

2.5D Printing

Bridging the Gap Between 2D and 3D Applications

Carinna Parraman

University of the West of England, UK

Maria V. Ortiz Segovia

Océ Print Logic Technologies SA, France

WILEY

This edition first published 2018

Registered Offices
John Wiley & Sons, Inc., 111 River Street, Hoboken, NJ 07030, USA
John Wiley & Sons Ltd, The Atrium, Southern Gate, Chichester, West Sussex, PO19 8SQ, UK

Editorial Office
The Atrium, Southern Gate, Chichester, West Sussex, PO19 8SQ, UK

For details of our global editorial offices, customer services, and more information about Wiley products visit us at www.wiley.com.

Wiley also publishes its books in a variety of electronic formats and by print-on-demand. Some content that appears in standard print versions of this book may not be available in other formats.

Library of Congress Cataloging-in-Publication Data

Names: Parraman, Carinna, author. | Ortiz Segovia, Maria V., author.
Title: 2.5D printing : bridging the gap between 2D and 3D applications / by Carinna Parraman, Maria Ortiz Segovia.
Description: First edition. | Hoboken, NJ : John Wiley & Sons, 2018. | Includes index. |
Identifiers: LCCN 2018006230 (print) | LCCN 2018012067 (ebook) | ISBN 9781118967331 (pdf) | ISBN 9781118967324 (epub) | ISBN 9781118967300 (cloth)
Subjects: LCSH: Three-dimensional printing. | Texture (Art)
Classification: LCC TA1560 (ebook) | LCC TA1560 .P37 2018 (print) | DDC 621.9/88–dc23
LC record available at https://lccn.loc.gov/2018006230

Cover design: Wiley
Cover image: © e-anjei/iStockphoto printed as a Woodburytype by Peter McCallion

Set in 9/11pt TimesLTStd by SPi Global, Chennai, India

10 9 8 7 6 5 4 3 2 1

For Grace

Contents

About the Authors **xi**
Series Editor's Preface **xiii**
Preface **xv**
Acknowledgements **xvii**
About the Companion website **xviii**
Introduction **xix**

1 Defining the Field of 2.5D Printing **1**
1.1 What is Texture? 1
1.1.1 How to Quantify Texture 2
1.1.2 How do Artists Convey the Appearance of Texture?. 3
1.1.3 How the Natural World Mimics the Appearance of Texture 4
1.2 Measuring Texture and Colour 6
1.3 Images, Pictures and Reproductions 9
1.3.1 The Anxiety of the Reproduction 11
1.3.2 Reproducing Images: Tools of the Trade. 12
1.3.3 Reproducing Images: Colour and Texture 15
1.3.4 Reproductions versus Forgeries versus Copies 15
1.3.5 Are Facsimiles and Replicas Important? 17
1.3.6 Copying from Old Masters 19
1.3.7 Technical Examination of Artworks 20
1.4 The Authenticity of the Image and Object and Perception of Things 21
1.5 Current Industrial and Mechanical Methods to Reproduce the Appearance of Texture. 24
1.5.1 2D Printing Methods 24
1.5.2 The Emergence of 3D Printing 25
1.6 Conclusion 28
References 29
Bibliography and Further Reading 30

2 The Past **35**
2.1 Introduction 35
2.2 Artists' Observations on the Appearance of Illumination 36
2.3 Artists' Conversion of Images into Relief 39
2.4 Artists' Exploration of Different Sculptural Relief 41
2.4.1 Examples of Degrees of Projection 45
2.5 Coloration of Relief Surfaces 51
2.5.1 Visualising and Reconstructing the Past 57

2.6 Examples of Artists' Approaches to Representation and Reproduction of Texture 58
References 66
Bibliography and Further Reading 67

3 The Present: Materials, Making, Capturing and Measuring 69
3.1 Introduction: Universal Knowledge 69
3.2 The Relationship of Digital Technologies, Knowledge of Materials and Skills 70
3.3 Methods to Capture and Measure Texture 78
3.3.1 Commercial and Specialist Scanners 78
3.3.2 Scanning for Cultural Heritage. 82
3.3.3 In Pursuit of Standards: Metric versus Quality 84
3.4 Methods to Represent the Appearance of Texture 85
3.5 Physical Material Libraries 90
3.6 Methods for 2.5D Printing 92
3.6.1 Analogue versus Digital 93
3.6.2 Materials for Printing. 96
3.6.3 Printing Functional Materials. 102
References 105
Bibliography and Further Reading 108

4 The Future 115
4.1 Introduction 115
4.1.1 The Future of Education in Art Design and Sciences 115
4.1.2 The Future of Materials and Manufacture 116
4.2 Circular Economy and Sustainable Manufacturing 117
4.3 Worldwide Print Connectivity 119
4.4 Mass Printing for One 120
4.5 Security Printing 122
4.6 Conclusion 123
References 123
Bibliography and Further Reading 124

5 Case Studies 127
Case Study 1: Nature Printing in the Nineteenth Century 127
Case Study 2: Wallpaper Design 133
Wallpaper Printing Processes 134
Case Study 3: 2.5D Printed Tactile Books and Artworks 142
Examples of Tactile Images that Incorporate Multiplatform and Digital Print Technologies 148
Case Study 4: Coins and Medals 151
Making Coins 152
Medals of Dishonour Exhibition British Museum 155
Case Study 5: Capturing Texture of Paintings for Museum and Heritage 157
Visualising Surface Texture Through the Combination of 2D and 3D Data 157
Workflow: 3D Scanning and Processing 158
Generation of Colour RTI Images. 158
Generation of Colour 3D Models 160
Case Study 6: Textiles 161

Case Study 7: *Trompe l'Oeil* 167
Case Study 8: Marble 172
Case Study 9: Gold 179
Printing Gold in the Digital Market Place 183
The Stafford Hoard and Gold Printing at The Jewellery Industry Innovation Centre, Birmingham Jewellery Quarter 184
Case Study 10: Exterior Decoration Tiles and Ceramics 185
Twenty-first Century Ceramic Tile Inkjet Printing 190
Case Study 11: Microstructural Texture 195
Case Study 12: Painting Machines 203
Computer Simulations for Representing Materials and Objects 204
Direct Manipulation 205
Translating Brush Strokes into Painting Machines 206
Methods of Converting Images into Brushstrokes 208
Brush Strokes and Painting Machines 210
Case Study 13: Analogue Printing Methods 214
Intaglio 214
Relief Printing 217
Lithography 220
Screenprinting 221
Case Study 14: Relief Woodblock Printing 223
References 230
Bibliography and Further Reading 234

Index **239**

About the Authors

Carinna Parraman's understanding of 2.5D printing has evolved through her training in fine art printmaking. She is Professor of Design, Colour and Print and Director at the Centre for Fine Print Research, University of the West of England, Bristol, UK, and has an in-depth knowledge of traditional colour mixing, colour printing and photo mechanical printing processes. She collaborates with many different sectors including industry, heritage and fine art print.

Maria V. Ortiz Segovia's understanding of printing has evolved through Electrical Engineering and Imaging science. She is the leading scientist of the colour and image processing activities of the Innovation Team at Océ Print Logic Technologies, France. She is in charge of conducting collaborations and partnerships between Océ and different universities, laboratories and research institutions worldwide.

If you are reading this book in paper format, please also be aware that the electronic copy will provide you with hyper links to a wide range of extra material.

Series Editor's Preface

2.5D, you ask? Why not 2D or 3D, or both? Well, there is nothing indecisive about picking 2.5 dimensions to choose as the focus of a book. In the rapidly changing world of additive manufacturing, custom manufacturing, and hybrid 2D and 3D printing, the concerns of ***2.5D Printing*** – texture, relief, colour, reflectance, opacity, etc. – become paramount. In this book, part of the ***Wiley-IS&T Series in Imaging Science and Technology***, Drs Carinna Parraman and Maria V. Ortiz Segovia address these subjects to facilitate the advancement of all dimensions of printing and additive manufacturing. While colour science in two dimensions (think photos and videos) is a relatively mature field, its extension to 3D is not clear and the pathway will not be simple. Here, 2.5D benchmarks and measurements for quality, consistency, texture and translucency, among other measures, will provide an intermediate step to the eventual benchmarks and measurements used on digitally manufactured 3D objects. In some ways, this is analogous to high school mathematics – we proceed from planar (2D) to surface (2.5D) to volumetric (3D) geometries and calculus. There's a reason so much time is spent on 2.5D, in mathematics and in printing. Don't skip this step!

This book demonstrates that 2.5D printing provides the relationship of materials, texture and surface. The extra half dimension places the reader halfway between visual and haptic sensing, asking her or him to distinguish amongst thousands of different materials, textures and colours. The book provides a means to define, measure and assess a 'materiality of surface', meaning a qualitative and/or quantitative evaluation of the aesthetic, informational and other qualities of a surface. Where does the rubber meet the road? Perhaps nowhere more so than in the description of texture. The authors see texture as 'the microstructural details that can be perceptually distinguished from one surface property to another', usually through sight or touch. We can tell a red ball from a blue ball readily with visual perception, but two blue balls with the same radius are usually distinguished by 'interactive haptics'; for example, picking them up to note a difference in density or squeezing them to note a difference in elastic modulus.

The authors, like their subject matter, bridge the gaps between many fields of interest to the modern printer, maker and architect. Dr Parraman has deep colour expertise in screenprinting to digital wide format, which has resulted in her developing colour palettes for inkjet artists and ink multilayering. Her ability to describe and provide the means to deploy colour science bridges the worlds of art, science and industry. An innate multidisciplinarian, Dr Parraman considers colours through both space (multilayering, 2D and 2.5D) and time (colour fading and conservation, print history), and how these are tempered by the medium (paper and other surfaces) and the method (paint charts, colour circles and colour models). Parraman is currently Professor in Design, Colour and Print at the University of the West of England (UWE) in Bristol, and Director of the Centre for Fine Print Research. Among her many funded projects and awards, one that stands out in context is her development of a plethora of printed materials and surfaces that were developed as textiles or applied to walls – these materials adapted to changes in the environment, such as light, emotion and temperature (in collaboration with Roland DG, the project received a Roland Creatives award). Clearly, Dr Parraman is well-versed with the '2.5D world' – no wonder, then, a book on ***2.5D Printing***.

Dr Ortiz Segovia, meanwhile, appropriately complements Dr Parraman with her colour science research focusing on image processing, image quality and document management. She also has studied printing and sensor forensics, which are image processing-driven technologies that can be 'backed off' to provide inspection, validation and measurement (e.g. of texture, material properties, colour, and reflectance). She is currently an Imaging Scientist in Research and Development at the Océ–Canon Group, and earned her PhD at Purdue University (Electrical Engineering) and her Bachelors from Pontificia Universidad Javeriana (Electronics Engineering). The two bring a powerful, broad repertoire of skills and experiences that (pardon the play on words) cover the gamut of 2.5D printing. That is, these two authors hybridize nicely on a subject that also hybridizes 2D and 3D printing. Enjoy the read!

Preface

The print industry in the twenty-first century is a vital, economic and global contributor, adding value to a diversity of consumer products and services. Studies emphasise the importance of this sector of adopting a value added position by identifying and responding to the technological requirements of niche and large markets. These include artwork reproduction and applications in creative industries, 2.5 and 3D printing, tactile maps, security tagging, biomedical, textile, packaging, signage, and direct printing on complex curved objects.

What is the value of the print industry? In the way printing adds value, these figures are increasingly difficult to define, especially for niche or specialist sectors. Digital printing has evolved to become a catch-all term that has moved so far from its traditional graphics and packaging background. As demonstrated in the medical sector, 3D digital printing is now being used to create training replicas for brain surgeons, bioprinting for tissue growth, and bespoke prosthetic design, which has enabled mobility for millions of people around the world.

It is certainly hard to apply an accurate value, and more than likely, these projected figures may well prove to be an underestimation as we move into the next decade. The growth of the functional and industrial print market revenues worldwide is estimated to have nearly doubled in value from $37.2 billion in 2012, to $76.9 billion in 2017, with a projected value of $114.8 billion in 2020 (https://www.statista.com). In 2015, the global 3D print market size was estimated to have been around $5 billion, and is set to increase to $26 billion by 2021 (http://www.wohlersassociates.com).

The current print industry demonstrates the buoyancy of the market, but there are also global uncertainties, threats and weaknesses. In order to maintain market share in an increasingly competitive marketplace, there is a shift in the way large companies are consolidating product portfolios. And in the context of the ongoing decline in the traditional printing sector, there is a greater strategic emphasis in a diversification of products and services alongside company mergers & acquisitions. As a barometer of change, and a need to gain a significant market share over the next few years, patent applications are on the increase from international corporations, indicating greater investment and research into additive printing and manufacturing.

According to a recent report by Smithers Pira (https://www.smitherspira.com) electrophotography is the major contributor to the digital market, however, inkjet printing is growing rapidly and, by 2019, is anticipated to overtake electrophotography. By 2024, inkjet printing technologies will account for 56% of the value and 53% of the digital print volume. Although figures have plateaued over the last few years, we have seen an impact on commercial print volume which, since 2010, has shown a decline. This has largely been due to 'new media' platforms–e-books, online, electronic and social media, and search engines, which have steadily replaced traditional high volume printed products, such as catalogues and telephone directories.

According to the BPiF (https://www.britishprint.com), the gross value of printing adds relatively more value to all but one other sector, which is the manufacturing industry. For example, in the UK, with a turnover of £13.5 billion ($18 billion), the print sector has a gross value added of £6.1 billion ($8.1 billion), employing around 122,000 people in 8,600 companies, thus, making the UK printing sector an important economic contributor and employer in all UK regions.

As consumers we have an insatiable appetite for images and pictures, things and products. It is almost impossible to imagine a world without printing. The evolution of the printing press - from Gutenberg's moveable type in the fifteenth century to the multi-platform printing press of today - has grown to become a highly effective method for mass communication. Since the industrial revolution, mechanisation and mass production have made a significant impact on the way products are made. A quick glance around our desk or room, and printing may well have played an important part in the manufacture workflow – even on a micro level such as a useby date, washing label, raised braille text, or a CE safetymark – playing a crucial role in the way we use, consume, and dispose.

Some sectors - from engineering and health to the hobbyist and home user - may be considered as niche and overlooked, and yet each contributes to the rich and varied landscape that we can describe as print. Compared to the fifteenth century, we as technologists, manufacturers, educators, makers and designers are shaping and contributing to a very different landscape today.

Acknowledgements

The preparation for this book has not been accomplished alone. We owe our debt of thanks to our partners, families, friends, colleagues and mentors. Our understandings on colour and print have certainly been informed and challenged by a highly knowledgeable group of scientists, technologists, artists, designers and practitioners. It has been shaped by discussion through European colour projects including CREATE and CP7.0 and conferences including IS&T, CGIV and AIC. We consider ourselves as observers and collectors, theorists and practitioners in this newly emerging technology. We will also draw from many different aspects of visual culture, heritage and traditional industries including crafts, design and applied arts, which has provided alternative insights for 2.5D printing applications.

This work is in collaboration with Xavier Aure, Teun Baar, Michaela Harding, Jesse Heckstall-Smith, Stephen Hoskins, David Huson, Peter McCallion, Paul O'Dowd, Melissa Olen, Theo Phan-Van-Song and Peter Walters.

About the Companion website

This book is accompanied by a companion website:

www.wiley.com/go/bridging2d3d

The website includes:

- Video links

Scan this QR code to visit the companion website

Introduction

What is 2.5D printing? And what is the half-a-dimensional quality that we are attempting to describe? Does it address surface, relief, texture, material? Or about perception, appearance, illusion? Or terms-of reference or taxonomy, or methods of capturing, measuring and modelling material appearance? Or is it about trends and new technologies? The simple answer is that it is all of the above and more. The primary objective of this book is to scope and identify the essential 2.5D qualities and benchmarks. The challenge is how to arrive at definitions and exemplars that – in this rapidly developing and changing technology – effectively reflects the current state of the art of 2.5D printing, and to provide insights into the future of printing and additive manufacturing.

As the title of the book suggests, there are two primary aspects to this enquiry: the dimensional – the need to gain insights and understanding of a surface that is neither 2D nor 3D, but is somewhere in between, and yet can effectively describe the micro and macro textures; the print – how an object, a scene or an image can then be captured, measured, recorded and printed as a physical reproduction. Furthermore, there is also an extra element, which could be considered as the illusive 'x' or an extra half dimension that is much harder to describe. For the purposes of this enquiry, and because it is highly significant for us, it relates to, firstly, the printing processes, materials and colours to create a textured surface, and secondly, its appearance; our human, emotional and perceptual relationship with the printed image.

The Internet has provided access to unlimited images and things we may never have anticipated or known about. Digital technologies have irrevocably changed and challenged the way we look at, construct and print images and objects. We work digitally and incorporate numerous digitally aided technologies as a part of our daily workflow. As we move from real-world texture to screen-based or printed representations of texture, our understanding and engagement is mediated by the screen or printed matrix. Furthermore, colour-printing technologies have evolved from coloured dots on paper to coloured dots on three-dimensional objects. What has remained unchanged are the thin-film CMYK process inks and colours by which images are printed. The next step in colour printing is a modification to the thickness of the film to create a new dimensionality or functionality.

We are also working on the idea that 2.5D printing demonstrates a materiality of surface, which could be described as the relationship of materials, texture and surface. The aim here is to explore ways in which an extra half-dimension can incorporate an idea of otherness, a physical delight, or a visual engagement to imply a mixture of aura, illusion and paradox. We could begin by suggesting there is a halfway point between conscious and unconscious seeing: how do we observe, recall and differentiate thousands of different materials and textures, and, in particular, coloured materials and textures? This materiality of surface can also imply whether we are likely to find a surface or material convincing and whether we are unconsciously drawn to their tactile qualities, or are physically repelled by surfaces, materials and colours.

We are presented with thousands of different textures and materials every day and at a glance we are able to determine each of their material characteristics. Moreover, we are attracted to natural things, and the material qualities of these natural objects, who is not tempted during a walk along a beach to pick up sea-smoothed pieces of glass, wet pebbles or shells? We still like to preserve the analogue – there is renewed interest in vinyl records, analogue film, black and white photography and the rustle of the pages of hardback books. More frequently, we do not have time to engage with everything we see, and therefore it seems that there is more effort to ask someone to stop and look. Walking through a gallery, looking through a picture book or a clever advertisement, how long do we look at an image? Seconds? The challenge therefore is to ensure that images have the potential to arrest the viewer and stop them in their tracks and maybe to take a second look.

As more stuff is created in our increasingly hectic world we need to ensure that the things we design and make are of benefit to our health and safety, as well as provide us with pleasure and comfort. We suggest it is increasingly important to gain an understanding of the entire design process: firstly, that a product is well designed and fabricated and, secondly, the design process does not finish at the manufacturing stage, but continues beyond its lifetime. For example, what happens to the artefact after it is made? Should the materials that we use have a sustained lifecycle and not simply be thrown away? Therefore, the second aspect of this search is to consider the materials that we use: can the printing materials be useful? Save lives? Be eaten? Can they be touched? Do they provide wellbeing, comfort or visual delight?

In preparation for printing we may also struggle to work with materials that are neither aesthetically pleasing nor pleasant to the touch or smell. These materials may have been developed based on the limitations and constraints of the hardware, for example: printers, plotters and cutting tools, and materials: paints and inks, nylons and polyesters. It is, therefore, of inestimable importance that the materials, the processes and tools we use today and in preparation for the future are: of high quality, are pleasing to the eye and the touch, offer new design solutions towards social impacts, provide comfort and wellbeing, for example, and, finally, can ensure a legacy for future generations.

The structure of this book is divided into five chapters. In Chapter 1, we investigate the relationship between material and texture and how each conveys a character or quality. We explore the physical, perceptual, linguistic, natural and artistic interpretations of these appearances. We also consider the relationship between images, pictures and printing, how appearances are reproduced, and the idea that as reproductions, a picture may convey some sort of embedded aura or emotion. In Chapter 2, we study the past: how artists have observed scenes and objects, and used different tools and materials to translate the appearance of textured objects into pictures, reliefs and artefacts. Chapter 3 considers the present, and the range of methods that are being used to capture different surface qualities, and how these are measured, categorised, reproduced and applied in the twenty-first century. In Chapter 4, we suggest future trends and the implications for the print industry. The aim of Chapter 5 is to explore, through different case studies, concepts and ideas about material appearance and methods for reproduction. The case studies are chosen based on our experiences and interests. These include day-to-day objects, artefacts, materials and surfaces, many of which are taken for granted and overlooked, but that in reality represent the sum of efforts in many different creative and technological fields. The case studies may also reflect on a span of history- from ancient to contemporary production. Some case studies may be considered as niche or outdated, but the aim is to demonstrate the enormous variables, materials and crafts skills that could be used as benchmarks when considering the specifications of a design and production workflow. It may be obvious by now, but our examples are focused in aesthetic applications as they display the capabilities and the full potential of relief printing as a digital tool. We have used these case studies to explore and consider how manufacturers have adapted methods to incorporate new materials, technologies, and responded to emerging consumer trends.

Before we can embark on the '2.5D-ness' of a surface, we would firstly like to consider what does 2.5D really describe or imply? In essence, the question is what do we mean by 2.5D? What are the qualities that define it? One can start with a negative and suggest that it is not about printing objects (3D) or about printing images (2D), but something between the two – or more simply put: what is the relationship between the textural attributes of a material and the object? The important factors in gaining an understanding of 2.5D are:

To identify the key terms,
To address assumptions,
To establish the relationship between perceptual and physical (visual, tactile and physical),
To look back to existing historical benchmarks,
To survey contemporary making,
To identify current weaknesses and problems,
To demonstrate current emerging ideas that could lead to further exploitation,
To set out a series of benchmarks and objectives in need of addressing.

The underlying motivation is to ask 'why, what and how?' The rationale underpinning the book *2.5D Printing* has evolved through personal interest and our research in working towards developing methods that: capture, measure and model the surface qualities of 3D and 2D objects, as well as those that represent and reproduce the appearance of surface, materials and textured qualities. Therefore, with cross-disciplinary understanding and insights in arts and technologies, this book could be considered as a confluence of ideas, methods and applications that investigates the relationship between two and three dimensions. The subject crosses and overlaps a range of fields including science, technology, art, conservation, perception and computer modelling. It is an engagement and discourse on what the perceptual and objective differences are between two dimensions and three dimensions. Our interests and ideas may well differ to yours, nor do we provide definitive answers, but the aim here is to open the debate.

1

Defining the Field of 2.5D Printing

1.1 What is Texture?

What is the relationship between textures we see in the real world and as a reproduction? How convincing are these? As more images are digitally reproduced and printed, standardised methods may result in a uniform ubiquity, whereby prints, posters, photos and reproductions of paintings are all printed in the same way, and likewise objects that are fabricated using the same materials, using the same layer upon layer fabrication methods. On the one hand, we can say that digital technologies have assisted in printing things faster, cheaper, bigger, bolder; on the other hand, we could suggest these things fall short of expected levels of quality. We could suggest that quantity may have been achieved at the cost of quality or ubiquity at the cost of diversity. In our manufacturing world, as more and more things are designed and created digitally, how do we bridge the gap between images we see on screen and how it is physically reproduced?

Texture can be described as the microstructural details that can be perceptually distinguished from one surface property to another. One could consider texture as multisensory; we only need to look at a surface to gain a quick understanding of its textural properties, which may then be reinforced by other senses (smell, taste and touch) (Klatzky and Lederman, 2011). Texture could be broadly categorised according to whether it is tactile or visual, it can be described by its appearance, as a noun or adjective and by comparison or difference.

Tactile texture. Tactile or physical texture describes the minute variations in the surface elevation created by the changes in orientation, density and distribution of tiny particulates of the surface. At our most primitive level and from birth, we engage with the world as a tactile experience and by using all our senses, we are compelled to touch materials, surfaces and objects to find out whether they are smooth or rough, soft or hard, or understand their material properties, for example, by handling a fabric to discern whether it is flexible or stiff or how it drapes. As a survival mechanism, it is highly important to recognise whether a food appears rotten (discoloured, wrinkled, mouldy) or whether an object may do us harm (prickles, barbs, spikes) (see Figure 1.1).

Visual texture. Visual texture usually refers to flat changes on the surface, a sort of drawing that demonstrates certain properties of periodicity and colour but does not present topographical

2.5D Printing: Bridging the Gap Between 2D and 3D Applications, First Edition. Carinna Parraman and Maria V. Ortiz Segovia.

Companion website: www.wiley.com/go/bridging2d3d

Figure 1.1 Texture is useful for recognising the difference between something fresh and rotten (courtesy: Grace Parraman).

changes. For instance, a plastic table designed to look like wood may appear to be wood at a first glance (visual texture) as the drawing or projection of a picture is presented to be real wood, but may not feel like wood to the touch (tactile texture). We may miss the minute ridges, the roughness/smoothness of the material or the feeling of warmth. On a perceptual level, we use our library of knowledge to quickly recognise and chose from the range of materials and textures, and from a priori experience we no longer need to touch, for example, tree bark, velvet, brick. We can begin to order materials, for example, according to roughness or gloss. We can differentiate between the properties of materials or discern their suitability for a particular application, for example, different roughness, weight or pliability of a paper, or whether a shiny material is made of plastic or glass. Likewise, from our knowledge of handling fabric we can begin to use these for different garments.

Appearance of texture. In order to represent different textures or demonstrate an object's material qualities, it is important to render with convincing likeness. Artists have sought to paint and draw realistic representations of material surfaces and objects for hundreds of years, and now, through computer modelling, textures can be created virtually, whereby objects can be rendered for a wide variety of applications (film, animation, games, architecture) and bodies can be dressed in garments that drape and move in a lifelike way.

Taxonomy of texture. Using our skills to compare different textures, we are able to sort, categorise and name the appearance of a wide range of materials; some terms may be specific to different trades, subject areas, cultures and countries. From this, we have developed a rich vocabulary of adjectives to describe the differences and nuances of appearance (rough, smooth, prickly, slippery, slimy) or to describe their chemical or mineral appearances (gold, copper, granite, limestone) and material properties (denim, silk, wood, leather).

1.1.1 How to Quantify Texture

How may one quantify the appearance of a texture – by its characteristics or material properties? Some terms have different connotations depending on subject specialisms and requirements. Analogous to the many colour models that have evolved to suit different requirements, specialists and industries have developed their own quantitative methods – scales, comparisons and measurements – to define the parameters of manufacture or appearance and assist in selection, for example, roughness through the International Roughness Index (IRI), surface qualities of metals (R) (Black and Kohser, 2011), measurement of different aspects of the gloss of a surface (Hunter) Gloss Units (GU) or the dynamic viscosity of a liquid (centipoise cP). In commercial specifications it is sometimes easier to predict the behaviour of a material by comparing its material properties to well-known materials. For example, a manufacturer of adhesives, in approximating the viscosity of different products to a client, might achieve this by suggesting different domestic fluids and foodstuffs as a comparison: at 300 000 cP toothpaste is a highly viscous material that does not move without being squeezed, whereas a syrup at 4000 cP can be poured or dripped, and water has a low viscosity at 1–3 cP, and can be easily poured.

1.1.2 How do Artists Convey the Appearance of Texture?

Since the beginning of art history and throughout the centuries practitioners craftspeople, artists, designers, engineers and technologists – have been fascinated by the material they use, and by selecting different materials and mediums are able to translate meaning and concepts into form. A visit to any museum may well demonstrate a diversity of historical artefacts, each created using different tools, and each designed to achieve a specific mark. Over time, makers have developed different craft skills and tacit knowledge of materials, tool and techniques, and this knowledge passed down through workshops and studios.

The relationship between two and three dimensions is a long and enduring area of interest for technologists and artists. Artists have been constantly fascinated by the pictorial representation of a three-dimensional world through the two-dimensional media of painting and drawing, and by employing drawing elements such as perspective, illusion, colour, texture, light and shade to create more convincing and immersive environments.

Over the last one hundred years, photography and photomechanical processes have also played an important role in recording and reproducing images for mass consumption. Over the last few decades, sophisticated computer graphical interfaces have transformed methods of image construction, so that now we can no longer easily detect which components of an image are computer generated. Now we are presented with images that are 'a plausible rendering of visual effects that create the illusion of life-likeness'. Surprisingly, this reference was made by art historian E.H. Gombrich (1909–2001) in his critique and analysis of the psychological aspects of image making (Gombrich, 1959, p. 246). Gombrich, most noted for his book *The Story of Art* (1950), also wrote prolifically during the latter half of the twentieth century on the arts and sciences. His writing demonstrates that our enquiry is certainly not new, but can be reappraised within this new digital context, namely of the development of surface fabrication and digital fabrication technologies.

So-called rapid prototyping and additive manufacturing technologies have shifted the technological focus from 2D inkjet printing to 3D digital printing, whereby a virtual object that has been generated and designed on computer can be exported and printed in layers to create a physical three-dimensional object. 2.5D printing, as an ad hoc and evolving technology, has borrowed elements from 2D wide-format inkjet and 3D digital printing systems. It has incorporated similar design workflow components to 3D printing, enabling printers to apply multiple layers of ink (and/or other materials) until a desired low relief elevation per layer, pixel or unit is achieved. The resulting surface could be considered either as a flat surface with some sort of topographic feature, or as a skin to wrap 3D objects.

Artists have created images using different mediums including stone, wood, paint and drawing as low-relief and two-dimensional narratives, scenes and pictures. From the perspective of appearance, materiality may also relate to the choice of material that an artist or a designer has chosen as a medium to transmit the experience of the image or artefact. In some cases, materials and media are selected for reasons of longevity, such as stone or wood, which may reinforce, challenge or confound the viewer's perceptions of the artefact. As illustrated in Figure 1.2, the Italian sculptor Giovanni Strazza chose a hard, solid block of Carrara marble, which we would argue is the most inflexible and hardest of materials to convey the appearance of a diaphanous veil. However, Strazza has captured a translucency and

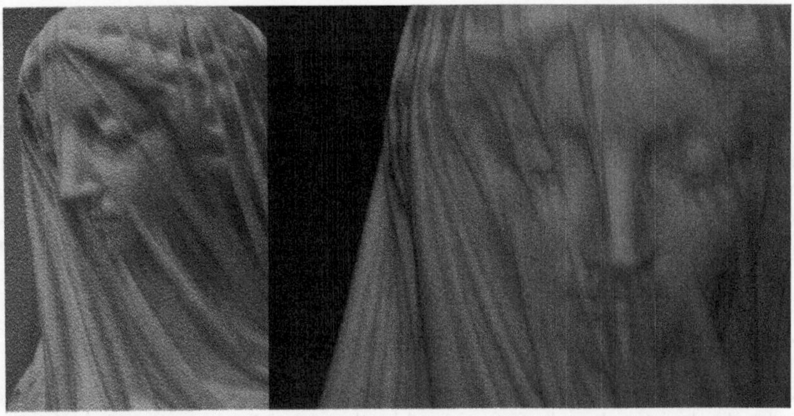

Figure 1.2 Giovanni Strazza (1818–1875) *The Veiled Virgin*, Presentation Convent, Cathedral Square, St John's, Newfoundland, a sculpture carved from a single block of marble (courtesy: Philip Chircop).

softness to the features of the woman's face and we can see through the close folds of the material and the braiding of her hair is also discernable. In this example, the hardness of the marble is at odds with the subject, and yet the artist has created a convincing appearance. Consciously we know the figure is not real, but by its verisimilitude we are moved by an aura of its emotional power.

1.1.3 How the Natural World Mimics the Appearance of Texture

We also find animals, insects and reptiles are excellent at using texture and colour to blend into their environment. Species around the planet depend on camouflage and mimicry for daily survival and communication. Animals have different ways of mimicking their surroundings. Depending on their needs, evolution has made some of them specialists in colour and others in physical texture. Chameleons may be considered the con-artists of the animal world, which can change the colour of their skins to blend into their environments, but there are many insects and sea creatures that are capable of even better illusions (Figure 1.3). For instance, the cuttlefish (not actually a fish but a cephalopod) is a master of

Figure 1.3 Cicadas in Japan blending with the textures and colours of their habitat.

disguise, despite assumptions of being colour-blind. Cuttlefish are known to have a diverse set of body patterns (colour, contrast, locomotion, posture and texture) that can switch almost instantaneously to adapt to their surroundings based on perceptual cues related to contrast, brightness, (micro) shape and visual and physical texture.

Humans have sought inspiration from them as a way of gaining insights into camouflage for re-creating patterns in fashion design or structures in architecture or biomimicry for electronic adaptive e-readers. In a study entitled *Biological vs. Electronic Adaptive Coloration: How Can One Inform the Other?* (Kreit *et al.*, 2012) developers of electronic reading devices of e-Paper have studied how cephalopods use subtle changes in their skin pigment to absorb or reflect available light. Technology based on 'emissive' light uses electrical fields in much the same way as cuttlefish.

These animals are able to mimic a wide range of colours and patterns that are highly convincing. However, we know when something appears right or wrong and are also very good at spotting differences and recognising patterns. Convincing texture reproduction and rendering tends to be more problematic, our visual system is able to pick out repeated elements (Figure 1.4) and we can discriminate

Figure 1.4 Left: sample image of a *Letratone* dry-transfer pattern of a wall traditionally used in graphic rendering. Right: an irregular pattern of a dry stone wall. Below a windbreak using a dry stone wall repeat pattern; note how the pattern is repeated for each panel.

the difference between natural and patterned texture. Likewise, a natural texture (grass, a brick wall, wood grain) appears homogeneous, but is infinitely random – each element is similar but remains unique. However, a patterned texture, although homogeneous, is composed of the same repeatable and recognisable elements (Hawkins, 1970). Furthermore, the effort in rendering surfaces with no discernable pattern structure comprising unlimited variations may result, as demonstrated by a computer-generated rendering, in exceptionally large file sizes. The new challenge for digital rendering is to investigate the relationship between the object and surface texture in a scene, and to render materials and objects where the textural attributes of the object are perceived to be convincing (Kim, Hagh-Shenss and Interrante, 2004; Adelson, 2001).

The complexity in the creation of a convincing textural render is essentially due to the enormous range of physical components that are required to incorporate all the nuances of a texture, such as colour, fibre, grain, reflectance, specularity, weave, hardness, softness, glossiness, fluidity and, as demonstrated in the previous list, the range of descriptive adjectives, cultural and specialist terms that extend these more subtle characteristics of a texture. Furthermore, these multivariables of textures tend to be stored as a visual taxonomy in the human memory, whereby subtle textures and surfaces can easily be identified and differentiated by our visual memory. In a real-world scenario, planed wood can quickly be distinguished from paper (grain, surface, flexibility), and animal fur from human hair (direction, colour, smoothness, curl).

1.2 Measuring Texture and Colour

How do we decide on a colour if a material surface is textured, and appears to contain a range of different subtle colours? How might we go about choosing a colour to paint a wall? For example, we may be inspired by bright yellow daffodils or wish to colour-match a plush red velvet sofa. Looking at the range of household paints in a DIY store highlights many subtle variations of colour, surface and texture. As consumers we are very exacting, we may leave the store with a pot of daffodil-yellow paint, yet when we apply it to a wall at home, the final appearance may be very different depending on the outdoor light, or artificial illumination in our room (Figure 1.5). The following image shows a simultaneous colour viewer (http://www.gtilite.com/products/scv-simultaneous-color-viewer), which demonstrates the different light sources: lighting in the home (such as tungsten, which is a warm white light), fluorescent lighting commonly found in offices and commercial environments, and average outdoor light (https://www.gtilite.com/pdf/Various-Light-Sources.pdf).

Choosing colour can be an expensive process for consumers, and marketing paint to a DIY market is fundamental for maintaining consumer loyalty. As exampled in a rebranding exercise by a UK household paint manufacturer (1998), the aim was to understand the change in the domestic paint market. Their key insight into its consumers' choices was that they did not want *paint* - they wanted help with matching colours. Based on the possibility of living in a painted room that may be hideous, often led to a 'non-purchase, delayed purchase or compromise. People needed help – colour help.' The slogan for the manufacturer's paint-matching advertisement was, 'You find the colour, we'll match it' (https://www.marketingsociety.com/the-library/2008-dulux-marketing-communications-case-study). This advertising campaign made a significant visual impact on British DIY consumers.

When presented with an array of colour swatches, it is surprising we can still be uncertain about making colour matches. In general, humans are fairly good at colour constancy (McCann and Rizzi, 2011). We are able to discount misleading information, such as modulations in colour caused by surface texture or effects (weave, bumpiness, grain), highlights and shadows. A brick wall that is painted white is still perceived to be white even if it is a bright day or a dull day. As demonstrated in Figure 1.6, although the pink jacket demonstrates a wide range of reflectances, highlights and shadows, we still perceive the colour to be pink.

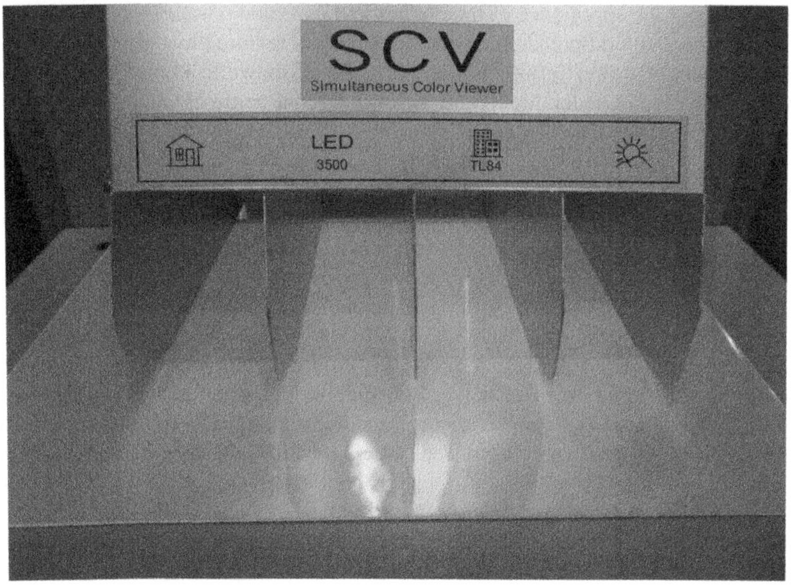

Figure 1.5 Illustrating the appearance of the same colour under different illumination. Simultaneous Color Viewer (SCV) produced by GTI Graphic Technology Inc.

Figure 1.6 Dulux advertising campaign 2008 (courtesy: AkzoNobel) (http://www.mischiefpr.com).

However, comparing the appearance of the clothes and the colour swatches of the advertising campaign (Figure 1.6), there is a difference between flat surfaces, textured materials, and materials, each demonstrating complex angles of reflection. Furthermore, as shown in Figure 1.7, although the coloured square is seemingly flat under direct light, by changing the angle of the illumination we perceive a range of bumps and textures; the coloured square is transformed into a noisy surface of highlights and shadows.

Therefore, in order to obtain an exact colour match from a textured material, by simply measuring a surface using a colour measuring device, such as a spectrophotometer, could lead to a range of results. By moving the spectrophotometer across different areas may lead to a variety of colours or tones, which may result in an unwanted colour reproduction and a potential DIY blunder (Figure 1.7, right).

In order to model the appearance of complex surfaces and geometries, scientists and technologists are now striving to find methods that accurately capture and measure low relief, texture, and surface qualities that incorporate quantitative and qualitative methods. The objective evaluation allows a quantitative validation of data acquisition methods and measurements of the real world. It provides information about the feasibility of reproducing details, colours and textures. On the other hand, the subjective evaluation provides a sense about whether a reproduction is pleasing or appealing.

We make use of different methods and techniques to judge the quality of 2D rendering systems such as printers, cameras and screens, but little work has been undertaken in the assessment of 3D reproductions (as described in Chapter 2). Not only the extension of 2D quality metrics appear to be the answer but also previous methods from the field of optics are just as suitable to estimate quality. The current lack of standardised methods in the 3D print industry can be compared to the colour printing industry of the 1980s. Instead of working towards a common standard to unify results and open the discussion about quality. each emerging 2D industries were developing their own file formats, hardware systems and workflow parameters.

It seems that no matter how accurately we can measure the properties of colour and textures, from the point of view of perception, the human visual system is adept at interpreting these two-dimensional visual clues to imagine three-dimensional spaces. The human visual system is also highly objective and critical in identifying when objects and spaces do not appear to be correct. Therefore, subjective and objective approaches are essential for comparison and improvement to evaluate the quality of the outputs.

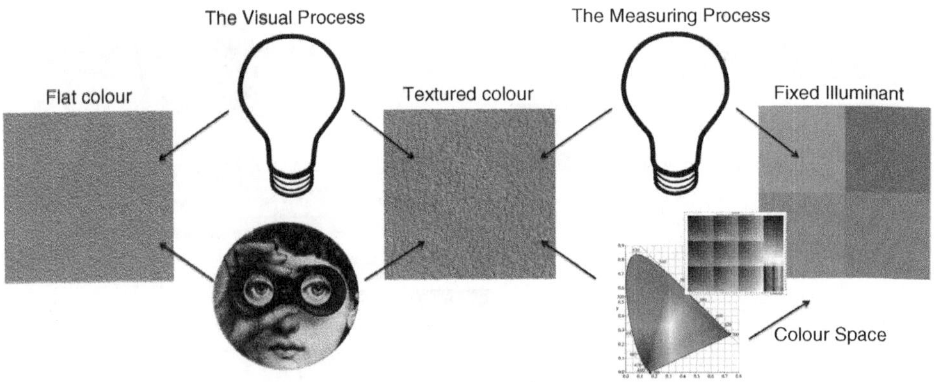

Figure 1.7 Human perception is able to discriminate an overall colour and discount texture (left). Measurement of the surface could lead to an unwanted reproduction of colour (right).

1.3 Images, Pictures and Reproductions

What is it about an image that is so compelling? And what is the relationship of an image to its reproduction, representation on screen or on paper? Or even a reproduction of a reproduction? Images in the twenty-first century are ubiquitous. Alongside our own old photos kept in shoeboxes or on hard-drives, there is the World-Wide-Web, with unlimited access to digital databases, *Picassa* and *Flickr*, commercial stock-imagery, as well as magazines, books, museums, galleries and archives, we are surrounded by pictures and images. A quick word search for 'image' on the Internet reveals about ten billion hits. Yet, as regularly demonstrated in news reports, images can still cause offence, are feared and venerated, and can even bring about violence and death. Images could be considered as wide ranging including something that is fixed to the more ephemeral and fleeting, including paintings, pictures, diagrams, graphs, illusions, photos, cartoons, ideas, memories, dreams, imagination, or our mind's eye. Images are 'notoriously ambiguous' (Mitchell, 2006), we rely on different images to distil complex information (info-graphics, pie-charts, bar-charts), provide accurate and specific information (city maps, sea charts, architects plans, car-repair manuals) or to tell a story (sketches, paintings, films). Each of these carries specific codes for interpreting or reading. A misinterpretation of a map or a road sign may even be life threatening. The following arrangement, by cultural thinker W.J.T. Mitchell, suggests how some images can be categorised according to different disciplines based on how we might use them, including arts, optics, perception and psychology (Mitchell, 1986).

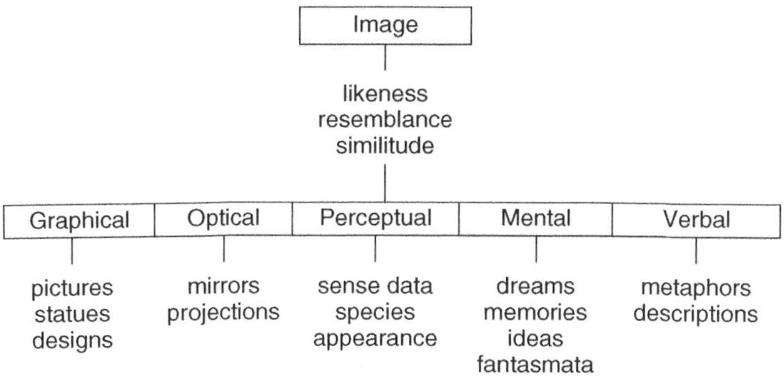

The most common way of looking at images is via photographs, on screen or as prints. Our ongoing fascination with the camera, is the ability to extend beyond the human eye. Likewise, spectacles, magnifying glasses, telescopes and microscopes are an aid to vision, which can change the scale of objects, bring objects closer and draw the viewer to things, surfaces or materials, and may assist in our understanding (Edwards, 2006). This idea is certainly not new. In the eighteenth century, a short story entitled *Micromégas* (1752) by Voltaire (François-Marie Arouet 1694–1778), describes Micromégas and his travelling companion who had evolved on different planets; both are giants by Earth's standards. As part of their quest to find new insights and different life forms, they have arrived on Earth, but because of their huge size are unable to see the human inhabitants. They assume that as they cannot see any signs of life, the planet Earth must be uninhabited. Micromégas' companion, who is from Saturn, by chance spots a speck swimming in the sea, and realises the life form is communicating. However, they conclude that because these life forms are so small, their brains must be incapable of holding reasonable thought, and therefore must be stupid. The story of Micromégas demonstrates that although our vision of the world seems effortless, there is a difference between looking, seeing and apprehending; thus, from an epistemological perspective, we ourselves define the limits of our knowledge. Likewise, the story demonstrates limits in technology; as we are able to see clearly or closely, we are able to gain new insights about the everyday things around us. Today, we have powerful methods of capturing the unseen and, furthermore, if extended to high-speed motion capture, high definition, X-ray, infrared or microcameras, then our ability to see beyond human vision creates new insights and challenges our own assumptions of our world. Traditional photography captures different intensities of light through the physical changes as a chemical reaction, or chemical solutions using light-sensitive chemicals. Digital image sensors capture the brightness of light, that in turn are converted into electrical signals and converted into a numerical code. Likewise, the mechanism for translating images into a code for the brain, visual information, is converted as neural impulses to the brain whereby the brain creates 'symbolic descriptions' that have no relationship to images or pictures but generate their 'own alphabet of nerve impulses' (Ramachandran, 2012, p. 47).

We continue to be fascinated by images. The photographic recording, such as the Hubble Telescope, has bought the furthest regions of our universe closer to us, and in turn, the process of looking at the world through a lens, such as photography, has become a medium that anchors us to our world. However, the downside is that we use the camera to mediate our experiences without properly looking, which can result in a secondhand experience. The image is not an image at all but a direct re-presentation of lived reality (Edwards, 2006, p. 69). Art historian James Elkins echoes Voltaire's message in his book *How to Use Your Eyes* (Elkins, 2000), who suggests that looking is based on memory, previous experience and day-to-day routine through familiar landscapes; we are not actually seeing but perceiving what we think it should look like. If we 'see' every detail it would be an exhausting experience. Elkins describes how familiar everyday things slip by unnoticed. He reintroduces the reader to a range of commonplace objects, including postage stamps, moth wings, sand, rainbows, and *Craquelure*. *Craquelure* are the tiny cracks that appear in the surface of paintings. These cracks and the formation of patterns can tell us about its provenance, how it was painted, varnished or repainted, conserved or even ill-treated.

The camera has changed these perceptions and impacted the way we look at things, 'Strange realism of the camera … is a compelling illusion' (Edwards, 2006, p. 2006, p. 69). The camera can reproduce pictures and via imaging software and the Internet, we can modify them -to any size, send files anywhere, and reproduce images for any purpose. The same image can be viewed on millions of different screens, accessed from thousands of different locations and surrounded by many different contexts. Originally a painting, such as *The Annunciation Diptych*, a small devotional piece by Jan van Eyck (1390–1441) (https://www.museothyssen.org/en/collection/artists/eyck-jan-van/annunciation-diptych), may have originally resided in a private house or chapel. The Assyrian carved panel (Chapter 2, Figure 2.5) was once part of the architecture in an ancient building. However, the meaning of the artwork no longer resides in the place where it was painted or installed.

According to English writer and painter John Berger (Berger, 1972), looking at and interpreting artworks, and the process of translating real life into artworks, is less spontaneous and natural than we assume, and is more to do with training, habit and convention. For example, the use of perspective is unique to European art, which focuses objects towards the gaze-appearances travel in and the lines of perspective tend to draw the eye to the centre of the picture, but as we walk about or travel, our perception travels with us. The photograph can do this perfectly: one can point a camera down a street lined with buildings and the camera produces a picture in perspective.

French philosopher and art historian Georges Didi-Huberman (B.1953) suggests our perception of an artwork may be changed as we move closer to a painting and celebrate its wonder. His suggestion is to not look at a painting but to see it. By getting closer we can analyse all the detail and arrive with an exhaustive description, we can see the brush marks, the defects, and the embedded brush hair. Under close scrutiny, we can find and describe all the microelements, but from a distance we understand its drama (Didi-Huberman, 2004, p. 229). A good example of this can be found in the paintings by Diego Velázquez (1599–1660), who was a master of different contrasts and textures. His work is intriguing:

when viewed at a distance the canvases present an opulent and detailed rendering of precious pearls, lace, metal, fur and feathers, and gold, silver metal work, and glittering brocade – all rendered in a seemingly high contrast and dynamic range. At a distance, these constructed compositions are a mimesis of materials, but close up, the micro details appear out of focus and gestural (Chapter 2, Figure 2.33). Velázquez celebrates his still-life subjects as a careful study of appearance through the rendering of paint - by bringing together his knowledge of materials, he is exploiting our assumptions that drive perception (Ramachandran 2012, p.51).

With the invention of photography and film, comparisons can be drawn to the early twentieth century, when artists began to capture the world mediated by the lens. These ideas eventually led to a logical progression of the moving image or cinema – a twentieth century synthesis of art and science. Photographers such as Eadweard James Muybridge (1830–1904) was able to capture the movement of animals and humans using trip wires and a row of cameras (http://100photos.time.com/photos/eadweard-muybridge-horse-in-motion), and Étienne-Jules Marey (1830–1904) invented a handheld rotating drum that could expose 48 plates in 72 seconds (https://www.youtube.com/watch?v=dU5llQM7lzs). Composition and movement are some of the aspects that provide more dimensions to a flat media and connect painting, film and photography. This photographic idea of movement was also captured in the painting. The Fourth Estate (1901), by Giuseppe Pellizza da Volpedo (1868–1907); by playing with transparencies and changes in light, the artist creates the impression of movement among a compact group of people (http://museodelnovecento.org/index.php/en/collection?giuseppe-pellizza-da-volpedo). This is similar to the work by Gaetano Previati (1852–1920), *Funeral of a Virgin* (1895), in which we have the sense that the crowd is moving so lightly as if they were ghosts (http://pinacotecabrera.org/en/collezione-online/opere/funerale-di-una-vergine).

1.3.1 The Anxiety of the Reproduction

Perceptually we know that images – paintings, prints, photographs – are visual statements, that by their flatness, are illusions of our world. We are experts in looking through or into a flat-land and understanding its spatial dimension. As exampled by the marble *Veiled Virgin* (Figure 1.2), these images and objects, and the stuff from which they are created, resonate on so many different emotional levels.

Cultural theorist of the early twentieth century, Walter Benjamin (1892–1940), stated in 1936 that an original image cannot exist without the knowledge of a reproduction. However, in the twenty-first century, as more images are viewed on screen, the aura of originality in the age of mechanical and screen reproduction is so far removed from its place of origin that even reproductions are now commonplace (Benjamin, 1936). To source an image in the twenty-first century, the art historian, curator, publisher, researcher or art lover has many opportunities to search for images. Although we have access to more coloured images, the issue of image quality on the Internet is questionable. Of course, the accuracy of printed colour reproduction has improved over the last hundred years, but a badly reproduced colour image might be considered too misleading, and a black and white image tonal image a more useful alternative. Can we rely on the Internet? As an example, the thumbnail images collected from a Google image search of a famous painting demonstrates the many variations of colour gamut and contrast (Figure 1.8).

The work by Doug Manchee, *20 Works of Art in the Age of Mechanical Reproduction* (2006), dramatises the variability of images we think of as reproductions. Manchee gathered reproductions of well-known artworks from books that were printed at different resolutions, sizes and colours, and which were often cropped to fit to the book format. Manchee scanned and superimposed them, resizing to align the primary elements of the composition. In the final composite image, he reveals all the layers. Here one can identify the different halftone processes, different colour gamuts, cropping and stretching, as well as realising that compositions are fitted to suit the demands and shortfalls of the publisher, rather than keeping the integrity of the image.

Moreover, looking at the range of colours in the screen grab (Figure 1.8), one could ask which is the closest to the original? This poses a problem for gallery curators wishing to maintain control over the works in their care, and, furthermore, in the current interest in the creation of textured paintings that

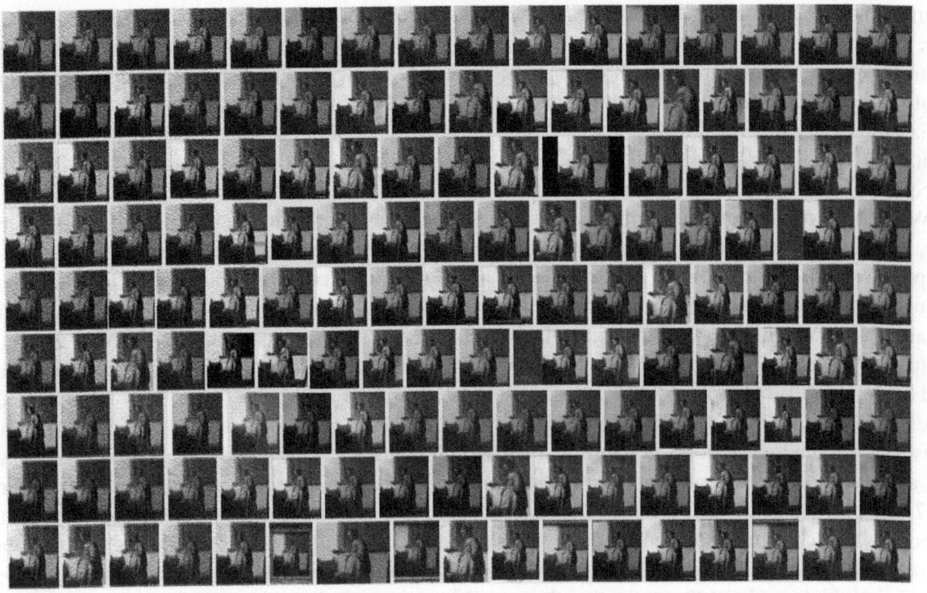

Figure 1.8 A screen-grab of a page of thumbnail images from an image search of Johannes Vermeer's *Woman in Blue Reading* (1664).

demonstrate brush strokes, does this add yet another dimension of complexity? What of the term original reproduction – an oxymoron?

1.3.2 Reproducing Images: Tools of the Trade

From the sixteenth century onwards, etching and engravings of paintings and drawings were highly sought after and prized in the print markets. A large number of chiaroscuro woodblock prints made in the mid-sixteenth century were designed solely for a print market. Prints were an important source of artistic production in circulating and expanding awareness of the original paintings, and through reproduction to generate income from an original painting. Often these prints were the primary source of access for the public and scholar; therefore, the reproduction needed to be accurate. The translation of brush strokes to printed marks, colour to tone, required skill and knowledge, and the transcription from a painting had to be undertaken in reverse on the plate. Our contemporary understanding of original and editioned prints (the artificial method of limiting the quantity of prints that are numbered sequentially) was less of a consideration, and blocks were printed in different colourways and often modified and reworked. The art works by Albrecht Dürer (1471–1528), made in the sixteenth century, were prized and highly valued, and remain so today. It is also highly unlikely that Dürer made his own prints; instead a studio comprising different skilled artists would have transcribed his drawings into line on to woodblocks, which were then carved and printed as a relief. Likewise, John Constable (1776–1837), who was painting gestural and highly evocative landscapes at the turn of the nineteenth century, also attempted to capitalise from the burgeoning print market and to commercially reproduce his paintings. David Lucas (1802–1881), working with the medium of mezzotint, used a burin to transcribe the painterly marks of Constable's brush strokes on to copper. Working from black to white, the fine process of mezzotint enabled the engraver to convert colour and texture into a full range of marks and tones. (see Case Study 13: Analogue Printing Methods). Lucas' skill could be considered as an equivalent to the photographer's approach to sharpening, dodging and burning, deepening the shadows in some sections and increasing the contrast in others (Wax, 1990, pp.120–122).

Speculation on whether optical devices were used by Old Master painters to accurately draw scenes and portraits has undergone considerable (and often heated) debate over the last few years. It seems that historians, scientists and artists are both passionate and divided about how artists make art: at worst regarded as cheats, who employ clever trickery using lenses and drawing machines, or at best, artists able to transcend any technology whereby artworks are solely achieved by eye and hand. Martin Kemp, writing on the history of the art and sciences, documented a significant body of apparatus that could assist the artist, including the drawing devices by Leonardo da Vinci (1452–1519), and perspective machines by Leon Battista Alberti (1404–1472). As recorded in *De Pictura* in the sixteenth century, Alberti describes how the camera obscura and drawing frames were important components for arranging and composing figures and objects. Likewise, in terms of describing the influence of illumination, colour and composition, and careful lighting, composition in their paintings, Kemp compares the contemporary photographer to Vermeer. Kemp explains that Johannes Vermeer (1632–1675) was not making a 'mindless or mechanical image' but the artist was making 'aesthetic choices at every stage', and 'like great photographs, can be regarded as utilising optical mechanics for highly individual effects' (Kemp, 1990, p. 195). Furthermore, Kemp's observations call to mind the work of artist David Hockney (1937–), who during his career has explored many different imaging-making technologies – contemporary to their time – including Polaroid cameras, fax machines, iPads (as sketch books) and digital camera-recorders.

Hockney is continually interested in the different ways to capture and record the landscape, using a range of media and technologies in a highly individual way. Hockney's own work and ideas originate in drawing and therefore he is expertly able to recognise drawing traits and characteristics by other artists who sketch and draw. It is possible that as Hockney was more responsive to technologies as used by Old Masters, this may therefore have triggered a recognition in marks made in a sketch by nineteenth century painter Jean-Auguste Dominique Ingres (1780–1867), and to view it with fresh insights and understanding. He compared the drawing marks made by Ingres to a drawing made in 1975 by Andy Warhol (1928–1987) who often used a camera obscura to project an image in order to trace its outline. This observation led to a lengthy investigation into drawing methods and apparatus. His suggestion that Old Masters used drawing machines resulted in a publication, *Secret Knowledge* (2001). The story behind artist David Hockney and experimental physicist Charles M. Falco's investigation into how artists used mechanical devices is described by Lawrence Weschler in his blog, *Through the Looking Glass, Further Adventures in Opticality with David Hockney* (http://www.believermag.com/hockney/lookingglass). Weschler explains that although not based on any hard science or fact, Hockney had collected photocopies of artworks and had pinned them across a wall, asking: 'When did an optical look first emerge?' Hockney's suggestion was that artists were not necessarily cheating, or that all Old Masters were using optical devices to make their art, but analogous to the twentieth century influence of photography, billboards of the 1970s, high dynamic range imaging of the twenty-first century, or objects rendered using CGI, we begin to recognise characteristics or trends in the way pictures appear to us. Artists may be influenced by a particular characteristic when creating images, and likewise may be drawn to a particular method; for example, artists working on computer drawing tablets may modify their methods simply by the constraints of the pen and the tablet or the software they are using. One can quickly recognise how images are filtered or the software that has been used to create a particular style or manner, for example, the depth of field as viewed through a camera obscura, or the tonal range as viewed through a *Claude Lens*, or an image that is filtered in Photoshop. As Elkins suggests, our unconsciousness is persuaded and mediated by the things that surround us.

From artistic to commercial exploitation and mechanisation, drawing devices demonstrate a bridge between reproduction by hand, to reproduction by photochemical, and then photomechanical means. In the eighteenth century, the silhouette apparatus enabled an artist to create profile portraits. The sitter was illuminated by a spotlight, and paper placed behind them. The artist could then draw around the shadow. In 1786, the French musician Gilles-Louis Chrétien (1754–1811) introduced the *Physionotrace* machine, and up to twelve images could be produced at the same time using an engraving needle. Similar to a *Pantograph*, images could also be created to a different scale. The Pantograph was used widely as an engraving tool for engraving into metal (Case Study 4: Coins and Medals). The English potter and manufacturer Josiah Wedgewood (1730–1795) was able to create silhouettes of his sitters using chemical methods (although unable to make the images permanent), by painting the background surface with light-sensitive chemicals and illuminating the sitter against the light-sensitive background. In the early nineteenth century, inventor Charles Babbage (1791–1871) described two other drawing machines called the *Prosopographus* and the *Corinthian Maid*. By means of a Pantograph and Camera Lucida he captured the likeness of the sitter.

Early photographers attempting to capture scenes on to paper using silver chloride were frustrated by the fugitive nature of these images. Other experiments conducted by William Henry Fox Talbot (1800–1877) and Alphonse Poitevin (1819–1882) worked with pigments and presses to make their images more permanent. Realising the fugitive nature of photography, and to obtain a light-fast image, printers sought to experiment with pigments instead of dyes, and with plates and presses instead of sensitised paper. Printers found that by combining print and photography they could not only produce an image that would last much longer, but the image could also be reproduced many times and at greater speed. Printers at the end of the nineteenth century were able to reveal an astonishing range of pigment and carbon processes. The most reliable, and possibly the most beautiful, was the Collotype (Thirkell, 2000) and Woodburytype (Atkinson, 2005; McCallion, 2017), novel processes that used the light-sensitive qualities of bichromated gelatine to convey detailed and subtle continuous tones.

Since the fifteenth century, the colour print has evolved through an incremental process of comparison and commercial expediency (Gascoigne, 1997; Benson, 2008). Printing has progressed through the development of techniques that emulate processes, which were perhaps too laborious and time consuming, through halftoning and colour, the objective was to provide a near facsimile of the previous technology, but remained appealing and captured the elements of the process it was emulating. This can be demonstrated, for example, in the use of the mezzotint to emulate the colours and brush marks of a painting, lithography to emulate the marks of a woodcut or a photogravure to emulate the qualities of a photograph. One could suggest that the attainment of such qualities through mechanical methods could be considered to be a compromise of quantity, and speed over quality.

In 1881, a commercial method for printing an image using three colours was made possible by Frederick E. Ives (1856–1937), which he later refined in 1885 using a halftone screen. His technique employed the same principle as James Clerk Maxwell's (1831–1879) theory, of exposing an image through three coloured filters. Working from Fox Talbot's photogravure process, Ives developed a method for breaking down an image into small dots, but Ives used a more symmetrical matrix. He made a series of ruled opaque lines on to glass, each line of the same equal width, which was then placed on top of a second ruled glass screen and positioned at right angles, to give a pattern consisting of small openings on an opaque background. This method of halftoning – the process of translating an image into a series of graduated dots – has provided the basis for the way we give thought to images and, akin to Hockney's search for the 'optical look', the halftone such as *Benday Dots*, *Letratone*, stochastic screens, the halftone pattern may be considered as the optical look of the twentieth century. In the latter half of the twentieth century, printed colour has been further advanced by electrostatic print, enabling colour, text and image to be printed in one pass. As the refinement process of inkjet has attained higher levels of accuracy, a range of processes are able to emulate all the marks, colours and qualities of both photographic and traditional printer techniques; it has widened the scope for artists to engage with a new medium.

The next steps in the twenty-first century are the technological developments in colour and inkjet printing, for example, inks that can be physically deposited to create optical nanostructures (Yakovlev *et al.*, 2016) or printing organic thin-film transistors (OTFTs) that will play an important role in electronic devices, for example, e-paper, antennas, sensors, batteries and solar cells.

1.3.3 Reproducing Images: Colour and Texture

In this fast-moving industry, commercial and industrial manufacturers, makers of hardware and software require cross-platform standards and benchmarks. In order to bridge how images are reproduced, we need to understand how different surfaces can be captured, measured and reproduced. There are many potential different industrial and commercial uses and applications. The print industry is beginning to incorporate novel materials, decorative printing inks, textures and embellishments to enhance the surface qualities of packaging and prints. Printed textures are considered acceptable where the surface is purely decorative (i.e. repeat patterns for wallpaper). For some applications a low-relief texture is applied to photographic images, however a correlation between image and a convincing naturalistic rendering of texture has proven to be difficult. A well-used traditional CNC method to create a photographic relief is to assign a range of heights according to the tonal range of the image. This works well for greyscale but works less well as a coloured image. The conflict between texture and image is more apparent where there are contrasts, edge contours or attempts are made to distinguish relief from a flat picture plane. The appearance of false shadows and edges tends to amplify these problems.

As humans we inherently know what appears to be wrong, whereas computers have no such powers of perception. As demonstrated by halftoning, an image is converted into dots and the process of printing a full colour image is achieved by making a colour separation of each colour channel, after which each channel is converted into a greyscale image. When reprinted, using the four-colour process, colours will optically mix to appear as a full colour image (Burch and Gamble, 1983). On inspection of a print under magnification the dots are still discernable but at an appropriate viewing distance we perceive a continuous tone image. In comparison to an image that is printed, for example, by photocopier, inkjet or offset lithography, the clustering of the halftone dots will appear to be very different, often resulting in very different reproductions. To create a halftone reproduction of a textured surface, texture is implied by shadow and highlights, which is converted to dense areas of dots to create shadows, and absence of dots to create highlights. To obtain a similar colour and tonal range in traditional painting there are different methods to use; for example, Rembrandt used many layers of thin translucent glaze to create a deep black while Jan van Eyck applied thick dashes of white to indicate a highlight. Figure 1.9 illustrates the difference between a brush stroke, the density of a colour that can be obtained by overloading paint, the appearance of a brush stroke under magnification and a brush stroke that has been electrostatically printed and then captured under the same magnification.

1.3.4 Reproductions versus Forgeries versus Copies

In Dafen Oil Painting Village (http://www.dafenvillageonline.com) in the suburb of Shenzhen in Guangdong province China, an army of artists produce paintings and drawings on a mass scale. Highly skilled artists copy from Old Masters to produce artworks that may either resemble an image originally painted by Picasso, Manet or Van Gogh, or may be commissioned to incorporate unusual subject matter and incorporating painterly styles, such as a portrait of the customer or a picture of their wedding or favourite pet. Arranged along the walls of a nearby seventeeth century country mansion house, serving now as a function hall for weddings and conferences, are arranged Old Master oil paintings in ornate gilt frames. On further inspection these paintings are simply inkjet reproductions on to canvas, upon which brush strokes of varnish have been added by hand to obtain a surface texture. However, to the persistent eye, the brush strokes do not necessarily relate to the edges or sections of paint, but strokes are applied randomly.

How are these two examples connected? They are linked by our preoccupation with reproductions and artworks, the use of texture that shifts from what are ostensibly copies to convey something else – a facsimile that enhances an existing collection, a pseudo relic of the past, a long-forgotten treasure or to simply to surround ourselves with materials that move beyond the flat poster. Curiously, one of the paintings in the Ashton Court Mansion House is a copy of the *Portrait of Sir Thomas Gresham*; the real

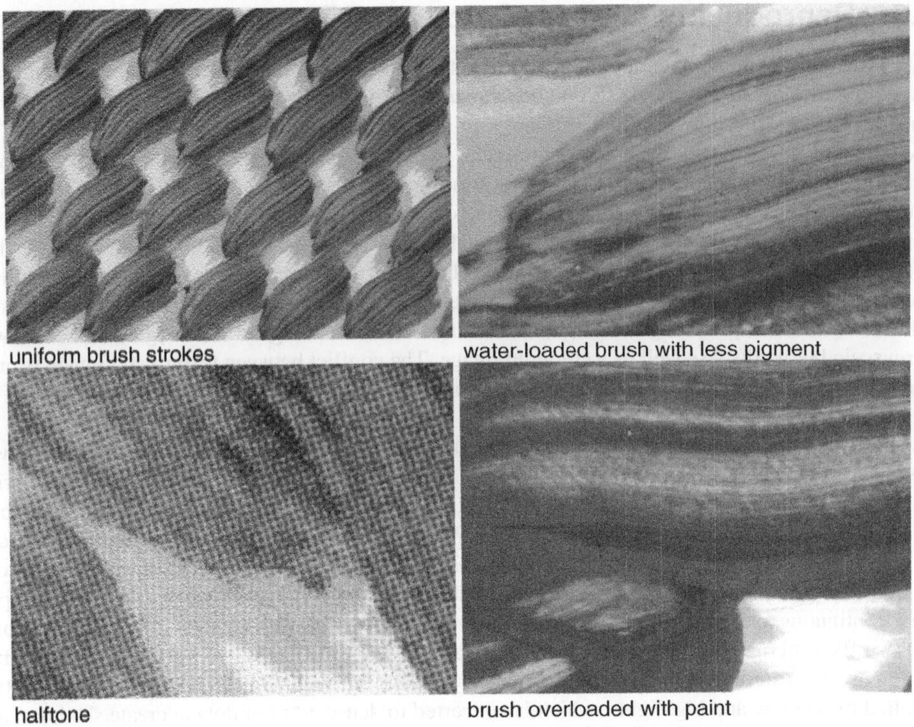

Figure 1.9 Darker brush strokes over lighter coloured uniform brush strokes, a water-loaded brush with less pigment, brush overloaded with paint, halftone at 40× magnification.

painting is exhibited in the Rijksmuseum and discussed in Chapter 2 (see Figure 2.30, Anthonis Mor (1517–1576), *Portrait of Sir Thomas Gresham*). Moreover, analgous to the artists working in Dafen Village, Figure 1.10 is a reproduction from a copy of the original painting. This reduced print version has lost much of the fine detail and mid-tones, and has more than likely faded, hence giving a greenish skin tone to Gresham's hands and face.

We are now highly accustomed to viewing works of art on screens, as postcards and printed reproductions. An established market for many museums and art galleries is the 2D print-on-demand fine art poster, enabling visitors to request any painting from collections that can then be printed at different sizes on different papers or canvas, one can even browse entire collections online without putting a foot inside a museum.

There is also a commercial initiative by museums to accurately reproduce paintings with colour and surface texture of brush marks (https://www.metmuseum.org/blogs/digital-underground/posts/2013/3d-printing). Museums have used high-tech scanning methods to minutely capture the *x-y-z* data of the surface (as described in Chapter 3, scanning methods to accurately capture the data and materials and methods to print them). When scanning paintings for conservation purposes, what then can be done with all the data? There remains a fundamental difference between paintings made by artists and a digital relief printer. In current digital relief printing a topographical height map – similar to the contours one would see on a map – a digital printer applies texture and then a layer of CMYK colour is added. The painter applies colours that are bound in mediums and materials to create texture, adds different directions to strokes, and layers of colour where needed. The aim of the artist is to strategically place brush strokes that direct the viewer's focus on particular objects or to generate an energy through the application and overlayering of marks, by orientating the brush marks in different directions to create perception of depth or specific regions.

Figure 1.10 A copy of a painting of *Portrait of Sir Thomas Gresham* in Ashton Court Mansion House, Bristol. The original is in the Rijksmuseum, Amsterdam.

There is of course a difference between forgeries, copies and reproductions. Forgeries are made to mislead or dupe buyers into believing they have purchased an artwork of value. The move from poster art and inkjet to more textured artworks may provide the consumer with a facsimile as close as possible to the original, which may even provide insights to the magic or aura of authenticity. The painted works produced in Dafen Village, although more than likely executed by a highly skilled painter, do not purport to be originals but are works that imitate a painterly style. It is interesting to note that copying, capturing and even studying the ways fakes and forgeries are made may provide insights into how painters paint, for example, how images are composed, the range of brushes and drawing tools that are needed to create different marks, and which may also provide insights as to whether a painting is authentic.

1.3.5 Are Facsimiles and Replicas Important?

Whether one considers copies and reproductions from a painting as art, fake or forgery depends on the intention of the maker, seller, dealer and collector. Are computer-generated brushstrokes from a digital scan or artists working at Dafen Painting Village realistic or more convincing than each other? These artworks sit in an uncertain position in the commercial art world: Are they art? Experiment? Entertainment? Do they hold any value? They are not unique and so what is their place to a museum or a collector? Today most consumers are eager for things and experiences that are authentic.

Museums have undergone a radical change since the 1970s, incorporating stunning new spaces by building in to traditional buildings or the creation of brand new buildings. In this climate of 'new museology' and through increasing social change and education, these are no longer just places for the elite, but museums are increasingly about openness and social inclusion that appeals across disciplines. Although museums are still in charge of caring and preserving a nation's cultural heritage, this new openness has enabled the curator to organise and present their cultural artefacts in new ways, by the 'breaking down of established boundaries … to some transgressing of the boundaries of specific subject disciplines' (Ross, 2004, p. 96).

Technology provides us with new insights as to how a painting or an artefact is made. 3D scans and computer-aided methods provide us with new methods to recreate them. They can also be used to improve our experience of exhibitions on multiple levels – popular/scholarly, aesthetic/critical – and also to be able to contextualise painting and artefacts in a new light (Freedberg, Jackson-Stops and Speer, 1987). Museums also incorporate facsimiles of artefacts by assisting the curator in completing their display or to create a narrative for existing objects. It is unlikely that the majority of the public are even aware that some artefacts are replicas, as it is in the skill of the artist to create realistic matches to existing sculptures and coins. Possibly the most well-known and iconic facsimile is the replica of the *Rosetta Stone*, which is in the Kings Gallery, British Museum, London. The original stone is just a few rooms away. George III officially donated the stone to the British Museum in 1802. Even in the 1800s, the preservation of the stone had always been a concern, and so it was placed behind glass. However, after complaints from visitors who wanted to touch this iconic stone, a compromise was found and a replica made, which is now in the Kings Gallery. Indeed, this replica has assumed its own significance, history or aura, and during its many years as a facsimile, has required regular attention and maintenance. It is interesting to note that even though it clearly states it is a replica, visitors still wish to photograph and touch it.

In the late 1980s, author and playwright Alan Bennett created a fictional conversation between two real-life characters – Queen Elizabeth II and Sir Anthony Blunt. In real life, Blunt was the Surveyor of the Queen's pictures, but was later exposed as a Soviet spy and stripped of his knighthood. The scene takes place in the Royal Picture Gallery in London, during which they discuss their fascination for forgeries. The Queen asks Blunt why the Vermeer paintings by Han van Meegeren, painted in the 1940s, were not exposed at that time as forgeries. It seemed inconceivable to the Queen character that these 'terrible daubes' were anything but fake. However, the Blunt character suggested that even the most brilliant forgery of their time, even though vetted by experts, they would eventually 'come to be seen as old fashioned'. Hundreds of works had been forged by van Meegeren and had been authenticated by important historians and critics of the time and caused a great deal of embarrassment and paranoia regarding forgeries. Alan Bennett was referring to the Dutch forger Han van Meegeren (1889–1947), who was able to imitate the works of the Old Masters including Frans Hals and Pieter de Hooch, earning him the equivalent of tens of millions of pounds (in today's money). He also claimed to be the forger of a piece initially authenticated to be a piece of work by Vermeer *The Supper at Emmaus* (1937), which had originally passed hands from van Meegeren via a Nazi banker and art dealer to Hermann Göring, and was finally discovered along with thousands of other looted works. Yet, as highlighted by the characters in Bennett's play, 'it is inconceivable to think it is not what it claims to be'. Artworks that are so readily thought to be the real thing in one time and context, that based on new knowledge gained through new technologies, can impact our perception as to how artworks appear today (Morris, 2009). These comments, although fictional, could be compared to Hockney and Elkin's observations that on seeing old artworks from a fresh perspective, as if for the first time new insights would be revealed.

In the UK, Tom Keating was regarded as an renowned forger of the twentieth century. Keating, who was also an art restorer, would have gained insights into how painters created their works and appropriated their methods to make new or minor art works and through the late 1950s and 1960s, placed his forgeries in auctions in the UK and with London dealers. He is reported to have forged over two-thousand paintings by around hundred major artists. An interesting addition to Keating's forgeries was his inclusion of an invisible signpost in his pictures. He started by daubing rude comments in lead white on the canvas before he started the painting, knowing that if the work was X-rayed, the lead white would appear. In 1987, John Drewe, responding to a small advert in the *Private Eye* magazine in 1987, conceived a plan to defraud the artworld with artist John Mayatt, who could turn his hand to painting many styles of nineteenth and twentieth century painters. Described as the 'the biggest art fraud of the twentieth century' by Scotland Yard in London, Drew and Mayatt created and sold around two-hundred forgeries through private dealers and auction houses (Battersby, 2014). Described as the 'one man Renaissance', Shaun Greenhalgh was jailed in 2007 and made hundreds of forgeries, including, paintings, ceramics and metalwork (Greenhalgh, 2015). He explained that he always put a 'tell' into the works, so that to a trained eye, would show they were not real. The work *La Bella Principessa* was attributed to be by Leonardo da Vinci, but has also been claimed by Greenhalgh to be his own work. The *Bella Principessa* is a portrait of a young woman in profile wearing a fifteenth century style dress. Greenhalgh claims the portrait was drawn by him in 1978 of a young woman called Sally, whom he had met in Bolton in 1975 (Jones, 2015). The provenance of the work remains contested.

One may wonder how many works are in museums and private collections that have passed as real. The long-running BBC programme entitled *Fake or Fortune* considered the science behind Old Master paintings. Using multispectral scanning or X-ray may reveal hidden clues and hidden information about a work. According to the BBC there are over 17 500 'unattributed' oil paintings – or by an 'unknown' artist – in UK public art collections. A *sleeper* is a term commonly used by art experts to describe a painting that has been miscatalogued or just overlooked, in a large collection (http://www.bbc.co.uk/programmes/b01mxxz6).

Fakes and forgeries remain in the public consciousness as works made by a roguish miscreant who steals from the rich and hoodwinks the experts, who dabbling in forgeries and, having served their time, are then invited on television programmes to reveal their secrets. The flip side is the impact on those who purchase forgeries, those who assumed they had purchased an investment only to find their purchase seized and destroyed thus losing thousands or millions when it was found to be a fake. The difference here is the relationship of art as a commodity that is bought and sold through dealers and auction houses or art that is enjoyed for the sake of its pictorial qualities that may be a copy or a reproduction.

Of course, throughout history there have been countless small workshops and large factories intent on profiting from any number of consumable articles, intent on the production of fakes and forgeries, including drugs, banknotes, handbags and artworks. In particular, inkjet printed artworks present a new opportunity for the forger. In order to address issues relating to reproduction of digital prints, one approach has been to tag artworks, whereby each print is allocated its own security code and DNA, which can then be linked to a certificate of authentication. In order to prevent attempts to copy the tagging many layers of complexity are added to the tag (www.tagsmart.com).

1.3.6 Copying from Old Masters

According to Josh Kaufman it only takes twenty hours to learn a new skill (Kaufman, 2013). Experts in a particular field may well disagree; asserting it takes a lifetime. However twenty hours may be enough to give the learner an idea about the complexities of the skill. Analogous to the way a mimic chooses to portray a famous character, they may only need a recoginsable attribute, for example, their voice, a gesture, or their posture. Today there is less access to learning from so-called Old Masters, but teaching is replaced with a myriad of how-to books and online teaching or just by searching YouTube. Likewise, we now have the opportunity to gain insights into how artists use different tools to create images, and how artists observe and reproduce different textures in paintings, prints and drawings. Secondly, we can also gain an understanding of artworks made by forgers and copyists, as they have studied brushstrokes and signature styles of the Old Masters. The objective for our study, is to gain an understanding of how forgers and copyists set about capturing the qualities that are important for a convincing likeness.

Copying from Old Masters was traditionally a well-established method for training the classical student and a direct way to learn the techniques of painting and drawing, understanding of tone, perspective and anatomy. Students began by observing and copying directly from their master, or as an alternative teaching aid by drawing from plaster casts of classical sculptures. Students in the Academies were only allowed access to life models once they could demonstrate proficiency in drawing from casts (http://www.ashmolean.org/departments/castgallery/about). Leonardo da Vinci, who had trained in the workshop of Andrea del Verrocchio (1435–1488) in Florence, learnt by observing and copying the work of his master. According to Vasari, Leonardo had painted one of the two angels in Verocchio's *Baptism of Christ* (1475). Verocchio was so overcome by da Vinci's painting that he declared he need never paint again. Following the same teaching style, similar evidence shows students' attempts in copying the work of Leonardo. A stunning example recently revealed at the Prado Museum is a copy of Leonardo's *Mona Lisa* that was undertaken by his apprentice. In the studios of established masters such as Velázquez and Rembrandt van Rijn (1606–1669), technicians were trained to perform the more laborious tasks including backgrounds, patterns of dresses, leaving the master to undertake the finer aspects of a painting. In the eighteenth century, in order to gain an understanding of the technique of drawing, J.M.W. Turner (1775–1851) set his students the task of copying from his own drawings. Stephen Farthing, curator of *Drawing from Turner* at Tate Britain (2006) explained that the importance of copying from Old Masters in the Academies was 'not to forge, but to try to understand' (http://www.tate.org.uk/whats-on/tate-britain/exhibition/drawing-turner/drawing-turner-copying-turner).

1.3.7 Technical Examination of Artworks

Scanning of artworks is a significant component for understanding the health of a painting, to assist in conservation and to understand how a painting was made. Galleries and museums are increasingly using digital scanning methods to measure the surface of paintings (Case Study 5: Capturing Texture of Paintings for Museum and Heritage), as demonstrated by Tim Zaman *et al.* (2014), who was working with the Mauritshuis (www.maritshuis.nl), the Kröller-Müller Museum (http://krollermuller.nl/visit) and the Rijksmuseum (https://www.rijksmuseum.nl/en) to undertake the scanning of the works by Rembrandt and Van Gogh. Then the Canon Océ Group made a series of high-resolution colour 3D prints (http://www.3dartistonline.com/news/2013/09/3d-printing-fine-art-reproductions). The digital scanning revealed the sculptural qualities of the brushstrokes and different brushes and tools employed by the artists (see Figure 1.11).

In 1968, The Rembrandt Research Project (RRP) (http://www.rembrandtresearchproject.org) was established in order to reduce the huge number of supposed Rembrandt forgeries. It used specialist scanning and art historical investigations to categorise and discover new facts about the works of Rembrandt. This was a mammoth undertaking, resulting in six volumes entitled *A Corpus of Rembrandt Paintings*, of which the sixth is a revision of the collection (http://www.springer.com/series/6494). As a result, through re-evaluating previously attributed works of art, the RRP

Figure 1.11 Textured reproduction of Van Gogh's *Flowers in a Blue Vase* (1887). The high-resolution 3D prints were made by the Océ Canon Group (courtesy: Tim Zaman).

became the authoritative organisation for the identification and authentication of Rembrandt's works.

The method of assessing its provenance was based on both technical examination using scientific aids, such as X-ray radiography, infrared imaging, alongside traditional and historical insights and connoisseurship using, 'judgements [...] largely based on intuitively applied criteria concerning style, brushwork and quality and taking into account every piece of evidence, that whilst not conclusive if taken separately provided a more compelling account' (White, 2015). Whilst the Rembrandt Database has incorporated a bulk of the RRP, the aim of the joint initiative of the RKD (Netherlands Institute for Art History) (www.rkd.nl) and the Royal Picture Gallery Mauritshuis (www.mauritshuis.nl) in The Hague, the Netherlands, is to work as a repository for all research undertaken into Rembrandt. In terms of documentation it lists a wide range of infrared, radiographic, other imaging such as stereomicroscopy, raking light, ultraviolet light, sample analyses and analyses of canvas and panels.

One could also ask, 'with all this data available, what could experts do with it?' As well as identifying forgeries, how could the data be useful to understand how Rembrandt painted? Did he use a particular brush stroke or geometry for creating a face? During a recent collaborative project entitled *The Next Rembrandt* (https://www.nextrembrandt.com) between Microsoft and ING, Delft University of Technology in the Netherlands, Dutch museums Mauritshuis and Rembrandthuis posed the question, "What would a new Rembrandt painting look like if we could use all the existing data?" According to the eminent Rembrandt expert, Ernst van de Wetering of the RRP, there is 'no such thing as a typical Rembrandt, each painting is unusual in its own way' (White, 2015). The goal of the project was to set about creating an entirely new picture based on the style of Rembrandt and using 168 263 Rembrandt painting fragments to create a new picture. Joris Dik of the Technical University Delft, when considering the methods used to create a new Rembrandt, started by using all the existing data and then designed an algorithm to identify and classify the most typical features. They also used data from the 3D scans to create texture that emulated the brushstrokes, 'We looked at a number of Rembrandt paintings, and we scanned their surface texture, their elemental composition, and what kinds of pigments were used. That's the kind of information you need if you want to generate a painting by Rembrandt virtually' (www.nextrembrandt.com). The data-generated file was printed using an Océ colour UV flatbed Arizona printer, reproducing all the colours and brush strokes that were digitally generated from the data. Thirteen layers were printed to create the texture with a coloured layer on top to add the detail. The result is startling and appears to be of a subject and style painted by Rembrandt. What might we learn from this, or is this just an interesting project? If a visitor to a gallery is unaware of the painting as a construction, how might they connect with the painting as an artwork *experience*?

1.4 The Authenticity of the Image and Object and Perception of Things

Throughout ancient civilisations, different cultures and groups have incorporated additional elements and dimensions to decorate utensils, jewellery, buildings and clothing. Learning skills and trades through manual and technical copying, trial and error, are examples of a transfer of traditional skills from maker to maker, material to object. Clothing was designed to be functional, beautiful and decorative, and using materials that were readily to hand (skins, fur, feathers, woven materials). A visit to a museum of artefacts, such as the Pit Rivers Museum in Oxford (https://www.prm.ox.ac.uk), reveals drawers and cabinets that are crammed with strange and curious archaeological and ethnographic objects from around the world. This museum could be considered as a rich resource dedicated to material culture, demonstrating how objects and artefacts were created with whatever materials were to hand; textured surfaces with bas-, mid-, and high-relief have been a constant denominator between cultures.

As demonstrated in Figure 1.12, 3D printing used for fashion, an eighteenth century protective waterproof coat made from animal skin or a cloak made from feather for rituals, each demonstrates a

Figure 1.12 (Left) Dress from *Voltage Haute Couture* (2013), Iris van Herpen in collaboration with Philip Beesley made from polyester microfibre and polyester foam. (Iris van Herpen is one of the leading fashion designers using 3D printing to create flexible 3D printed dresses. Her collaborations include material collaboration with Philip Beesley, Neri Oxman of the MIT Media Lab, as well as Keren Oxman and Professor Craig Carter of MIT with Stratasys, and architect Julia Koerner with Materialise.) (Middle) Gut Parka (circa 1980) waterproof coats made from the intestines of sea mammals. They were originally associated with feasts and rituals and considered as garments that possessed the power to protect the wearer from misfortune and protect them from bad weather, Alaska. (Right) Cloak of feathers from Kiwi birds (circa 1914), New Zealand.

different material and tactile quality for different applications. In contemporary culture, designers continue to work with many different materials and methods for fabricating objects. A visit to a contemporary design museum, a couture house, or end of year student show will demonstrate the many ways artists, craftspeople and designers are drawn to combinations, richness and varieties of colours, materials and textures. How do high-value items become translatable into objects that are designed for reproduction and mass consumption? Does mass consumption require a cheapening in quality? What are the materials and surfaces that we are drawn to? Or would wish to keep? A quick look around a typical office raises the question: can we improve on the things that surround us? Modernity (printing, photography and cinema) has enabled the mass reproduction of objects, eliminating the authenticity and authority they initially had. In technical processes we use cameras and scanners that would allow us to see things that were invisible to the naked eye and that change inevitably the perception of the actual object. Technical reproductions such as lithographs, printouts and videos would place the copied object in a remote context far from the original (Benjamin, 1936).

It seems that culture has moved to a place where the average is the commonplace, uniqueness has been replaced by universal equality and authenticity ceases to be applicable to artistic expressions. Do reproductions lack a sense of aura? Are these just empty containers without a place in material culture or tradition? If this is the case, is there any way of finding anthropological value in reproductions?

If we work under the prior assumption that authenticity can be culturally constructed, that is of a socially intangible value, we may then be able to bring to life reproductions by involving entire communities into their creation. The sense of accomplishment attached to reproductions adds value to them (social relationships with objects) and it could possibly validate their authority or generate a new aura (ACCORD is an Archaeology Community Co-Production of Research Data, Stuart Jeffrey, https://accordproject.wordpress.com). Our visual system has been trained to recognize materials based on their appearance and even though we may be easily deceived by *trompe l'oeil* and optical illusions, we are able to compensate for our weaknesses readily. Just by looking at objects we form ideas in ourselves of their material composition. For instance, we quickly make estimates of different material properties such as uniformity (rough, porous, coarse, smooth, grainy, bumpy, corrugated, spiky, slippery), luminosity (cloudy, opaque, matte, brilliant, shiny, reflective, transparent), hardness (soft, stiff, flexible, elastic, brittle), viscosity (dense, solid, liquid, jelly, oily, firm), materiality/substantiality (silky, flaky, fibrous, feathered, interwoven, crystalline) and temperature (warm, hot, frosted, icy), among others.

We are capable of detecting many complex stimuli from an object, but our brain can resolve just a certain number of them. Elements such as perspective, translucency and colour, texture and shape-metamers can play tricks on our perception. Artists have long known this and have challenged the laws of physics to take advantage of our confusion (Case Study 7: *Trompe l'Oeil*). By mixing pigments, materials and painting techniques, they have been able to recreate the sensation of velvet, the depth of a room, the lightness of veils or the influence of forces such as gravity. How do we find the right balance before it becomes vulgar or cartoonish? Maybe that is what separates artists from amateurs.

Given the inability of controlling viewing environments outside our laboratories or workshops, we need to have a wide range of parameters in our creations that are still pleasing under different viewing conditions. For example, velvet still needs to be perceived as velvet no matter what the lighting source or the viewing position is. Achieving such a wide bandwidth of perceptual attributes in our creations demands a high level of understanding of our limitations.

Preliminary studies have shown that human observers are willing to make compromises regarding print quality in 2D prints but are stricter when it comes to the evaluation of relief or 3D prints. It seems that the closer we get to representing reality the more demanding and pernickety observers become. When we look at 2D representations of the real world, we know (consciously or unconsciously) that they are just that: representations. Therefore, we seem to be more forgiving and more ready to make concessions with a lack of detail and misrepresentation, as they are practically just a 'caricature' of our reality. On the other hand, we are more susceptible to artefacts in representations made in higher dimensions.

When we are called to judge the accuracy of the reproduction of scenes and objects, we tend to mix our understanding of them with our association and memories. Are they linked to a pleasant experience? Were they part of our childhood? Do they bring us back to a bad souvenir? Our understanding of things is based on our quotidian experiences with them. Nevertheless, we can still perform and rely on psychophysical experiments for evaluation as there are ways to avoid these limitations to a certain degree. A great deal of research has concentrated on making recommendations about test images, viewing conditions and the kind of questions that need to be asked in order to have successful psychophysical test results with a very low variability.

Intuitively, reproducing shapes and forms gives the impression that it is easier (from a technological point of view) than reproducing appearances but, as one can imagine, mathematical (objective) accuracy in reproduction does not guarantee a successful reproduction. This can be due to the mathematical part not being mature enough to quantify the subtleties of surface perception or because measuring perception of reproduced materials is so convoluted that technical accuracy will never measure what the subject entails.

We argue that accuracy is not necessarily required for most reproductions, as a certain degree of interpretation of things may be more valuable and more welcome than a cold replica. A clear case where interpretation is fundamental has been investigated by Studios Durero and their Didu project (http://www.estudiosdurero.com/estudios_durero/opencms/Presentaciones/didu.html?idioma=en), which attempts to make Old Master paintings accessible to visually impaired people through relief prints. Their approach could be considered as semantic – what is the salient information that is important? Their designers work along with art specialists and a group of visually impaired people to either exaggerate textures already present in a painting or create new ones that would be visually relevant but need to be simplified to be understood by blind participants.

1.5 Current Industrial and Mechanical Methods to Reproduce the Appearance of Texture

In the age of industrial reproduction, printers have struggled to represent natural surface textures that can already be achieved so creatively by artists. From a technological point of view, our task is to observe the behaviour, gestures, tools and techniques artists have mastered across many centuries and to programme our machines (printers, displays) to develop similar tricks to transform their resources into surfaces that are perceptually analogous. The interest is no longer in reproduction of fine art prints but using similar concepts involved in such creations to be able to reproduce the appearances of different materials.

As part of the course of (technological) evolution of humans across different fields, technology is beginning to study ways of reproducing and communicating material perception in the same way that nature and artists have previously mastered. To some extent, resources in printing are still highly limited as the robot is unable to emulate the movement, nuances and autographic gestures of human fingertips, hands, wrist and arm that is adept at holding a pencil or brush. However, for the human body to learn a craft or a skill may take many years – as exampled in the art of Chinese calligraphy, where children start to learn at school and then move to being an apprentice to a master. The human body may become adept to almost every circumstance given time and practice. However, to emulate craft through reproductive methods, technologies, substrates and inks may be developed to achieve similar results. Something that plays to our advantage is that as our senses are equally limited in finding the right recipe for the blend is the motivation for current and future research.

Therefore, why try to mimic existing textures industrially? What is industrial reproduction good for? Parallels can be made with the transition from analogue to digital printing. In this new era packed with personalised items and easy access to technology we can anticipate more and more creative opportunities and new methods for artistic and decorative applications. Technology and art now have common grounds that morph rapidly, making it difficult to keep up to date. New trends come and go, but what remains is the idea of being able to recreate existing stuff based on or with other existing stuff.

Artists took centuries to master dimensionality and transparency and one could wonder how long it will take us to replicate that, and, looking to the future, how long it will take us to master material appearance reproduction in contemporary reproductions. The process is somewhat overwhelming, maybe due to the huge amount of variables involved, maybe due to the flow of virtual of information available everywhere in the planet or perhaps because of the different technologies available to play around and experiment with in a more creative sense.

Nevertheless, current 3D print researchers have taken a more pragmatic approach, leaving behind the aesthetics to produce functional pieces that were typically constructed by costly industrial processes. Printing technologies are now moving away from the reproduction of entire objects to concentrate on macrotextural elements and some are concentrating on microtextural aspects that will allow printing to conquer the appearance of reproduction fields in the future. For the moment, there is no magical solution to mimic relatively simple but typically hard-to-reproduce materials such as velvet and hair, not only in printing but also in computer graphics where the images on screen still look synthetic.

1.5.1 2D Printing Methods

In planar printing we can try to emulate the artists' approach to texture reproduction by controlling the way the ink interacts with the media. The printing industry has made an effort to provide more dimensions for creators by moving beyond the cyan, magenta, yellow and black process inks, and proposing distinct colours, translucency levels, and gloss options. Moreover, the usage of media with different rugosity properties is currently supported by many printing processes. However, most artworks have a dimensionality that cannot be conveyed with a flat 2D print. To go beyond planar printing, we can print different layers of ink to control the surface and subsurface scattering of light and therefore generate different visual effects, similar to the ones seen in artists' conceptions, and even decide to either keep or discard the effect of the supporting substrate.

For centuries most of the first printing technologies were based on the use of plates and blocks that contained the information to be printed in relief. Such plates were covered with ink and pressed against the media to produce the printout. Japanese techniques such as the tonal woodblock printing of Ukiyo-e would even allow the use of different colours in the same print, where the colour range of the final print depended on the choice of the printer and the complexity of the image. As newer alternatives for printing appeared, an interest in the printing of special effects developed. After graphic design consolidated as a profession during the beginning of the twentieth century, digital techniques to create different textures such as embossing emerged. The packaging industry (offset printing) started to use textured zones in the wrapping of products to make them more appealing because tangible materials leave a deeper imprint in the brain as it involves more emotional processing. Presses composed of several stations are capable of adding different effects by gluing foils, marking the media and printing varnish.

The advantages of digital printing versus analogue printing mainly consist of improved versatility, personalisation and better interfacing with other technological processes. The importance of continuing the development of digital machines that handle multiple printing materials and techniques is to keep these advantages of the digital world to cover different aspects needed for material reproduction in one (or a few) steps.

Printers and printmakers certainly appreciate the fine balance between image, matrix, ink and paper that may impact the surface dimensions, either by the translucency, opacity, gloss or viscosity of the ink, and thus the 2.5D qualities of the ink and paper relationship. From the three essential print processes (relief, intaglio and graphic), there has developed an enormous range of manual and subsequently mechanical print processes, including screenprinting, rotogravure and flexography. Each demonstrates different textural qualities, albeit on a microscale. Furthermore, the impact of the matrix or plate, such as the edge of a lithostone or embossing from a woodblock, letterpress or engraved copper plate, can also create added dimensions. Although inkjet printing is assumed to be a non-impact process, adding a nanolayer of ink to the paper by a skilled digital printmaker will change its surface qualities (Case Study 13: Analogue Printing Methods).

1.5.2 The Emergence of 3D Printing

Thanks to new technologies to render objects in multiple dimensions we have the potential to show micro details, texture, and different illumination settings to reproduce the qualities and characteristics of the originals. Regardless of the recent advances in the 3D market, the issues of printing high-quality details, colours and reflection effects is far from being solved. Many assumptions applied to 2D technologies may well have hampered the development of 3D printing, as the complexities of the 3D workflow have certainly been underestimated. Not only have printing techniques begun to languish since their creation, but also the development of higher-dimensional displays and scanners have been impacted and therefore the entire workflow has suffered an anachronistic evolution.

In a way, the printing industry has always served as the protagonist in historical, social and political revolutions, and we are about to experience a significant change in the way we look at, make and consume products over the next decade.

Today, the maturity of 2D printing technologies has enabled a switch in research priorities to 3D printing methodologies. The market of 3D print has found a place in the areas of rapid prototyping and additive manufacturing and the technological advancements have concentrated on building functional objects where aspects such as colour, detail resolution and other important material appearance properties have been neglected. Despite its aesthetic limitations, additive manufacturing provides the technology needed to unite materials and media.

The history of additive manufacturing dates back to the late 1950s and early 1960s when the first patents describing printing 3D objects were submitted (Munz, 1956). The US Patent issued to John

Munz in 1956 described a process that he called 'photo-glyph recording' to produce three-dimensional objects using a light-sensitive emulsion. In the 1950s and 1960s, the DuPont Company gained a number of patents for their experiments using two laser beams of differing wavelength to solidify a photo polymer resin. More than 20 years passed before the process gained any commercial interest when Charles W. Hull commercialised his invention called stereolithography in the 1980s (the term 'stereolithography' was introduced by Chuck Hull in his patent in 1986) and established 3D Systems. Other startups based on similar 3D printing techniques were launched (mainly in United States and Japan) but just a few have survived until the 2000s (Stratasys, ZCorp, 3D Systems). The lack of user-friendly applications and high costs of materials and processes seem to have influenced the initial delay in the development of this technology.

It was not until 2009 that an ASTM (American Society for Testing and Materials) Committee, called 'F42 on Additive Manufacturing Technologies', was established to join efforts in the standardization of terminology, design, materials and processes. In 2011, the ASTM F42 Committee and the ISO Technical Committee 261 on Additive Manufacturing announced an agreement of cooperation to avoid duplication of tasks. The involvement of two standardization communities shows the interest of the industry and encourages 3D print practitioners to continue their work. Regarding the colour and more general reflection aspects, the International Color Consortium (ICC) has also recently contributed by proposing a new way of thinking (iccMAX) about the typical colour management undertaken in 2D printing by including reflection properties that were less important before. Its adoption by the industry is still yet to be seen as not only the format of colour profiles change but also other modules in the classic colour pipeline or colour management system (CMS). Despite the efforts made until now, a universal file format and standard workflow to communicate reflection and geometric properties has not been established. Printers have either adapted a stereolithography CAD format created by 3D Systems called STL, or have their own proprietary format that works on a handful of machines. STL files describe only the surface geometry of a 3D object without any representation of colour or texture. A newer digital 3D file type that includes colour is commonly known as the Virtual Reality Markup Language (VRML), designed mostly for Internet and virtual reality applications. In 2005, the consortium Web3D proposed an update of VRML called Extensible 3D (X3D) but since then big players have left the consortium and created their own formats, namely, Intel with U3D, Microsoft with XAML and Dassault Systèmes with 3DXML. The amount of options regarding file formats, file compressing protocols, colour/features transformations and the variety of parameters available in the 3D field is overwhelming, not to mention the fact that quality standards to evaluate performance have not yet been put in place. Past experiences in printing, such as colour management back in the 1980s and 1990s show that the implementation of standards helps to benchmark products and processes as well as to communicate with customers and other makers. A full adoption of 3D printing and other 3D technologies may be hindered by the delays in putting together a clear standard workflow and quality targets. On the other hand, relief printing has been able to remain closer to what is already known in terms of file formats and standards, but the integration of 3D technologies into a relief workflow (or even the merger of both technologies) is a reality and we need to be aware of the weaknesses in the area.

Meanwhile the 3D printing industry continues to grow; since 2010 new and financially stronger companies have emerged with a range of cheaper printers (for example, Makerbot, Bits from Bytes) that produce more appealing content creation environments, whilst being user-friendly enough to attract a good percentage of the consumer market for easy upload to the leading 3D print bureaus (for example, Shapeways, Make XYZ, Sculpteo, Ponoko). Traditional players in the printing industry such as Roland, Canon and Hewlett-Packard offer subtle additions that can change the surface of prints, for example, to build surfaces with contrasting visual effects made of varnishes or special inks (metallic, translucent, opaque). In 2014, Roland, Canon and Hewlett-Packard announced their intention of moving into the 3D printing field. In 2014, Roland DG introduced a desktop ARM-10 3D Printer that could be used in conjunction with the SRM-20 desktop milling machine. In 2015 Canon entered into an agreement with 3D Systems and in 2016 HP launched their HP Jet Fusion 3D 3200 Printer. New markets have started to develop around the concept of 2D+ (2.5D, 3D) printing as an option to diversify their production and integrate different technologies into one single product. For example, in 2016, Ricoh launched the AMS5500P Printer, which can print using different materials to build functional parts for product testing.

On one hand, we would like to have more options of 2.5D and 3D printing materials but, on the other hand, there is the question about sustainability and ethical usage of toxic substances, petrol-based derivatives, and their biodegradability and recyclability. These concerns go beyond the printing materials as most 3D printers are 'energy hogs', as concluded by scientists at Loughborough University (2009) (http://sustainabilityworkshop.autodesk.com/blog/environmental-impacts-3d-printing), if compared to traditional casting, moulding, milling or machining. Moreover, the waste generated from the support materials in 3D printing is also more important than what is produced in classic industrial processes. Unhealthy air emissions due to plastic melting, solvents and toxic by-products, and possible contact with food or skin, are also significant factors.

New biodegradable printing materials are slowly becoming available, but their usage is not yet widely adopted. In 2014, 3D systems launched the Ekocycle Cube Printer, a machine that works with filament that is composed of 25% recycled material. An open source initiative called RecycleBot (http://www.thingiverse.com/thing:12948) converts household plastic waste into usable material to be fed into small-scale 3D printers. There is also a project called 3D Print Canal House (http://3dprintcanalhouse.com) that aims to 3D print an entire building with Macromelt, a printing material made of 80% vegetable oil. Despite these few efforts to use biodegradable options in printing, these alternatives may be less desirable, as these 'biodegradable' options are usually corn-based and other essential food sources that also impact on poor countries where knock-on effects such as food price rises and soil erosion can occur as a result. Furthermore, there remain other health, ethical and legal aspects yet to be addressed including: regulations, licensing, counterfeiting, bioprinting, gun legislation and drug abuse.

Aside from the complex issues surrounding environmental sustainability, there is also discussion about economic sustainability. Taking into account that the cost of 3D printers range from a couple of hundred to several hundred thousand dollars, printing materials also range from US$15 to US$800 per kilogram, and finished objects printed by a third party commercial vendor are an added cost. For example, to print a 3D object with dimensions of 15 cm × 4 cm × 4 cm (with a printed surface of 272 cm^2 and volume of 240 cm^3), could significantly range in cost depending on the type of material used. The following table provides a comparison guide.

Material	Process	Cost
Nylon (polyamide)	Selective laser sintering (SLS)	$7
ABS (acrylonitrile-butadiene styrene) or PLA (PLA (polylactic acid)	Fused deposition modelling (FDM)	$60
Mineral powder	Inkjet 3D printing or drop-on-powder	$220
Epoxy resin	Stereolithography (SLA)	$350
Alumide (polyamide powder and fine aluminium particles)	Selective laser sintering (SLS)	$420
Photopolymer resins	Polyjet matrix printing	$2000
Stainless steel	Direct metal laser sintering	$3500

This cost may increase depending on the finish, the addition of colour and post-processing. Although makers can benefit from no hardware costs and short runs, there is also the question of margins and profitability. In 2012 Amazon opened a 3D store called 3DLT so that customers could upload 3D files and buy their products, but it was closed a few years after. Around the same time Threeding and Pinshape emerged to facilitate the trade of 3D models, similar to the service that was expected from 3DLT. Another business model has been developed by Materialise and Prodways who specialise in additive manufacturing to produce prototypes for industrial partners or to provide the software to others with additive manufacturing capacities to produce prototypes for their own customers. Sculpteo and

Shapeways offer a range of online 3D printing services and a wide variety of printing materials. Other printing providers such as Project Eiger (http://www.projecteiger.com) and Estudios Durero (http://didu.estudiosdurero.com) provide 2.5D printing for a specialised range of applications.

We can find more examples in the open source community that provide repository and file exchange services to their customers. However, the printing applications and opportunities still feels underexplored and it seems for the well-established companies the actual printing exploration is left to hobbyists and geeks. Political leaders around the globe have added 3D printing as one of their priorities in the belief that there are opportunities for change and impact on current manufacturing practices. Barak Obama announced in 2013 that 3D printing had 'the potential to revolutionize the way we make almost everything'. In 2014, he hosted a *Maker Faire* in the White House to promote Fab Labs (fabrication labs). Neil Gershenfeld, the creator of Fab Labs, defines Fab Labs as places where we can make 'almost anything' that will bring us to 'a new digital revolution'. These labs are equipped with computer-controlled tools such as CNC, 3D printers, scanners and laser cutters. To illustrate his point he makes an analogy between Fab Labs and the history of computing. By the time personal computers were released into the home computing market, the World Wide Web had already been developed by networks of scientists to share their research on minicomputers many years before. This phenomenon resulted in almost no delay in integrating the two technologies and develop their applications. In the same way, Fab Labs emulate today what personal fabricators will eventually do in the future and 2.5D/3D printing is an important part of that. However, in order to understand 2+D printing as a part of the future of sustainable manufacturing, both from an environmental and economic perspective, the different technologies must support new models of sustainable businesses (circular economies) and supply chains and versatile pipelines.

Regardless of the experience gained during recent years, the many issues of quality in 3D printing are far from being solved. Formal studies into the different constraints that exist in 3D printing have only taken place in the last twenty to thirty years. During the recent explosion and hype in the press regarding the future of 3D printing and its applications, we have overlooked the important details of reality in exchange for fast production, functionality, costs and probably an easier expansion and acceptance of a printed 3D world. The macrogeometry of things has taken over the reproduction of objects, forgetting that our senses are used to seeing the real world with a pinpoint degree of resolution. For the moment, the printing industry has concentrated on making objects and prototypes that convey more aesthetic qualities and is just beginning to develop materials for printing objects that are functional.

Even though 3D printed objects are commonly limited to one colour, which is the same as the printing material, some technologies offer a reduced colour palette that can be used in a per-area basis. If more realistic colours or surface effects are required, manual finishing processes are used to achieve the desired appearance. Moreover, undesired rough surfaces also need to be used at a polishing/finishing stage. Besides the lack of quality (aesthetically speaking), the manual processes involved in the final step make a full colour 3D or polished object an expensive and requires the skills of the technician or craftsperson. Mastering the technical aspects of 3D printing is not enough for the production of artefacts, especially if related to cultural heritage, where a high degree of accuracy is required. Creating accurate replicas of historical objects is fundamental to ensuring the cultural, historical and archaeological context and as part of the narrative when curating museum objects.

1.6 Conclusion

In this introduction we have explored the relationship between images, pictures and reproductions. We have begun to explore different ways to describe 2.5D, the relationship between texture, material and object, our emotional and perceptual relationship with materials, how artists convey the appearance of materials and how materials can be measured and quantified. One can start by a negative and suggest that it is not about printing objects (3D) or about printing images (2D), but something between the two – or more simply put, the relationship between the surface and the substrate, and textural attributes of a material and the object.

2.5D printing has not yet abandoned several aspects borrowed from 2D printing but it has adapted them to the new challenges. Likewise, 3D printing is not ready to implement what 2.5D brings to the field as it adds another layer of complexity to the already complicated field of 3D. Therefore 2.5D printing remains for the moment as a standalone technology that may merge in parallel to 3D in the future for the sake of multimaterial, multifunctional, multiapplications, multidimensional, multiplatform printing.

If we consider 2.5D or relief printing as the bridge that connects 2D and 3D printing, then, on the one hand, we have 2D printing, which lacks enough degrees of freedom to cover the complexity of reality and, on the other hand, we have 3D printing with an overwhelming amount of variables and unable to concentrate on more subtle aspects. We as humans are complex subjects and our reactions are not only physical but also psychological. Our memory of shapes, colours, textures, our previous experiences, our cultural background, as well as our physical skills and limitations play important roles in the overall perception of an object. 2.5D or relief printing concentrates on studying the external layer(s) or skin of objects as this is the part that is more in contact with our senses. This superficial layer contains the microstructural information needed to understand material properties such as texture and reflection. Emulating the fine details contained in surfaces is the key to bridging the gap between an artificial and a natural reproduction.

References

Adelson, E.H. (2001) On seeing stuff: The perception of materials by humans and machines, ins *Human Vision and Electronic Imaging*, VI (eds B.E. Rogowitz and T.N. Pappas), The Society for Imaging Science and Technology.

Atkinson, A. (2005) The creation of digital photographic fine art print through the Woodbury type model, PhD thesis, University of the West of England, Bristol.

Battersby, M. (2014) The talented John Myatt: Forger behind the 'biggest art fraud of 20th century' on his criminal past – and how he went straight, *Independent*, 28th November 2014. [Online]: http://www.independent.co.uk/arts-entertainment/art/features/the-talented-john-myatt-forger-behind-the-biggest-art-fraud-of-20th20th-century-on-his-criminal-past-and-9889485.html.

Benjamin, W. (1936) *The Work of Art in the Age of Mechanical Reproduction*, Penguin, London.

Benson, R. (2008) *The Printed Picture*, Museum of Modern Art.

Berger, J. (1972) *The Ways of Seeing*, Penguin Books, London.

Black, J.T. and Kohser, R.A. (2011) *DeGarmo's Materials and Processes in Manufacturing*, John Wiley & Sons, Inc., Hoboken.

Burch, R.M. and Gamble, W. (1983) *Colour Printing and Colour Printers*, with a chapter on Modern Processes by W. Gamble and Introduction by Ruari Mclean, Paul Harris, Edinburgh.

Didi-Hubermann, G. (2004) *Confronting Images: Questioning the Ends of a Certain History of Art*, Pennsylvania State University Press.

Edwards, S. (2006) *A Very Short Introduction to Photography*, Oxford University Press, Oxford.

Elkins, J. (2000) *How to Use Your Eyes*, Routledge, London.

Freedberg, S.J., Jackson-Stopzs, G., and Spear, R.E. (1987) On 'Art History and the "Blockbuster" Exhibition'. *The Art Bulletin*, 69, 295–298.

Gascoigne, B. (1997) Milestones in colour printing, 1457–1859: With a bibliography of Nelson prints, in *Sandars Lectures in Bibliography*, Cambridge University Press, Cambridge.

Gombrich, E.H. (1959) *Art and Illusion*, Phaidon Press, Oxford.

Greenhalgh, S. (2015) *A Forger's Tale, Confessions of the Bolton Forger*, Allen and Unwin, London.

Hawkins, J.K. (1970). Textural properties for pattern recognition, in *Picture Processing and Psychopictorics* (eds B.S. Lipkin and A. Rosenfeld), Academic Press, New York.

Hull, C. (1986), Apparatus for production of three-dimensional objects by stereolithography. US Patent 4,575,330.

Jones, J. (2015) This is a Leonardo da Vinci? The gullible experts have been duped again. *Guardian*, 30th November 2015. [Online]: https://www.theguardian.com/commentisfree/2015/nov/30/leonardo-da-vinci-experts-painting-la-bella-principessa.

Kaufman, J. (2013) The first 20 hours: How to learn anything … fast!, Portfolio.

Kemp, M. (1990) *The Science of Art: Optical Themes in Western Art from Brunelleschi to Seurat*, Yale University Press, New Haven and London.

Kim, S., Hagh-Shenas, H., and Interrante, V. (2004) Conveying shape with texture: Experimental investigations of texture's effects on shape categorization judgments. *IEEE Transactions on Visualisation and Computer Graphics*, 10, 471–483.

Klatzky, R.L. and Lederman, S.J. (2010) Multisensory texture perception, in *Multisensory Object Perception in the Primate Brain* (eds M.J. Naumer and J. Kaiser), Springer, New York.

Kreit, E., Mäthger, L.M., Hanlon, R.T., *et al.* (2012) Biological versus electronic adaptive coloration: how can one inform the other? *Journal of the Royal Society Interface*. [Online]: Available at: http://rsif.royalsocietypublishing.org/content/early/2012/09/22/rsif.2012.0601.

Manchee, D. (2006) *20 Works in the Age of Mechanical Reproduction.* [Online]: http://www.dougmancheeprojects.com/20w.html.

McCallion, P. (2017) The development of methods for the reproduction in continuous tone of digitally printed colour artworks, PhD thesis, University of the West of England, Bristol.

McCann, J.J. and Rizzi, A. (2011) *The Art and Science of HDR Imaging*, John Wiley & Sons Ltd, Chichester.

Mitchell, W.J.T. (1986) *Iconology: Image, Text, Ideology*, The University of Chicago Press, London.

Mitchell, W.J.T. (2006) *What Do Pictures Want?: The Lives and Loves of Images*, University of Chicago Press, Chicago.

Morris, E. (2009) Bamboozling ourselves, *New York Times*, Blog 17th June 2009. [Online]: http://morris.blogs.nytimes.com/category/bamboozling-ourselves/.

Munz, J. (1956) Photo-glyph recording. US Patent 2,775,758 A, December 25, 1956.

Ramachandran, V.S. (2012) *The Tell-Tale Brain Unlocking the Mystery of Human Nature: Tales of the Unexpected from Inside Your Mind*, Windmill Books, London.

Ross, M. (2004) Interpreting the new museology. *Museum and Society*, 2, 84–103.

Thirkell, P. (2000) The integration of digitally mediated imaging techniques with 19th century continuous tone printing processes, PhD thesis, University of the West of England, Bristol.

Wax, C. (1990) *The Mezzotint: History and Technique*, Thames and Hudson, London.

White, C. (2015) The Rembrandt Research Project and its denouement. *The Burlington Magazine*, CLVII, 71–73.

Yakovlev, A.V., Milichko, V.A., Vinogradov, V.V., and Vinogradov, A.V. (2016) Inkjet color printing by interference nanostructures. *ACS Nano*, 10, 3078–3086.

Zaman, T., Jonker, P., Lenseigne, B., and Dik, J. (2014) Simultaneous capture of the color and topography of paintings using fringe encoded stereo vision. *Heritage Science*, 2, 1–10.

Bibliography and Further Reading

Alexa, M. and Matusik, W. (2010) Reliefs as images. *ACM Transactions on Graphics*, 29 (60) 1–7.

Ayres, J. (1985) *The Artist's Craft: A History of Tools, Techniques and Materials*, Guild, London.

Benjamin, W. (1936) *The Work of Art in the Age of Mechanical Reproduction*, London, Penguin, London. See also in Arendt, H. (ed.) (1970) *Illuminations,* Jonathan Cape, London.

Charney, N. (2015) *The Art of Forgery*, Phaidon, London.

Dougherty, W.R. and Reitnauer, A. (2001) Overview of industrial digital printing, in *Proceedings of the IS&T NIP17 International Conference on Digital Printing Technologies and Digital Fabrication*, Society for Imaging Science and Technology.

Elkins, J. (1996) *The Object Stares Back*, Simon and Schuster, New York.

Evans, B. (2012) *Practical 3D Printers*, Springer.

Faken, H. (2002) Inkjet printer and a method of printing a receiving material. US Patent 2,020,180,819 A1, OCE-Technologies, B.V., Netherlands.

Filip, J. and Haindl, M. (2009) Bidirectional texture function modeling: A state of the art survey, *IEEE Transactions on Pattern Analysis and Machine Intelligence*, 31, 1921–1940.

Ford, S.J., Routley, M.J., Phaal, R., and Probert, D.R. (2014) The industrial emergence of commercial inkjet printing. *European Journal of Innovation Management*, 17, 126–143.

Gershenfeld, N. and Vasseur, J.P. (2014) As objects go online. *Foreign Affairs*, 93, 60–67.

Gilmour, P. (1978) *The Mechanised Image*, London, Arts Council of Great Britain.

Gombrich, E. H. (1959) *Art and Illusion*, Phaidon Press, Oxford.

Gregory, R.L. and Gombrich, E.H. (1973) *Illusion in Nature and Art*, Gerald Duckworth, London.

Haindl, M. and Filip, J. (2013) *Visual Texture. Advances in Computer Vision and Pattern Recognition*, Springer-Verlag, London.

Heynderickx, I. and Kaptein, R. (2009) Perception of detail in 3D images. *Proceedings of SPIE 7242, Image Quality and System Performance VI*, 72420W.

Hiller, J. and Lipson, H. (2009) STL 2.0: A proposal for a universal multi-material Additive Manufacturing File format, in *Proceedings of theSolid Freeform Fabrication Symposium (SFF'09)*, Austin, Texas, 2009, pp. 266–278.

Hochberg, J. (1979) Some of the things that paintings are, in *Perception and Pictorial Representation* (eds C. Nodine and D. Fisher), Praeger, New York.

Hoskins, S. (2017) *3D Printing for Artists, Designers and Makers*, Bloomsbury Visual Arts, London.

Hoskins, S. and McCallion, P. (2013) Continuous tone colour printing in two and a half dimensions through a combination of 19th century analogue methodologies and 3D printing. *Proceedings of the IS&T NIP 29 International Conference on Digital Printing Technologies and Digital Fabrication*, Seattle, USA.

Hoskins, S. and Thirkell, P. (2003) The relevance of 19th century continuous tone photomechanical printing techniques to digitally generated imagery. *Proceedings of SPIE PICS Conference 5008*, pp. 390–398.

Huang, P., Hu, Q., and Chiang, F. (2002) Double three-step phase-shifting algorithm. *Applied Optics*, 41.

Ji, W., Pointer, M.R., Luo, R.M., and Dakin, J. (2006) Gloss as an aspect of the measurement of appearance. *Journal of the Optical Society of America A*, 23 (1), 22–33.

Johnson, D.H., Johnson, C., Klein, A.G., *et al.* (2009) A thread counting algorithm for art forensics. *Digital Signal Processing Workshop and 5th IEEE Signal Processing Education Workshop*, Marco Island, Florida.

Jürgens, M.C. (2009) *The Digital Print: The Complete Guide to Processes, Identification and Preservation*, Thames and Hudson, London.

Kirchner, E., Müller, P., and Njo, L. (2013) The influence of high luminance levels on perceived colour and texture differences, in *AIC Colour 2013, 12th Congress of the International Colour Association* (eds L. Macdonald, S. Westland, and S. Wuerger), Newcastle upon Tyne, UK, International Colour Association, pp. 1073–1076.

Kirsh, A. and Levenson, R.S. (2000) *Seeing Through Paintings: Physical Examination in Art Historical Studies*, Yale University Press, New Haven.

Korn, P. (2014) *Why We Make Things*, David R. Godine Publisher, Boston.

Koudelka, M.L., Magda, S., Belhumeur, P.N., and Kriegman, D.J. (2003) Acquisition, compression, and synthesis of bidirectional texture functions, *ICCV 03 Workshop on Texture Analysis and Synthesis*, Nice, France.

Lambert, S. (1987) *The Image Multiplied*, Trefoil, London.

Landy, M. (1996) Texture perception, in *Encyclopedia of Neuroscience* (ed. G. Adelman), Elsevier, Amsterdam).

Landy, M.S. (2013) Texture analysis and perception, in *The New Visual Neurosciences* (eds J.S. Werner and L.M. Chalupa), MIT Press, Cambridge.

Larson, J. (2013) *3D Printing Blueprints.*, Packt Publishing Ltd.

Le Blon, J.C. (circa 1726) *Coloritto, or the Harmony of Colouring*, London.

Lipson, H. and Kurman, M. (2013) *Fabricated, The New World of 3D Printing*, John Wiley & Sons, Inc., *Indianapolis.*

Lipson, H., Moon, F.C., Hai, J., and Paventi, C. (2005) 3-D printing the history of mechanisms. *Journal of Mechanical Design*, 127, 1029–1033.

Marr, D. (1982) *Vision – A Computational Investigation into the Human Representation and Processing of Visual Information*, W.H. Freeman, New York.

Marr, D. and Hildreth, E. (1980) Theory of edge detection, in *Royal Society of London, Series B, Biological Sciences*, Royal Society of London, London.

McCann, J.J., Parraman, C., and Rizzi, A. (2014) Reflectance, illumination, and appearance in color constancy. *Frontiers in Psychology*, 5, 5.

Michalopoulos, K., Tsakalakis, M., and Bourbakis, N. (2013) Detecting texture paths and patterns in aerial images, in *Fourth International Conference on Information, Intelligence, Systems and Applications (IISA)*, IEEE.

Nassau, K. (1998) Fundamentals of color science, in *Color for Science, Art and Technology* (ed. K. Nassau), Elsevier Science, Amsterdam.

Nodine, C. and Fisher, D. (eds) (1979) *Perception and Pictorial Representation*, Praeger, New York.

Pappas, T.N. (2013) The rough side of texture: Texture analysis through the lens of HVEI, in *IS&T Human Vision and Electronic Imaging XVIII* (eds B.E. Rogowitz and T.N. Pappas), p.12.

Parkinson, R B., Diffie, W., Fischer, M., and Simpson, R.S. (1999) *Cracking Codes: The Rosetta Stone and Decipherment*, University of California Press, Berkeley,

Parraman, C. (2001) Preserving the identity of digital prints, in *IMPACT 2*, University of Helsinki, Finland.

Parraman, C. (2013) Colour mixing in the 21st century, in *Colour in the Making* (ed. P. Stubbs), Black Dog Publishing, London.

Parraman, C. (2013) Colour in the age of digital reproduction. *Art in Print*, 3, 28–33.

Parraman, C. (2014) 2.5D printing, in *All About Imaging: Transactions*, Royal Photographic Society, University of Westminster.

Parraman, C., Walters, P., Reid, B., and Huson, D. (2008) Specifying colour and maintaining colour accuracy for 3D printing, in *Proceedings of SPIE 6805, Three-Dimensional Image Capture and Applications 2008*, The Society for Imaging Science and Technology.

Pedersen, M. (2011) Image quality metrics for the evaluation of printing workflows, PhD thesis, University of Oslo.

Rogowitz, B.E. and Rushmeier, H.E. (2001) Are image quality metrics adequate to evaluate the quality of geometric objects, in *Human Vision and Electronic Imaging, 2001*, The Society for Imaging Science and Technology, pp. 340–348.

Ronchi, L.R. (2005) On the color-texture interaction, in *10th Congress of the AIC*, Granada, Spain, pp. 879–882.

Ross, M. (2004) Interpreting the new museology. *Museum and Society*, 2, 84–103. p. 96.

Sorensen, R. (2008) *Seeing Dark Things, The Philosophy of Shadows*, Oxford University Press, Oxford.

Stafford, B. (1994) *Artful Science*, MIT Press, Cambridge, MA.

The Marketing Society (2012) Dulux, Marketing Communications 2008 – Case Study 'Colour Me Beautiful', in *The Marketing Society*. [Online] Accessed 2016: https://www.marketingsociety.com/the-library/2008-dulux-marketing-communications-case-study#BKhb7ljgrt6pFarj.97.

Turner, R. (1995) *Inventing Leonardo, The Anatomy of a Legend*, Papermac, London.

Weschler, L. (2001) Through the looking glass: Further adventures in opticality with David Hockney. *The Believer*. [Online] Accessed 2016: http://www.believermag.com/hockney/lookingglass.

Weverbergh, R. (2014) A 3D scan of Van Gogh to create the perfect replica [Online]. *Whiteboard*. [Online] Assessed January 2014: http://www.whiteboardmag.com/3d-scan-van-gogh-perfect-replica-formart.

Wijshoff, H. (2006) Manipulating drop formation in piezo acoustic inkjet, in *Proceedings of the IS&T NIP22 International Conference on Digital Printing Technologies and Digital Fabrication.*

Wynne, F. (2006) *I was Vermeer*, Bloomsbury, London.

Zeki, S. (1999) *Inner Vision: An Exploration of Art and the Brain*, Oxford University Press.

2

The Past

2.1 Introduction

In order to deliver high-quality materials, processes and methodologies that assist the twenty-first century designer/user in obtaining the desired look, feel and aesthetic qualities, the primary motivation of this chapter is to examine appropriate contemporary methods and tools towards the creation of novel 2.5D printed surfaces. To do this we pick up from the introductory chapter by looking back to historical methods and artistic techniques of surface relief for inspiration and benchmarking. By benchmarking we mean, what are the qualities (materials, methods, perceptions, tools) that frame our understanding of what constitutes 2.5D printing? To gain an understanding of how artists have interpreted and represented our world in 2.5D, as pictorial relief and as cultural artefacts, merely requires a visit to a museum, a gallery or an ancient building. Today, many ancient artefacts have been removed from their place of origin to museums in order to preserve them from ravages of pollution, theft or war. They have enormous significance in the historical evolution of ancient art. The craftsperson's primary objective was to translate stories into two and three dimensions. The important task was to choose the tools and materials that were most suitable to help them in their narrative. Their second objective was to create the most impact by reducing a complex scene into a simplified message. This may well have been informed by the types of materials to hand, the relationship of the artwork to illumination (candle, sunlight, natural light), context (exterior or interior, house or church, public or private) and intention (to inspire through worship or aid contemplation, to instruct or provoke, to create visual delight or unease).

There are so many books that provide a narrative on the history of art, its social contexts and art techniques. This chapter will aim to capture the briefest contextual background to ancient methods of 2.5D creation and representation, that provides the reader with suitable knowledge and insights as to the state of the art and qualities for twenty-first century requirements.

This chapter considers illumination, colouration and texture and its relation to the production of relief, as well as the conversion of 2D to different degrees of relief. In order to understand how planar images can be converted into relief, it is important to keep in mind the relationship between objects that are illuminated from a single point (paintings and drawings) and how these might be understood from many dimensions (relief and sculpture). The following section explores the different two-dimensional representational approaches that are used to create the appearance of texture and relief. It could be argued that 'flat art', such as paintings, drawings and prints, does not contribute to 2.5D printing.

2.5D Printing: Bridging the Gap Between 2D and 3D Applications, First Edition. Carinna Parraman and Maria V. Ortiz Segovia.

Companion website: www.wiley.com/go/bridging2d3d

Materials, methods of application and artistic devices are used to create an illusional appearance of texture and the use of illumination is an important aspect when considering how to compose images in preparation for textured or tactile surfaces. This is particularly relevant when converting images, for example, tactile pictures for the visually impaired (Case Study 3: 2.5D Printed Tactile Books and Artworks). The important criterion here is: what are the essential details in order to convey the narrative?

2.2 Artists' Observations on the Appearance of Illumination

Artists have been analysing the world through pictorial form for thousands of years. They have been observing and recording from nature, and are able to interpret and synthesise texture and objects in a highly convincing way. As visitors to galleries or those looking at picture books, observers are highly tuned to converting these illusory artworks such as paintings, drawings and photographs into three-dimensional scenes (Gibson, 1950, 1971), but the suggestion is that these methods of conversion are still a cognitive process (Arnheim, 1964, 1970). However, as observers, we are not necessarily tricked into believing that what we are seeing is real, but the imitative abilities of the artists' plausible illusion of nature. Through 'illusion and expectation' the task of the artist is to translate a scene as a pictorial composition, and through editing, to draw the viewer into the picture and bring a picture to life.

The aim for artists studying the genre of landscape painting was to tell a story as well as convey a sense of scale and depth in their works. Many painters worked on a grand scale, which included paintings on ceilings and walls, thus inspiring awe and wonderment. On the other hand, a minute observation as exampled by Canaletto (Giovanni Antonio Canal, 1697–1768) invited a more intimate relationship with the viewer. Mannerist painting was a particular style of art in sixteenth century Europe that originated in Italy, and artists including Michelangelo (Michelangelo di Lodovico Buonarroti Simoni, 1475–1564) and Jacopo Tintoretto (1518–1594), used many of the stylistic methods. These styles were adapted by other artists across Europe, including El Greco (Doménikos Theotokópoulos, 1541–1614) in Spain and Hans Baldung Grien (1484-1545) in northern Europe. Figure 2.1 is an example by Claude Lorrain (1605–1682), who was highly influential on landscape painters of the eighteenth and nineteenth centuries, including J. M. William Turner (1775–1851). Lorrain created figurative artworks set within stylised and imaginary landscapes. He composed his landscape scenes according to three visual fields, analogous to how we might perceive a landscape: the colours in the background are rendered as tints (light colours) and the information (mountains, horizon, ruined buildings) is hazy as if influenced by atmospheric conditions. In the middle ground the information and colour are more distinct, whereas in the foreground, the colour, the figures and objects are in focus, containing more detail and rendered with richer tones and colours.

Contrary to expectation, most artworks are remarkable in their lack of detail, or as Gombrich explains, where 'there is less paint there to explain and disturb' (Gregory and Gombrich, 1973). This is exemplified in the work by Leonardo da Vinci, who created paintings without lines or boundaries, or *Sfumato*, which softened the overall appearance of a scene, and is particularly apparent in the *Mona Lisa* (c. 1503), where the blending of her face is so translucent that the viewer's gaze glides across her features. In comparison to the so-called photo-real paintings of the 1970s and 1980s that dazzle the eye with distorted polished surfaces, interreflections, highlights and colours, artists were able to manipulate oil and acrylic paint with just enough information or illusion of detail to simulate a scene with startling photo-like qualities (Letze, 2013).

The subject of Charles Bell's painting *Paragon* (Figure 2.2) cleverly capitalises on the surreal artificial environment of the pinball machines, the highly saturated colours, coloured lights and

Figure 2.1 Claude Lorrain, *Harbour at Sunrise* (c. 1637–1638), oil on canvas (74 cm × 98.3 cm) (courtesy: Rijksmuseum, Amsterdam).

Figure 2.2 Charles Bell, *Paragon* (1988) oil on canvas (127 cm × 244 cm), Louis K. Meisel Gallery, USA (courtesy: Louis K. Meisel Gallery).

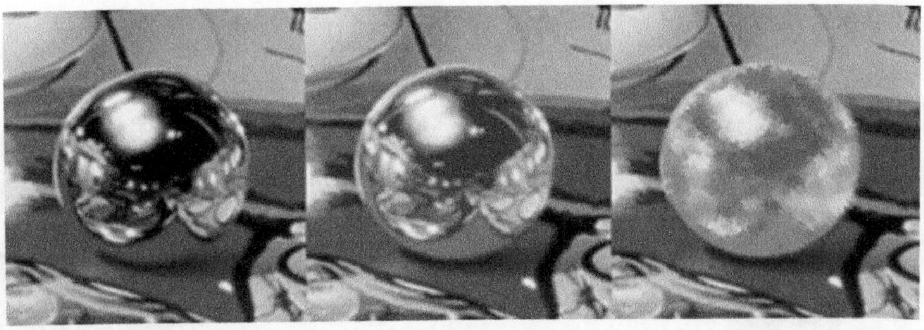

Figure 2.3 Example of simulated different rendering effects of Bell's pinball.

interreflective surfaces. The highly reflective pinball demonstrates the qualities in material perception, namely 'rich surface perception' that refers to the appearance of complex surfaces in the real world, viewed in the complex illumination of everyday life (Adelson, 2008; Fleming, 2013). Figure 2.3 illustrates how the original pinball in Bell's *Paragon* might appear if rendered to be more matte or textured.

Likewise, Richard Estes has captured the appearances of urban and artificially lit environments and has rendered the interreflective surfaces of glass and metal, the saturated warm illumination of low cast sunlight, the cool blue hues of a midday sun, or the artificial glow of neon lights. Such examples include *Telephone Booths* (1968), an oil on canvas (121.92 cm × 175.26 cm), or *Broadway Bus Stop, Near Lincoln Center* (2010) an oil on canvas (73.66 cm × 111.76 cm). Although these works span over forty years of Estes' painting career, they demonstrate the closely observed, illusory and the photo-real. Light is reflected across different surfaces that modify and reflect it in different ways: shiny metal has a high key with sharp lines, contours, interreflections and bright sharp glints of artificial light. The overall effect of Estes' work is what could be termed 'uniformly modulated' and has a similar appearance to a spatially balanced HDR image.

A different example of Estes' work, as shown in Figure 2.4, is *Times Square* (2004). The painting is of a street scene just after a rainstorm, and a man is hurrying across a gleaming road. The day must be warm or humid because people are wearing shirts. The light appears uniform and colours desaturated. There is no bright light or reflected sunlight, as in *Telephone Booths*, or artificial lights shining against a nearly dark sky, as in *Broadway Bus Stop*. The reflected lights on cold the wet pavements of the *Broadway Bus Stop* suggest a very different atmosphere, and we can clearly see a different rendering of appearance in of the patches on the damp road and deeper puddles from the defrosting snow. Returning to *Times Square,* the large window to the right both transmits and reflects images, which adds to the ambiguity as to whether we are looking at the reflection in the window of buildings and pedestrians. Likewise, in *Broadway Bus Stop* the interior lights from the stationary bus are transmitted through the glass of the bus shelter, whereas the lights from the bookshop are reflected in reverse on the same glass.

When looking at an object or a scene, as so strikingly painted by Estes, it is an enormous task to accurately convey all of the textural elements the observer can perceive, and so the artist has to choose the essential features and qualities, in order to decide what to draw and to leave out. This is also important when considering what is important for benchmarking in 2.5D printing. Illumination plays a vital role for both Estes and Bell. The angle of illumination on reflective surfaces and their interreflectances emitted light and the shadows that are created by objects and their illumination. Both Estes and Bell have explored many shiny surfaces and yet the relationship of the use of illumination to create large-scale environments, such as *trompe l'oeil* for architecture, is a particular challenge, which is explored in more detail in the case study on illusions (see Case Study 7: *Trompe l'Oeil*). Likewise surfaces with complex geometries, such as fabrics and textiles, will be explored in further detail in Case Study 6: Textiles and in Chapter 3 on texture and material reproduction.

Figure 2.4 Richard Estes, *Times Square* (2004), oil on canvas (94 cm × 163 cm), private collection (reproduced with permission of Marlborough Gallery, New York).

2.3 Artists' Conversion of Images into Relief

Most sculpture can be broadly distinguished as either a free-standing three-dimensional sculpture that is 'in the round', or as a relief sculpture that is attached to a background or surface ground. There are varying degrees of classification and projection that span the two. Reliefs can be categorised as one of three different kinds of sculptural artworks that demonstrate varying degrees of projection (low, medium, high) and how they relate to a surface or ground (Rogers, 1974, p. 2). Both approaches were practised concurrently for about 20 000 years, but do not appear to have begun at the same time. Evidence has suggested that relief could not have developed until three-dimensional works and drawing were established and resolved (Sanders, 1969).

The most elementary form of relief is as a progression from drawing by either cutting (incision, engraving, etching, cutting or abrasion) into a hard surface (stone, wood or metal) or where the outline is built up, or material is added-to a background, so that the outline is higher than the background (malleable materials such as clay or wax).

Based on the physical constraints of the placement of the surface relief inside or outside of a building, the objective for the carver was to create a degree of contrast between the background and the figure. The figures were carved to either protrude from the surface ground – whereby the carver had to remove the background, or by using an intaglio approach or sunken relief – whereby the carver incised the figure so the background was higher than the figures.

Illumination also plays an import role in creating a contrast, evoking an atmosphere and, depending on the illumination, creating movement and dynamism. The interplay between the relief and contrast is also an important factor, for example, contrast created by the edge of the stone and an increased or decreased contrast through the use of raking light. The position of the angle of illumination inside a temple could greatly enhance the drama of the story – one can imagine the effect of a flickering torch illuminating a scene of moving horses and figures engaged in a battle and the kinetic effect of the shadows cast by the arms and legs in relief. Figure 2.5 demonstrates what appears at first to be a confusing profusion of legs and heads; with the addition of flickering light these animals suddenly appear to be animated as if walking and nodding their heads.

Figure 2.5 Palace of King Tiglath Pileser III (728 BC) Neo-Assyrian gypsum wall panel relief (142 cm × 96 cm), British Museum, London (reproduced with permission: British Museum, London).

The elongated walls of the Sumerian temples were once adorned with figures of animals, lions, bulls, mystic trees and sphinxes. The subjects are of three kinds: hunting scenes, ritual scenes and the capture of a city. It is difficult to imagine the scale and impact of these panels and carvings, especially when removed from their original site. However, under museum conditions and the exact positioning of a raking spotlight, we can observe the wide range of intricate decoration and texture at close quarters, and we can begin to understand and appreciate the attention to detail that was applied to these surfaces. For example, dresses were incised with tiny decorative elements to represent embroidery, detail of the fleece on the goats and sheep, and leaves on the trees (Figure 2.5), all of which are difficult to see without close examination.

Furthermore, the artisan, knowledgeable of the effect of light on the surface of the stone, also created a sense of tone through the use of different heights. For example, the figures of the sheep are cut from the background to appear in high contrast. The effect of raking light plunges and delineates the edges of the figures in sharp relief. The decoration of the fur or the leaves on the trees are incised with a shallow contrast and therefore a softer appearance. If all texture and outline were cut in the same way, the appearance would be highly contrasted. In keeping with a linear perspective, the figures of the animals are depicted in profile. As illustrated in Figure 2.5, by arranging the sheep and goats in a line, the animal in the background is carved to appear to be behind the animals in the foreground, and which seems to be leading the group, as in a more conventional depiction. This solves the technical problem for the carver to assign an appropriate contrast between the rump of one and the head of another. Likewise, small adjustments to the composition, such as lifting or raising the head creates the impression of the appearance of animals as a dense herd.

In all, the panel is quite beautifully complex, all the bodies and legs are carefully included, contributing to the convention of the relationship between light and movement. 'A mastery of naturalistic representation in relief entails mastering all the problems that confront the draughtsman. In his tools and manner of working this artist may be closer to a sculptor in the round, but he must also have resolved the intellectual problems of drawing. A two-dimensional pattern lies embedded in the two and a half dimensions of the relief. The first steps towards relief were taken at much the same time that engraving was gaining assurance' (Sanders, 1969, p.19).

2.4 Artists' Exploration of Different Sculptural Relief

Relief carvings are classified by their degrees of projection from the surface ground, the spatial relationship between planar to full in-the-round and to describe the interaction and relationship between figures in relief. The first series are classified as haut sculpture relief (high relief or *alto-rilievo*).

In order to appreciate the varying degrees of projection, as illustrated in Figure 2.6, by looking sideways-on, one can imagine a figure stepping backwards into a smooth curtain of falling water, where the outer edge of the waterfall resembles the surface (back) ground. As the figure continues to step backwards by increments from high relief (left), eventually the body is just protruding from the surface ground (low relief or *basso-relievo*). By incremental means of reduction, the modelling of the figure would be flattened until some of the background elements are represented as no more than an etched line. The first degree of projection shows a relief (high relief or *alto-rilievo*) while the second series of relief shows a low relief or *basso-relievo*, whereby the creation of a relief is compressed into a shallow space and no part of the body is detached from the surface ground. The third series of relief is a combination of the first two, where the figures can be both compressed or in the round. In this instance the background is flattened into a low relief. The central figures are distinguishable by a higher relief; the heads of the background figures are modelled with a higher relief than their bodies.

There are other varying degrees of projections, which range from a hollow, intaglio, sunken or concave relief (*cavo-relievo*), where the carving is made in reverse and the illusion of a rounded convex figure or portrait is perceived. This is similar to the hollow face illusion, shown later in Figure 2.11. There is also a crushed relief (*relievo sticciato* or *schiacciato*) where the lines hardly project from the surface. A range of these parameters is also considered when making dies for coins and medals (Case Study 4: Coins and Medals).

Figure 2.6 Moving from in the round (left) to varying degrees of projection to achieve a combination of high relief, low relief and sunken relief.

The sculptural artist has at their disposal a variety of tools and materials for modelling. In fifteenth century Florence, the masons, carvers and carpenters belonged to the same guild of sculptors, but were separated from those who engraved into copper. To make a sculptural relief, the sculptor combines different degrees of projection to create a perceptual effect of depth and space through a limited physical depth and their degrees of projection from the surface ground. 'Western perspective corresponds to certain objective facts of human perception and is not merely conventional. Within certain limits it provides either a true record of a view of a segment of a real world or a convincing construction of an imaginary view of what could be a segment of a real world' (Rogers, 1974, p.50). In essence the sculptor had to accommodate as many elements as possible within a shallow space and to create the impression of depth whilst providing a narrative. One such example is the so-called *Gates of Paradise*, situated on the east side of the Baptistry of Saint John in Florence, an intricate and masterful piece of work by Lorenzo Ghiberti (1378–1455). The panel in Figure 2.7 can be found at the bottom right corner of the Gate, which depicts the scene of King Solomon and the Queen of Sheba. Ghiberti used a range of sculptural techniques including a flattened relief (*rilievo schiacciato*) and figures in-the-round, placed within an architectural setting. Giorgio Vasari described them as 'undeniably perfect in every way' and

Figure 2.7 Images captured of the same panel of the *King Solomon and the Queen of Sheba* at different times of the day. From *The Gates of Paradise* (1452) by Lorenzo Ghiberti (1378–1455).

Michelangelo described the doors as suitable to be the 'Gates of Paradise' (*Porte del Paradiso*) (Giusti and Radke, 2012).

Figure 2.7 compares one panel from *The Gates of Paradise* by Ghiberti, to photographs of the same panel, which has been captured at different times of the day (as well as using different cameras). As weather and sunlight changes, these images highlight the interplay of light and shadow and illustrates the relationship between the composition and degrees of projection. The figures in the foreground are created in a high relief relative to the architectural archway in the background. Ghiberti has used perspective to create a long gallery space, in which the figures are presented in the foreground in two rows. The elevated row are standing on a balcony and arranged in a diagonal with the two figures of King Solomon and the Queen of Sheba taking up a central position. Unlike the reliefs in temples and museums, where the direction of illumination is constant, these panels appear to be more dynamic. In miniature and as exampled here in the panel of the King Solomon and the Queen of Sheba, the drama of illumination is heightened as the shadows lengthen and contrast increases. It is interesting to compare the drawing (Figure 2.8) by Fedor Ivanovich Kalmyck, which he made from Ghiberti's panel of

Figure 2.8 Fedor Ivanovich Kalmyck (1763–1832), after Lorenzo Ghiberti, *The Queen of Sheba before King Solomon*, c.1798, drawing, British Museum, London (reproduced with permission: British Museum, London).

King Solomon and the Queen of Sheba. Kalmyck has drawn the figures in the foreground with shading and contrast as if illuminated from the left. The drawing is rendered in chiaroscuro and describes the degrees of projection using lines, tones and shadow to obtain definition. For example, the foreground is described with greater detail and contrast, whereas the buildings appear lighter and with less contrast (Giusti and Radke, 2012).

Physical methods of creating depth and tone through layering of materials include nineteenth century print processes such as the Woodburytype, which utilised the thickness of coloured gelatine to indicate different tones, whereby the thickness of the relief surface correlated to the depth in tone. The Woodburytype, although often mistaken to be a photograph, falls into a category known as photoplastography (McCallion, 2017), which was a continuous-tone photomechanical printing process that 'translated the two-dimensional tonal gradation present in a photograph into a three-dimensional model in gelatine' (Hammond, 1989, p. 174). Whilst permanent prints could be made using the Woodburytype process, it was not possible to create full colour images. Due to the complicated method of production, these traditional photoplastographic processes remained monochrome. Researcher Peter McCallion (2017) has explored plastographic variations in detail and has developed a novel method using 2.5D digital relief printing to obtain a multicolour Woodburytpe image. There also evolved a family of ceramic processes that either utilised the translucency of the stepped thickness of porcelain, for example Lithophanes, or photoceramic relief, which used the same stepped thickness but included a translucent glaze applied to the surface to create stepped ceramic or photoceramic relief (Huson, Hoskins and Thirkell, 2005, pp. 39–42) (Case Study 10: Exterior Decoration and Ceramics).

Cameotypes are similar in production methods to coins and medallions (Case Study 4: Coins and Medals). The Neoclassical Jasperware (Figure 2.9) by Josiah Wedgewood (1730–1795) resembled ancient cameo glass and was influenced by Greek and Roman subjects (http://www.wedgwoodmuseum.org.uk). The cameos were created by pressing the white porcelain clay in moulds and applied as sprigs on to solid coloured bases. Similar to the relief in Ghiberti's *Gates of Paradise*, the fine relief of the white clay of the cameo and the interplay of light are important components of the illusion of sculptural qualities of the object.

The astronomer and physicist David Brewster (1781–1868) observed how the appearance of a cameo can change from convex to concave when light falls across a relief as it moves direction: 'The illusion … is the result of the operation of our own minds, whereby we judge the forms of bodies by the knowledge we have acquired of light and shadow … where depressions become elevations. These shadows can be used to gain textural information and to measure far away surfaces such as the moon or surfaces that are difficult to measure such as paintings.' (http://biodiversitylibrary.org/page/24899400)

Brewster's observation (Figure 2.10) and the example of the Wedgewood trinket box (Figure 2.9) illustrate how the appearance of the cameo can change depending on the angle of illumination. The images are taken using the same illumination and the same angle of illumination, and yet the shadows created by the relief cause a dramatic change in the appearance. The illusion of whether surfaces appear to be concave or convex is also demonstrated by the rotating hollow mask and by understanding the connection between how the illusion is made and how we perceive it. According to cognitive psychologist Richard Gregory (1970), the *Hollow Face* illusion explains the power of the top-down processing performed by our brains to recognize faces (see Figure 2.11). As the hollow face rotates between convex and concave, our brains are hardwired to perceive the mask as convex, the way we have learned to 'see' faces since we were little. We disregard the shadows and shading produced by the rotating mask to the point of inverting these viewing queues, even at the point of flipping between the two. No matter how many times the face rotates and how much we know the illusion, the mask appears to be convex. The convex–concave approach has also been employed by painters and sculptors where the eyes of portraits are significantly darkened in the painting or carved below the surface to create a reverse relief that gives the sensation of having the eyes of a person following us when we move in front of the painting.

Figure 2.9 Wedgewood Jasparware trinket box. The cameo relief is photographed under raking light from different angles.

2.4.1 Examples of Degrees of Projection

It is more often the case that sculptors employ both first and second degrees of projection, or simply manipulate and arrange the elements in the pictorial space to gain a convincing balance between narrative and spatial arrangement of the figures within the composition. As demonstrated in Figures 2.12 and 2.13, the artist employed different degrees of projection and, by increasing the relief (and contrast) to draw the viewer's attention to the central figure, thus enhanced the drama of the scene.

The large Battle of Britain monument (https://bbm.org.uk/the-monument/) in central London features a sculptural relief by the British sculptor Paul Day (b.1967), who is well known for high-relief sculptures in terracotta, resin and bronze. Day uses a creative and unusual approach to perspective and depth, using a high degree of detail. In 2002, he was commissioned to honour the achievements of the fighter pilots during the Battle of Britain in 1940 (see Figures 2.14 and 2.15). These sculptures encompass a wide array of scenes, the most prominent one showing a group of pilots *scrambling* away from the centre piece of the relief to run to their imaginary aircraft

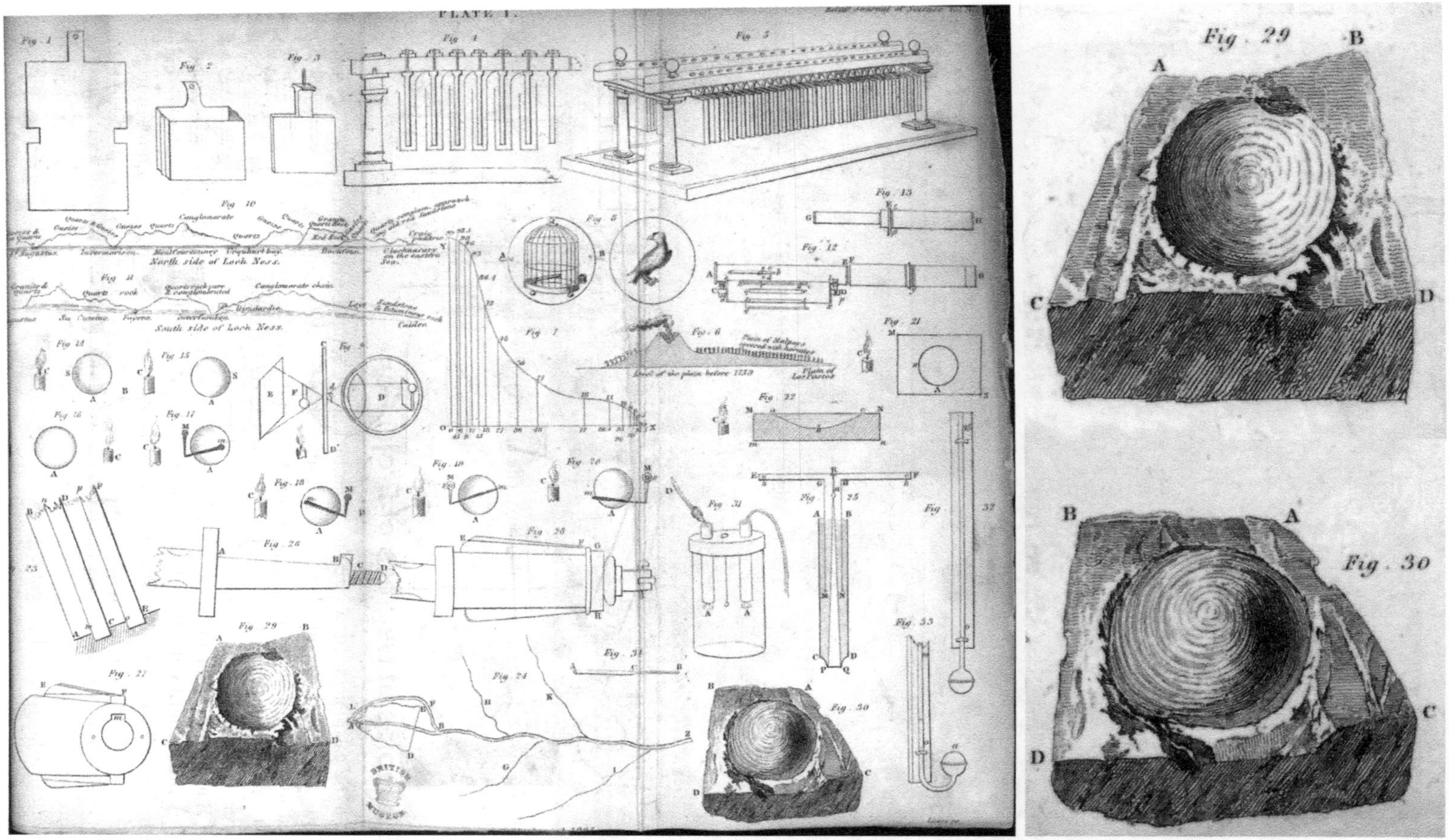

(a) (b)

Figure 2.10 Illustration from David Brewster's *On the Optical Illusion of the Conversion of Cameos into Intaglios, and of Intaglios into Cameos* (1824) and detail of a concave and convex hemisphere (right) (http://biodiversitylibrary.org/page/24899400).

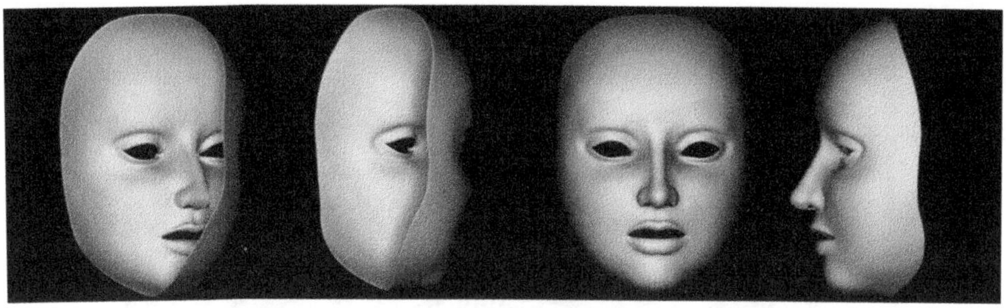

Figure 2.11 The hollow face illusion showing concave and convex masks that are continually rotating (courtesy: www.echalk.co.uk) (https://www.youtube.com/watch?v=sKa0eaKsdA0; see also http://www.echalk.co.uk/amusements/OpticalIllusions/hollowMask/hollowface.html).

Figure 2.12 John Collier, *Our Lady of the Bluebonnet* (60 cm × 70 cm) bas-relief. In Collier's vision of this Madonna and Child they are in a stable with Texas Longhorns and has included Bluebonnets, which is the state flower (courtesy: John Collier, www.hillstream.com).

Figure 2.13 *The Battle of Cape St Vincent* by Musgrave Watson and William F. Woodington. Bronze relief on the west face of the plinth of Nelson's Column Trafalgar Square, London (courtesy: Wikimedia/Creative Commons Credit: Eluveitie) (https://commons.wikimedia.org/wiki/File%3ANelson's_column_Battle_of_Cape_St_Vincent_relief_(Musgrave_Watson).jpg).

Figure 2.14 Paul Day (b.1967), *Mechanics and Armourers* (2002–2005), bronze and resin (courtesy: Paul Day, www.pauldaysculpture.com).

Figure 2.15 Paul Day, *Scramble* (2002-2005), bronze and resin (courtesy: Paul Day, www.pauldaysculpture.com).

(Figure 2.15). It is interesting to note these textured sculptures have been highly successful for blind visitors to the monument.

We can advance the sensation of depth even further with the use of interreflection and lighting, or so-called infinity mirrors. In the artwork *Endlessly Repeating Twentieth Century Modernism* (2007), the artist Josiah McElheny creates the illusion of an infinite number of silver vases in a glowing glass box. In reality, bottles are not made in silver but blown glass and they are not placed in a glass box. The bottles are actually placed within a sequence of mirrors and the ensemble is housed in a two-way mirrored box that allows us to see inside without being reflected in the bottles. Likewise, the artist Yayoi Kusama creates life-sized infinity rooms, entitled *Infinity Mirrored Room – Filled with the Brilliance of Life* (2011). As the visitors enter the dark, seemingly endless space, they experience thousands of pinpricks of coloured light that are reflected from all directions. The walls, ceilings and floors are mirrored and LED lights are suspended from the ceiling and flash on and off in different colour configurations (http://www.tate.org.uk/whats-on/tate-modern/exhibition/yayoi-kusama). These artists are able to achieve the visual illusion of a space that extends beyond the parameters of the existing space.

A different use of mirrors, called anamorphic art, combines a cylindrical mirror that is placed on a distorted image. The mirror transforms the flat deformed image into a 3D undistorted image on the side of the mirror (http://www.boredpanda.com/anamorphic-cylinder-art). This is demonstrated by Jim Sharpe with *The Illusions Parade* in Liverpool (https://news.liverpool.ac.uk/2015/08/21/liverpool-welcomes-visual-perception-experts/) (Figure 2.16) and his study of optical illusions that include hidden images, anamorphic distortions (Figure 2.17) and lenticular lenses. Some of his examples require red screens or lenticular decoders. Figure 2.16 is an example from Sharpe's *Camera Phone Illusion Book*, as a flat page the image is distorted until one page is tilted at right angles to the other, and the picture is captured with a camera (Pinsharp 3D Graphics; www.pinsharp3D.co.uk).

Anamorphic art was traditionally used in the late sixteenth century as a way of hiding pictures, as paradoxical art or playful demonstration through illusions (see also Holbein's skull hidden in Figure 2.26 later). Some anamorphic art does not require mirrors, for example, anamorphic projections painted on roads *pop up* as a driver comes closer; these are used to great effect for traffic calming.

Figure 2.16 Anamorphic distortion by Jim Sharp, *Illusions Parade*, Liverpool (2015) (courtesy: Jim Sharpe).

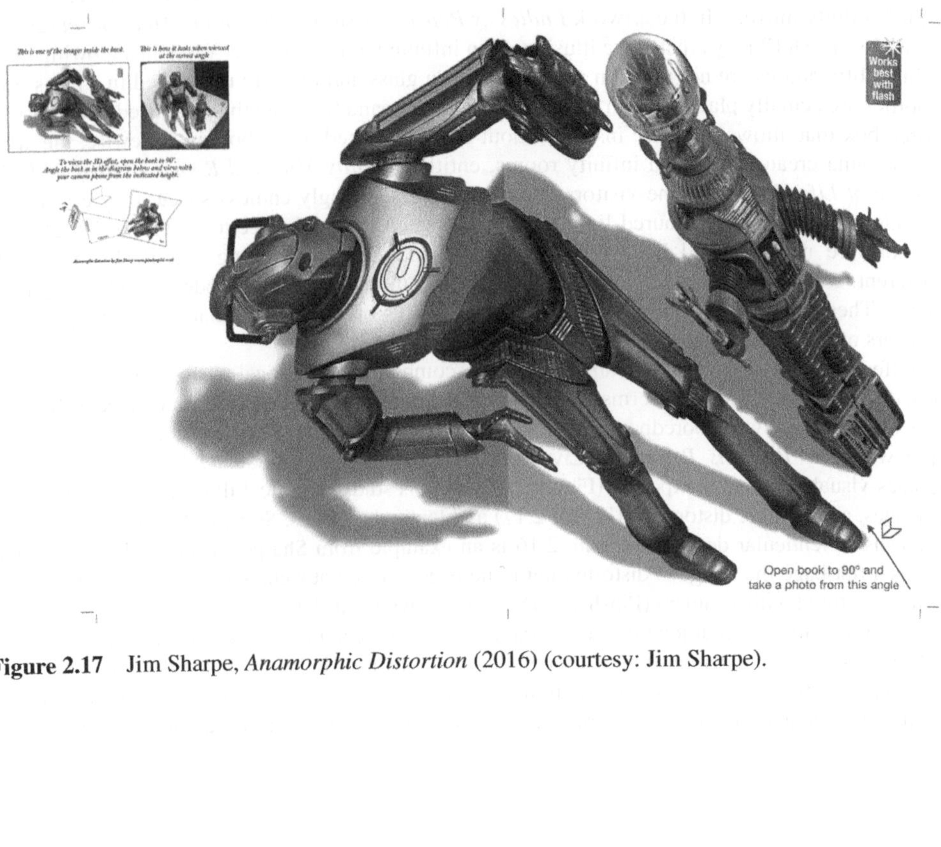

Figure 2.17 Jim Sharpe, *Anamorphic Distortion* (2016) (courtesy: Jim Sharpe).

2.5 Coloration of Relief Surfaces

There are many stunning examples of polychromic objects and surfaces. These range from the incorporation of naturally occurring coloured materials, such as marble and mosaics, as well as materials and surfaces that are enhanced through the addition of colours, such as glass, stained glass, frescos and ceramics (Case Study 11: Microstructural Texture) tiles and ceramics (Case Study 10: Exterior Decoration and Ceramics), metal, stone and wood. Colour and texture are also used to highlight areas or to differentiate from one plane to another, for example in the mural made from different types of wood, as exampled in the wooden mural at Kew Gardens (Chapter 3, Figure 3.3) and case studies, such as marble (Case Study 8: Marble).

The stunning mosaic example demonstrated in Figure 2.18 highlights ancient sensitivities for colour and tone through the use of small tiles or stone tesseraeto create permanent pictures that are integral to floors and walls of buildings. The most complex example is a mosaic made from tesserae in St Peter's Basilica, Vatican State, Italy. These tesserae are so small that the final work could easily be mistaken for a painting. The tiny squares are only discernable when close to the painting (Figure 2.19).

An example of planar reliefs can be found in the decorative detailing in the lacquered, intricately layered and coloured ebony *Coromandel* screens. Screens and cabinets were called Coromadel after the coast of the same name in India, from where they were exported to Europe in the seventeenth and eighteenth centuries by traders of the English and French East India companies. The examples in Figure 2.20, carved in China and displayed in the Ashmolean Oxford, demonstrate a range of surface pattern and decoration. Trees and flowers are carved or incised with chisels using minutely repeated elements, resulting in the textured appearance of foliage or fronds of the trees. In the bottom left of Figure 2.20, the artist has cut a sinuous and jagged blue rock, where the darker lines are in relief and the background is filled with a blue and white pigment. Parallels could be drawn with wood engraving or relief cutting and arguably could be considered as a crossover with engraving for woodblock printing (which is described in more detail in Case Study 14: Relief Woodblock Printing). This process was known as *Kuan Cai* (incised polychrome), where panels of wood are applied with a thick layer of ash, a mixture of powdered brick, pig's blood and lacquer. Several layers of black lacquer are then applied. The designs are carved using fine sharp chisels, and, as demonstrated in Figure 2.20, the patterning is determined by the profile shape of the chisel. For example, the craftsman used a rounded chisel to

Figure 2.18 The *Gypsy Girl* mosaic, Gaziantep Museum of Archaeology (courtesy: WikiCommons/ Creative Commons 3. Credit: Nevit Dilmen) (by Antep_1250575_cr.jpg: Nevit Dilmen (talk) derivative work: Durova (Antep_1250575_cr.jpg) [Public domain], via Wikimedia Commons).

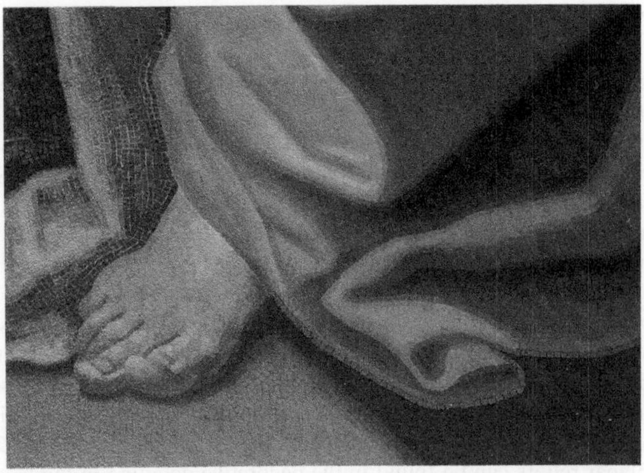

Figure 2.19 Detail of mosaic of the *Immaculate Conception*, St Peters Basilica, Vatican State, Rome, Italy (courtesy: WikiCommons/Creative Commons 3. Credit: Adrian Pingstone/Arpingstone) (https://en.wikipedia.org/wiki/Tessera#/media/File:St.peters.basilica.tesserae.closeup.arp.jpg).

obtain the blossom-shaped elements (top left), whereas a V-shaped chisel would have been used to create the fronds of a weeping willow tree (top right).

Another example of coloration of relief are the wooden friezes in the courtyard of City Hall (Rådhuset) in Oslo, which were created by painter and sculptor Dagfin Werenskiold (1892–1977). The sixteen wooden polychromic panels depict scenes from Norse mythology, of which Figure 2.21 shows *Thor* (Tor) in his chariot drawn by two goats through storm clouds, over rooftops of a farm and towards the sun.

Although many ancient buildings and statuary are seen today as a neutral bare stone or white marble, there is now convincing evidence that ancient monuments and statuary were polychromic. There is much debate as to what proportion of ancient sculpture was painted (https://edition.cnn.com/style/article/gods-in-color-ancient-world-polychromy/index.html); and over centuries the majority of sites and artefacts have returned to their natural state, leaving only tiny traces of paint that are hidden in crevices. A chance discovery of the head of a statue of a young woman in the Basilica Noniana at Herculaneum, in Italy in 2006, provided compelling evidence of polychromic statuary and the subtlety and naturalist skill employed by the painter (Figure 2.22). Gareth Beale, whilst undertaking a PhD at Southampton University on the subject, explained that very few examples of statue painting from this period and location survive, so it is difficult to compare this statue to other examples in terms of the painting quality and level of skill. The paint on the surface of the marble was analysed using Fourier transform infrared spectroscopy – a method using the infrared spectrum to analyse materials – and X-ray fluorescence (XRF) – a method used to analyse the elemental composition of pigments. Beale found the pigments contained traces of egg yolk, which was traditionally used as a binding agent. According to Beale, unlike other binders, egg does not yellow or change colour, and therefore the pigments remain bright and clear. In order to achieve dense layers of paint, a pigment was mixed with egg to create a tempera, which were applied in thin layers. This technique is still widely used in contemporary practice and experts have considered the possibility that Roman statue painters also used this technique. Tempera works well on hard non-flexible surfaces and it dries to a tough film, but over time it also becomes brittle and delaminates. In these instances, sculptures would have required regular attention, probably by a master skilled in the art of retouching.

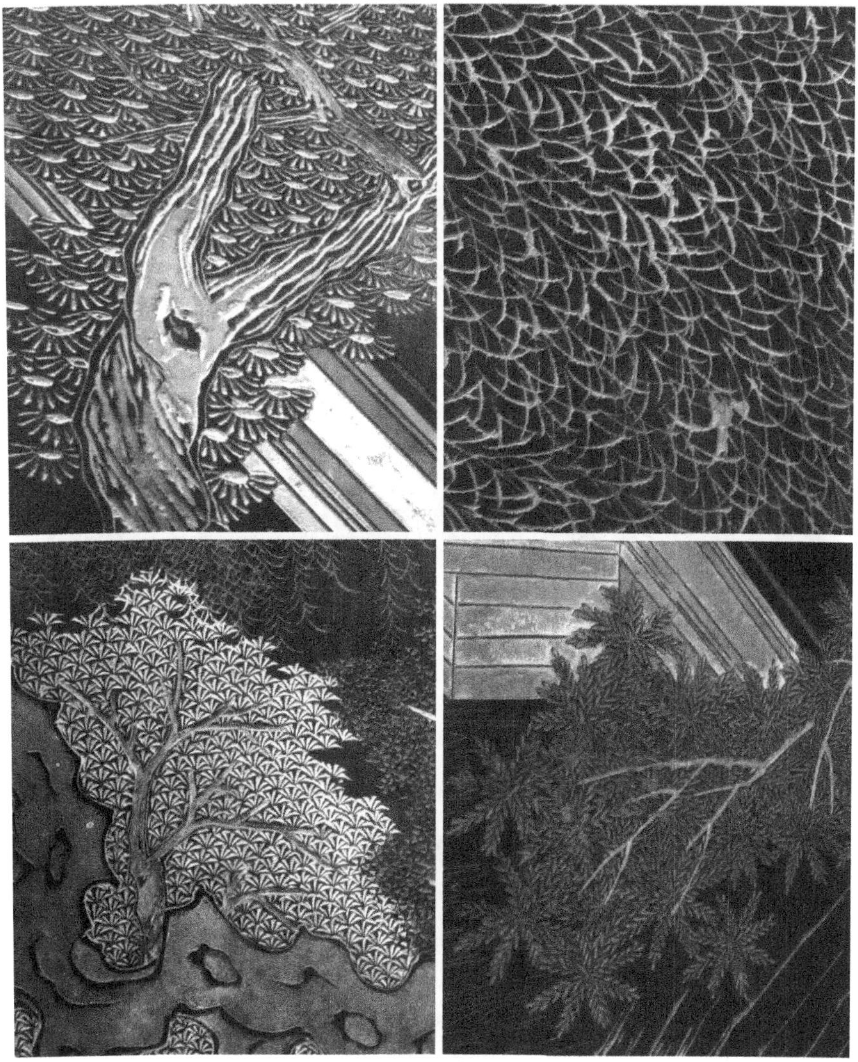

Figure 2.20 Details of carving from a twelve folding lacquer screen Coromandel Screen, China (1600–1700), which is decorated with flowers, landscape, trees, and animals. Lacquer on wood (reproduced with permission: Ashmolean Museum, University of Oxford) (http://jameelcentre.ashmolean.org/collection/8/per_page/25/offset/0/sort_by/date/object/17562).

Furthermore, study into the wall paintings at Pompeii, as presented by Sapienza University in Rome, has indicated that the so-called *Pompeiian Red* frescoes (Figure 2.23) were originally painted yellow, but as a result of the heat from the eruption of Vesuvius in AD 79, the yellow ochre walls were transformed to red (Omarini, 2012, pp. 61–67). Sergio Omarini, when presenting the Institute's findings, highlighted that as a result of their study the increase in number of walls originally considered to be yellow has doubled in number from 57 walls to 138. He said, 'The discovery allows us to rethink

Figure 2.21 Dagfin Werenskiold (1892–1977), *Tor is Driven by His Goats*, polychromic woodcarving relief.

Figure 2.22 Coloured pigment is still discernable on head of the statue, Herculaneum (Ercolano), Italy (courtesy: Richard and Phyl King).

the original appearance of the city in a radically different way from how we are used to – in which red, indeed Pompeiian Red, has been prevalent' (https://www.theguardian.com/science/2011/sep/22/pompeii-red-yellow).

Figure 2.23 The *Pompeiian Red* walls and pillars at the Herculaneum (Ercolano), Italy (courtesy: Richard and Phyl King).

Many ancient civilisations also painted their temples, buildings, statues and walls with a highly coloured finish, but due to many external impacts, such as weathering, building, war and pollution, many of these painted surfaces have been eroded. The water colour paintings and hand-tinted photographs made by Adela Breton (1849–1923) whilst she was travelling around ancient sites in Mexico in the nineteenth century represent a highly detailed record of polychromic wall paintings and painted reliefs of temples. As part of a nineteenth century Victorian gentlewoman's education Adela would have been taught drawing and painting by private tutors. By this time, women were also able to gain an income from their arts. In 1894 Adela travelled to many of the major ancient sites in central America and in her sketch book made many paintings that included the locations, dates and detailed notes. It may have been during this time that she came into contact with the British archaeologist Alfred Percival Maudslay; it is reported that she had asked if there was any work that she could do (Giles and Stewart, 1989). He had given her the task to travel to *Chichén Itzá* to check on the accuracy of his drawings and to make copies of paintings that were already showing deterioration. From 1900 to 1908 she continued to return to Mexico to make copies of the wall paintings, which she made to the same scale, and to accurately record the colours. In her diary entries she describes her frustration in obtaining the right level of accuracy. 'An occasional small bit of colour might be left out and I know by experience how many times one has to go over these complicated things to make sure that every scrap of colour is left in' (16 February 1916) (Giles and Stewart, 1989). The watercolour section of the *Acancéh Frieze* by Adela (Figure 2.24) is the only surviving record of the dazzling polychromic stucco frieze. Exposure to rain has partially destroyed the paintings. Adela was very careful about not adding to her watercolours if sections of a panel were missing, as reflected in the photographic records that she made. Figure 2.24 shows a series of relief carved animals and creatures in stepped frames with a red coloured background. This includes the long green feathers of a *Quetzal* bird (left), the central figure is a seated opossum, a monkey (above right) and a yellow and blue snake (right). She also added shadows, indicating the depth of the relief, which can be seen more distinctly in her hand-tinted photographs (see Figure 2.25).

Figure 2.24 Adela Breton (1849–1923), *Acancéh Frieze*, Panels 17–21, watercolour on paper (2130 mm × 1100 mm) (courtesy: Bristol Culture/Bristol Museums, Galleries & Archives/ Photographer Dan Brown).

Figure 2.25 Photograph of the *Acancéh Frieze* (left) and a hand-tinted photograph (right) by Adela Breton (courtesy: Bristol Culture/Bristol Museums, Galleries & Archives Photographer Dan Brown).

The paintings by Adela represent a large body of work that systematically records the wall paintings from ancient Mexican temples, which were already damaged and rapidly disappearing during her time of recording and now are lost. Her paintings are the only surviving archive and are also at risk of degeneration. It is interesting to note that visitors now have access to Breton's watercolours, which have been the subject of a digitisation project and with plans to reproduce the paintings and indeed to recreate sections as polychromic relief panels (https://artuk.org/discover/stories/digitisation-of-the-adela-breton-collection-at-bristol-museum-art-gallery). This highlights many interesting questions about how we use existing fragments of physical evidence to interpret and reconstruct ancient sites: How can we recreate the aura of experience through textures, material colours, scale and replicas? Are real materials used using traditional building and craft methods? Are reproduction methods digital fabrications?

2.5.1 Visualising and Reconstructing the Past

An essential part of archaeology is the discovery of decorations and buildings as they reveal the degree of development, the sense of aesthetics and the values attached to a given society. Usually, the presence of relief shows the wish to portray depth and perspective to either communicate something (such as hierarchy) or embellish objects.

Archaeologists work on the premise that their observations of a given artefact or excavation correspond to a particular scene in time. In order to give life to their theories a considerable number of drawings, annotations, diagrams and models are built. Such representations look to convey to the spatial-temporal context that better fits an archaeological finding. To discuss results about excavations, drawings are made where a complex set of lines (polylines) serve to communicate distinct measurements. When they work in the field they are compelled to propose an hypothesis at every stage of the excavation to indicate the different paths the work could take. It is an iterative process that has to be revaluated in vivo.

Usually, archeologists have a limited or a single opportunity to access ancient sites to examine and collect data. Budget, time and structural constraints expedite the decision-making process, meaning the recording process needs to be highly precise and focused. Even though new technologies such as 3D visualisation aids have started to assist they have not been widely integrated because of their costs, unfriendly environments, time delays and the lack of annotation tools. Archaeologists may have the impression that 3D renderings do not accurately relate to their work, as the tools they need to communicate may not be embedded in the models. The risk they may face is the loss of the in-situ interpretations.

Moreover, archaeologists are taught to create 2D drawings that are composed of shapes, lines and annotations to underpin their work. Despite the benefits of 3D technologies for visualization and archiving, in order to publish their studies, it is necessary for archaeologists to return to 2D graphical representation. For the moment the benefits and opportunities for the 3D community is overshadowed by the effort, training time and cost in 3D technologies. Therefore, a lack of understanding as to what is important in this application has highlighted a gap between technology and traditional methods.

Maybe looking only at the seemingly important advantages of 3D renderings (visualization, reproductions, etc.) is not the most complementary way of collaboration in between '3D-ers' and archaeologists. The efforts made in 3D technologies should be redirected to the actual needs of this field. For instance, their hypothesis about the positioning of objects in a scene and their purpose could be linked to their material, their colour and even the reflection they were emitting with respect to the illumination of the scene, but traditional 3D printers may well be fail to capture this. The final print may well be considered as just another model or mock-up, but not as a close-to-real representation of the artefact or building. In models, the level of details is important to reveal a range of visible archaeological features.

To limit the destructive nature of excavations, the impact on the environment, for example, planning for ditches and structures that may collapse during excavation is crucial. To model this, we need a fair understanding of the strata and their physical characteristics; for instance, a flat, slippery stratum cannot be printed as a rough surface. On return to the office, and to be able to accurately review and extrapolate excavation data, photorealistic reproductions, containing surface texture, patterns and decorations, colour and reflection properties are important to building a complete picture.

Attempting to reconstruct ancient buildings, artefacts and prehistoric people, is a balance between the excavated artefacts and fragments, and the extra unknown dimension - informed conjecture, hypothesis, that using 2.5D printing technologies may well provide significant contribution to communicating about ancient history. A real contribution will be made (which will expedite the adoption of these technologies) when we can achieve 2.5D and 3D reproductions that can incorporate different materials, details, structures, colours and other reflection properties. The aim is to enrich and inform archaeological understanding by creating accurate measurements, alongside subjective interpretation.

In the twenty-first century, it is difficult to gain a complete understanding when presented with a collection of abstract broken figures and ruined buildings that may have once formed citadels or major cities. For some ancient sites, it may be impossible to gain access because of political conditions, the fragility of ancient sites or simply that sections of walls may be widely dispersed between collections, in different countries, or may be in private hands. Moreover, there is a continued imperative to preserve any existing sites from impacts such as pollution and robbery. Based on these problems and with the benefit of computer graphics rendering, the alternative is to reconstruct these cities and then to piece together existing artefacts and virtually rebuild the lost sections, repaint the walls and replace all the objects. Using this approach, the digital historians can then demonstrate the appearance of all the different materials, textures and colours. These computer-generated environments are based on artistic interpretations, analogous to the watercolour interpretations of the nineteenth century archaeologists, but they also provide the historian, and more importantly the visitor to the museum, with deeper insights of social, cultural and historical heritages. The Institute for Visualisation of History hosted a project entitled REVEAL that has gathered other similar virtual archaeological projects (http://www.vizin.org/projects/nimrud/gallery.html); for example, the Learning Sites project (http://www.learningsites.com/NWPalace/NWPalhome.html) built a virtual reconstruction of the Northwest Palace of King Ashurnasirpal II of Assyria. In the mid-nineteenth century, Austen Henry Layard conceived the illustrations of the palace and was on the site during the excavation. His interpretation of scale, representation of the structure, the décor and wall patterning has largely gone undisputed until now. However, Snyder and Paley of the REVEAL project were able to piece together a more comprehensive environment by gathering all the available data, as well as the drawings made by Layard, the records and illustrations prepared by the later British excavations, Polish documentation and the Iraqi excavations. By using a Virtual Reality environment, the group has collated all the drawings and photographs of the bas-relief and have reconstructed spaces and applied décor patterns, textures and colours to furniture. They were also able to experiment with different decoration, bas-relief, colour, three-dimensional renderings, and even to place figures in the space and move around the environments as if on site.

2.6 Examples of Artists' Approaches to Representation and Reproduction of Texture

Texture is a visual attribute that enables us to distinguish the differences between materials, defined as substances or a substance out of which a thing is or is made. Texture also allows us to identify the structure and shape of objects and discriminate edges of objects in a complex pictorial scene. Materials may contain many microelements, which may be viewed under raking light or complex illumination, will reveal many different angles of reflection and will prove to be more difficult to accurately measure, model and reproduce (see Section 3.2, Figures 3.9 and 3.10 and Case Study 5: Capturing Texture of Paintings for Museum and Heritage) (model in Sketchfab: https://sketchfab.com/models/4737e8b449ba4d4eb10d43b4ca2c652e).

From a computational perspective, in order to translate between the objective and subjective, to extrapolate numerical data from natural objects, and to present ways that most people can understand is certainly a challenge. Sophisticated capture methods and mathematical models have been developed, but there is an element of ambiguity, subjective and comparison. Furthermore, to include every element or detail is highly costly both in terms of computing energy and time.

The evolving question is: What are the essential elements of paintings produced by artists that capture the qualities of texture, grain, reflection, translucency and absorption of a material, and those, through the application of coloured brush marks, that demonstrate a convincing likeness of the material qualities of wood, metal, glass and fabric? In order to gain an understanding of creating a verisimilitude of texture we can study colour, dynamic range, drawing and painting techniques used by artists (Bayer *et al.*, 2004; Hollman and Tesch, 2004; Constable, 2007; Jordan and Cherry, 1995) alongside scientists working on human vision and texture perception.

This section aims to look at the relationship between texture, objects and artists' approaches to reproducing texture in their art. It attempts to address the essential characteristics of a real-life world and how these qualities can be transcribed into a synthetic scene (Parraman, 2014).

The painting *The Ambassadors* by Hans Holbein the Younger (1497/8–1543) demonstrates not only a mastery of portraiture, but his deft skill in capturing the appearance of a range of textures, including metalwork, silk, fur, wood, paper and leather. Holbein was able to convey an experience of the weight of heavy textiles, such as the green *Jacquard* curtain in Figure 2.26 (see Case Study 6: Textiles) or, in Figure 2.27, the lynx-lined black fur trimmed coat worn by Jean de Dintville (left) or the dark brown floor length coat of velvet (right). Annotations on colour and texture are frequent in Holbein's preparatory portrait drawings. He also excelled in monotone areas of colour but still managed to convey the essence of silk, fur or velvet (Figure 2.26) (Foister, Roy and Wyld, 1997, p. 70).

Figure 2.26 Hans Holbein the Younger (1497/8–1543), *The Ambassadors* (1533), oil on oak panel (207 cm × 209.5 cm), National Gallery, London (reproduced with permission: National Gallery, London).

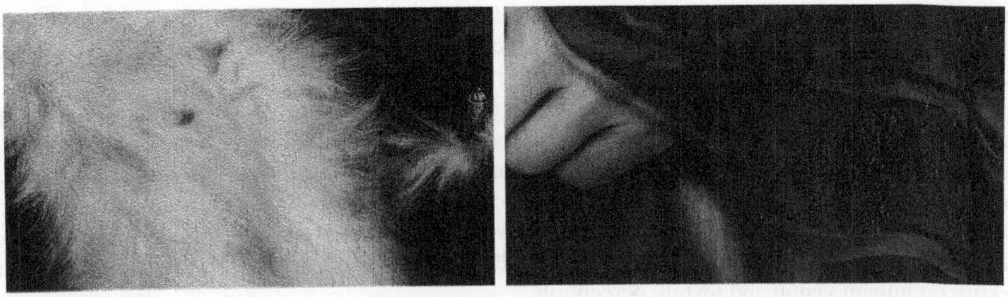

Figure 2.27 Details showing the relationship of fur and silk, by Hans Holbein the Younger, *The Ambassadors.*

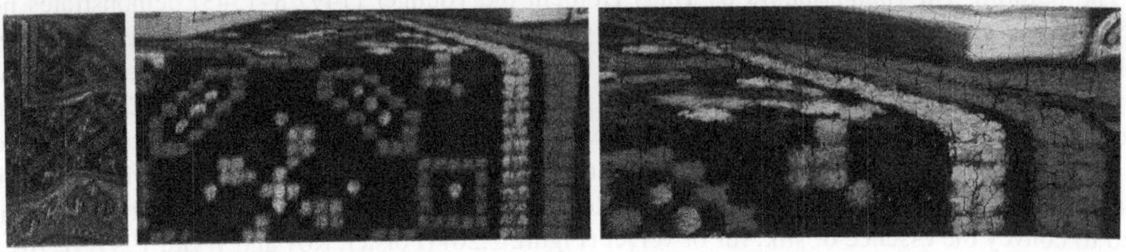

Figure 2.28 Details showing the knotted carpet, by Hans Holbein the Younger, *The Ambassadors.*

When work began in 1993 at the National Gallery in London, to clean and restore *The Ambassadors*, a series of cross-sections were taken to investigate Holbein's painting technique, which clearly showed how the artworks were layered. According to Wyld in the technical Bulletin (Wyld, 1998; see also https://www.nationalgallery.org.uk/paintings/research/technical-bulletin/the-restoration-history-of-holbeins-the-ambassadors), Holbein's approach was to build up the surface with a complex layering of different coloured pigments. For example, for the carpet that plays a prominent part of the composition, a black is applied to a chalk and mid-grey ground, followed by an intense red consisting of vermillion, red lake and white on top. Although seemingly complex, the pattern is composed of small squares of red, yellow, blue and white, with grey to represent the knotted threads (Figure 2.28). Looking at the high-resolution images (a highly useful resource of world-wide art collections and archives, which includes some high-resolution images of popular artworks; http://www.google.com/culturalinstitute), one can gain a better understanding of the optical complexity of the brushwork to convey the appearance of rug fibres and threads.

In the fifteenth century Jan van Eyck's observant eye and meticulous attention to detail was a key element in his highly accurate representation of materials. The essential factor of Jan van Eyck's art is perhaps the 'keen observation of the way things and phenomena manifest themselves in the visible world' (Kemperdick and Lammertse, 2012, p. 92). Visual comparisons can be made between Jan van Eyck's training in illuminated manuscript painting, and his meticulous attention to detail in his panel paintings. Works such as the *The Annunciation* (Figure 2.29) demonstrate Van Eyck's innovative approach to modelling the folds of clothing, the transparency of glass, colour and dynamic range, and his ability to transmute the medium of paint into gold, precious stones and pearls, and without the use of gold paint, 'he no longer simply painted the proper colour of an object, he captured the mutability of its appearance' (Kemperdick and Lammertse, 2012). The tiny flecks and dabs to describe the gilt stitching or the gleam of pearls is rendered in pale yellows and ochre in contrast to the deep reds and umbers in the background to create a high contrast. He also recorded the appearance of different materials, mimicking wood, marble and stone to create sculptures standing in shallow recesses.

Figure 2.29 Jan van Eyck (c.1390–1441), *The Annunciation* (1434/36), oil on canvas transferred from panel painted surface (90.2 cm × 34.1 cm), Andrew W. Mellon Collection (courtesy: The National Gallery of Art, Washington).

Vasari, writing in the sixteenth century on the lives of painters, described Titian (Tiziano Vecellio, 1480/5-1576) as demonstrating, almost to the point of abstraction, a sophisticated understanding of material applied through loose brushwork, and with an expert knowledge as to which components could be left out (Vasari, 1550). This is particularly highlighted in *Portrait of Pope Paul III Without Cap* (1543) in The Capodimonte National Museum, Naples, Italy (https://www.museocapodimonte.beniculturali.it/portfolio_page/collezione-farnese/), the *Mozzetta* – the large expanse of his red velvet cape is carefully rendered to convey the weight, texture, and highlights of the thick pile velvet. The rendering of materials that contain a solid body colour, or where different materials of the same colour presents a particular technical challenge for the artist.

In the twenty-first century, a similar problem is encountered when selling clothes of a single colour. Moreover, the ability to show the detail of a texture of a material is all the more crucial, and can be solved by allowing the customer to scroll over an image to enlarge the detail. However, there are parameters that cannot be controlled, such as the accurate representation of a colour between viewing platforms and printed catalogues. Online retailers have avoided this problem by introducing more pattern and embellishments to clothing, and digital textile designers have adapted their palettes to ensure the colours they choose on screen can be reproduced on fabrics (Henry, Westland and Cheung, 2013).

Figure 2.30 Anthonis Mor (1517–1576), *Portrait of Sir Thomas Gresham* (c.1560–1565), oil on panel (90 cm × 75.5cm) (courtesy: Rijksmuseum, Amsterdam).

Likewise, the portraits of *Sir Thomas Gresham* (Figure 2.30) and *Anne Fernely*, by Anthonis Mor (1517–1576) in the Rijksmuseum, Amsterdam, show the sitters dressed in black, only punctuated by white ruffs around the neck and wrists. As a study in light and dark (chiaroscuro) and local contrast, Mor has been able to create the appearance of a smooth silky shine of the black padding, compared to the appearance of a matte velvety dark trim along the bottom of the tunic and pleated capped shoulders. Areas are enhanced as highlights to the creases on the sleeve and slight indications of geometric stitching. Mor has paid particular attention to detail, including slashed ribbons at the shoulders, cross-stitching along the hem and velvet-covered buttons. As observers of the portrait we may be aware of the quality and the luxuriousness of the materials. Likewise, we may begin to make comparisons with real materials, for example the jacket could be made from silk or satin. Furthermore, the cap on his head conveys a slightly softer diffuse appearance, suggesting it may be made from velvet, moleskin, leather or felt.

Using a different medium of engraving, Hendrick Goltzius (1558–1617) demonstrates similar attention to detail in *Officers in Peascod Doublets* (Figure 2.31) by intermixing engraved lines of different thicknesses and varying the modulation of lines to suggest the physical structure and movement of different types of materials. For example, the ornamental metal work of the sword's hilt is rendered in detail, but, in contrast, the folds and tumbling fabric of the flag suggests that the material, whilst opaque, can be easily lifted by the bearer. The material is conveyed in a shimmering, crisp lightness that is in contrast to the smooth surface of the doublet using a sparse rendering of tone but conveying all the necessary information of ribbons, stitches, pleating and slashes (see also Case Study 13: Analogue Printing Methods).

The sixteenth century also witnessed the emergence of collectors of natural artefacts, and the growth of the empirical science based on observation and study of natural phenomena. Artists across Europe began to explore material qualities of objects that had not been considered in depth since classical

Figure 2.31 Hendrick Goltzius (1558–1617), *Officers in Peascod Doublets* (1587), engraving (28.7 cm × 19.3 cm) and (right) detail (courtesy: Rijksmuseum, Amsterdam).

antiquity. Observation of shells, insects, flowers and paintings were regarded as precious objects, and meticulous attention was given to even the most common of foodstuffs such as Antonio de Pereda's (1611–1678), *Still Life with Walnuts* (1634). Juan Sánchez Cotán (1560-1627) used the precise placement and lighting of objects, such as suspended fruit. He also included a three-dimensional setting, whereby the vegetables were placed or suspended in a niche, thus providing a sense of depth, space and *stillness*. The objects placed by Juan van der Hamen y León (1596–1631) on different steps appear to be precariously positioned (Figure 2.32), as if about to tip off the edge, or an overladen basket is waiting to topple. Each object and their material qualities of, for example, translucency, gloss, or roughness, are minutely observed.

These observations, as in the term *still life*, appear as life has been placed in suspended animation. Of course, these images were painted over a prolonged period of time and therefore the results are highly deceptive. The fruit and flowers would have rotted, the insects may well have either been already preserved or would certainly have flown away. These artworks therefore are careful observations of material qualities, surface reflections, light and shade, and three-dimensional modelling to create a complex composition.

During the seventeenth century, which is regarded as at the height of naturalist realism, artists were considering the illusory relationship of light on objects they were interested in, depicting convincing representations through close attention to material attributes of objects. Furthermore, artists did not limit themselves to a generic representation of wood or metal or fabric, but explored the difference in appearance of how materials were manufactured, for example as shown in Figure 2.33, Diego Velázquez's (1599–1660) *Old Woman Cooking Eggs* (1618). Although one of the Velázquez's earliest documented works, he already demonstrates a difference in appearance beaten and polished copper, or the pottery vessels that are glazed and unglazed. And through the shadows we are invited to explore the different tactile qualities of hard and soft: a muslin headscarf, a glass pitcher, and a woven basket. Interestingly, he also signals how materials have changed and decompose over time or through constant human use, such as the bruised melon, the chipped jug, or the wooden spoon that is roughened and darkened through use (Figure 2.33).

Figure 2.32 Juan van der Hamen y León (1596–1631), *Still Life with Sweets and Pottery* (1627), oil on canvas (84.5 cm × 112.7 cm), Samuel H. Kress Collection (courtesy: National Gallery of Art, Washington).

Figure 2.33 Diego Velázquez (1599–1660), *Old Woman Cooking Eggs* (1618), oil on canvas (100.5 cm × 119.5 cm) (reproduced with permission: National Gallery of Scotland).

Our appreciation of his work is based on his knowledge in the way he applied paint and his understanding of the optical appearance of materials at a distance. He demonstrates a sophisticated understanding of a range of surfaces including human skin, fur, metals and textiles. His highly gestural approach to brushwork, when viewed close up, appears abstract and impressionistic but seems to coalesce at a distance, 'The degree of understanding of the optical act of viewing had never before been so spectacularly demonstrated'(Keith, 2006, p.81). One hypothesis is that he used long-handled paint brushes to create the overall painting and a short brush to add the highlights and detail (Sonnenburg, 1971). X-ray photography of his work demonstrates very little overworking of his paintings and therefore suggests careful preparation and composition. His primary painting method is thick and direct, with limited earth pigments, and black and white. In order to undertake the rather laborious rendering of costumes, Velázquez attempted to systematize a method of painting so that his workshop could copy his handling technique. A study into the differences (Keith, 2006) between paintings made by Velázquez and his workshop demonstrate imitative methods of copying from the master (see also Section 1.3.4, Reproductions versus Forgeries versus Copies). Although the same materials and painting methods are used, the copied result appears a little too laboured and less fluid. This may be an interesting factor when understanding the relationship between the tacit mastery of painting as demonstrated by Velázquez versus proficient knowledge, and in the twenty-first century, the relevance of comparing machine-based methods of creating brush stokes (Case Study 12: Painting Machines).

Other examples can be compared such as Samuel van Hoogstraten's (1627–1678) mastery of *tromp l'oeil* (Case Study 7: *Trompe L'Oeil*), and still life compositions by Willem Claeszoon Heda (1594–1680), and his exploration of reflection, and the surface decoration of gold (see Chapter 5, Case Study 9: Gold, Figure 5.62).

Can parallels be made between paintings by Old Masters and artists working in the twenty-first century? There remains significant links between the still life paintings by Old Masters and hyper-real methods developed by artists working in the twentieth and twenty-first centuries (Letze, 2013). It is

Figure 2.34 Robin Eley (b.1978), *Immaculate* (2012), oil on Belgian linen (51 cm × 40 cm) (courtesy: Robin Eley; www.robineley.com).

interesting to note that painters are highly adept in capturing the appearance of materials, and yet for the computer graphics designers this has taken much longer. Visual comparisons can be made between, for example, the dullish fur trim on the sleeve of Jean de Dintville, and the matted fur that is computer generated by Pixar for *Monsters University* (RenderMan® is a software produced by Pixar that can simulate extremely realistic textures, hair, lighting and 3D animation; https://renderman.pixar.com/view/renderman).

Robin Eley (b.1978) is a contemporary artist who demonstrates a highly sophisticated understanding of the illumination and rendering of interreflective and complex surfaces such as crumpled paper, foil, polythene and coloured plastics. He also wraps human figures in the plastic and is interested in the differences in rendering the appearance of skin and hair as seen through semi-translucent materials. In the twenty-first century, there are greater opportunities to gain insights into the practical methods of art making – for example YouTube – and Eley demonstrates his method for rendering aluminium foil (Eley, 2011). Eley begins by sketching out the foil in pencil, he then applies light washes of colour and finally overworks the surface using a wider palette. The completed work is startling in its complexity and has a convincing appearance as the rendered material (Figure 2.34).

Contemporary artists are interested in creating life-like images, which on close inspection are still created using oil paint. These works demonstrate minute and detailed brush work and certainly lend themselves to a closer inspection. The range of twenty-first century material subjects are often highly challenging; even to capture using photography presents problems in colour rendering, tonal range and gamut. The twentieth and twenty-first century artists demonstrate their mastery of highly saturated colours, glossy or interreflective surfaces, different illumination points and colour temperatures, and highly textured or complex materials that contain a high degree of specular activity.

References

Adelson, E.H. (2008) Image statistics and surface perception, in *Human Vision and Electronic Imaging XIII* (eds B.E. Rogowitz and T.N. Pappas), SPIE-IS&T Electronic Imaging.

Arnheim, R. (1964) *Art and Visual Perception*, University of California Press, Berkeley.

Arnheim, R. (1970) *Visual Thinking*, Faber and Faber Limited, London.

Bayer, A., Gregori, M., Kemp, M. *et al.* (2004) *Painters of Reality, The Legacy of Leonardo and Caravaggio in Lombardy*, The Metropolitan Museum of Art, New York and Yale University Press, New Haven and London.

Constable, M. (2007) The painted photograph: Technical commonality between the digital composite and the pre-modern painting, in *World Conference on Educational Multimedia, Hypermedia and Telecommunications*, 25 June 2007, Vancouver, Canada. Association for the Advancement of Computing in Education (AACE), Chesapeake, VA, pp. 1121–1126.

Eley, R. (2011) *Process – The Painting of Aluminium Foil*, no. 1: https://www.youtube.com/watch?v=7tAPQ15kkig.

Fleming, R.W. (2013) Visual perception of materials and their properties. *Vision Research*, 94, 62–75.

Foister, S., Roy, A., and Wyld, M. (1997) *Making and Meaning Holbein's Ambassadors*, National Gallery Publications, London.

Gibson, J.J. (1950) *The Perception of the Visual World*, Houghton Mifflin, London.

Gibson, J.J. (1971) The information available in pictures. *Leonardo*, 4, 27–35.

Giles, S. and S, J. 1989. *The Art of the Ruins Adela Breton and the Temples of Mexico*, Bristol, City of Bristol Museum and Art Gallery

Giusti, A.M. and Radke, G.M. (2012) *The Gates of Paradise: From the Renaissance Workshop of Lorenzo Ghiberti to the Modern Restoration Studio*, Giunti Editore, Florence.

Gregory, R.L. (1970) *The Intelligent Eye*, Weidenfeld and Nicolson, London.

GregoryY, R.L. and Gombrich, E.H. (1973) *Illusion in Nature and Art*, Gerlad Duckworth, London.

Hammond, A.K. (1989) Aesthetic aspects of the photomechanical print, in *British Photography in the 19th Century* (ed. M. Weaver), Cambridge: Cambridge University Press, Cambridge.

Henry, P.M., Westland, S., and Cheung T.L.V. (2013) Colour selection strategies in colour design, in *AIC Colour 2013, Twelfth Congress of the International Color Association Proceedings*, pp. 1157–1160: https://3d3research.co.uk/student/becky-gooby/.

Hollman, E. and Tesch, J. (2004) *Trick of the Eye, Trompe l'Oeil Masterpieces*, Prestel, Munich.

Huson, D., Hoskins, S., and Thirkell, P. (2005) Photo ceramic relief imaging, in *NIP and Digital Fabrication Conference*, 2005, IS&T, pp. 39–42.

Jordan, W.B. and Cherry, P. (1995) *Spanish Still Life from Velázquez to Goya*, National Gallery, London.

Keith, L. (2006) *Velázquez's Painting Technique Velázquez*, National Gallery Publications, London.

Kemperdick, S. and Lammertse, F. (2012). Painting around 1400 and Jan van Eyck's early work, in *The Road to Van Eyck* (eds S. Kemperdick and F. Lammertse), Museum Boijmans Van Beuningen, Rotterdam.

Letze, O. (ed.) (2013) *Photorealism 50 Years of Hyperrealistic Painting*, Hatje Cantz Verlag, Ostfildern.

McCallion, P. (2017) The development of methods for the reproduction in continuous tone of digitally printed colour artworks. PhD thesis, University of the West of England, Bristol.

Melcher, D. and Cavanagh, P. (2011) Pictorial cues in art and in visual perception. in *Art and the Senses* (eds F. Bacci and D. Melcher), Oxford University Press, Oxford.

Omarini, S. (2012) Notes on colours and pigments in the ancient world. *Journal of the International Colour Association*, 8, 61–67.

Parraman, C. (2014) The visual appearance and surface texture of materials according to the old masters, in *Electronic Imaging*, IS&T/SPIE, San Francisco.

Rogers, L.R. (1974) *Relief Sculpture*, Oxford University Press, London.

Sanders, N.K. (1969) *Prehistoric Art in Europe*, Pelican, Harmondsworth.

Sonnenburg, H.V. (1971) The technique and conservation of the portrait. *The Metropolitan Museum of Art Bulletin*.

Vasari, G. (1550) *Lives of the Most Excellent Painters, Sculptors, and Architects, from Cimabue to Our Times*, Torrentino.

Wyld, M. (1998) The restoration history of Holbein's *The Ambassadors. National Gallery Technical Bulletin*, London.

Bibliography and Further Reading

Bergen, J.R. and Adelson, E.H. (1988) Early vision and texture perception. *Nature*, 333, 363–364.

Blay, M. (1995) Lumière et couleurs newtoniennes, in *La Lumière au Siècle des Lumières et Aujourd'hui* (ed. J.P. Changeux), Odile Jacob, Paris.

Boime, A. (1987) *Art in the Age of Revolution 1750–1800*, University of Chicago Press, Chicago.

Brewster, D. (1824) On the optical illusion of the conversion of cameos into intaglios, and of intaglios into cameos, with an account of other analogous phenomena. *The Edinburgh Journal of Science, VII*, 99–108.

Bryson, N. (1990) *Looking at the Overlooked, Four Essays on Still Life Painting*, Reaktion Books, London.

Caffin, C.H. (1899) Pneumatic power in stone-carving. *The Artist: An Illustrated Monthly Record of Arts, Crafts and Industries (American Edition)*, 26, xlv–xlvi.

Callen, A. (1994) The unvarnished truth: Mattness, 'primitivism' and modernity in French painting, c.1870–1907.*The Burlington Magazine*, 136 (1100), November 1994, 738–746.

Cardaci, D. (2007) *Drawing Made Easy Realistic Textures: Discover Your 'Inner Artist' as You Explore the Basic Theories and Techniques of Pencil Drawing*, Walter Foster Lake Forest, California.

Da Vinci, L. (1910) *The Notebooks of Leonardo da Vinci*, Duckworth, London.

Dawkins, R. (1998) *Unweaving the Rainbow, Science Delusion and the Appetite for Wonder*, Penguin, London.

Earl, G., Beale, G., Happa, J. *et al.* (2009) A re-painted Amazon, in *Electronic Imaging in the Visual Arts Conference* (EVA), London.

Earl, G., Beale, G., Martinez, K., and Pagi, H. (2010) Polynomial texture mapping and related imaging technologies for the recording, analysis and presentation of archaeological materials, in *International Archives of Photogrammetry, Remote Sensing and Spatial Information Sciences*, Commission V Symposium WG VI/4, Newcastle upon Tyne.

Elkhuizen, W.S., Zaman, T., Verhofstad, W. *et al.* (2014) Topographical scanning and reproduction of near-planar surfaces of paintings, in *Proceedings of SPIE 9018, Measuring, Modeling, and Reproducing Material Appearance*, 901809.

Gombrich, E.H. (1995) *Shadows. The Depiction of Cast Shadows in Western Art*, National Gallery Publications, London.

Gregory, R.L. (1998) *Eye and Brain. The Psychology of Seeing*, Oxford University Press, Oxford.

Guse, E.G. (1981) *Reliefs Form Problems: Zwischen Malerei und Skulptur im 20 Jahrhundert*, Westfälisches Landmuseum, Münster.

Hakala, J., Nuutinen, M., and Oittinen, P. (2011) Interestingness of stereoscopic images, in *IS&T/SPIE Electronic Imaging, Proceedings of SPIE 7863, Stereoscopic Displays and Applications*, vol. XXII, 78631S.

Heaven, D. (2014) ButtonMasher: photorealism takes gaming deeper. *New Scientist*. Accessed at: https://www.newscientist.com/article/dn25860-buttonmasher-photorealism-takes-gaming-deeper/

Hochberg, J. (1979) Some of the things that paintings are, in *Perception and Pictorial Representation* (eds C. Nodine and D. Fisher), Praeger, New York.

Schele, L. (1985) Color on classic architecture and monumental sculpture of the Southern Maya Lowlands, in *Painted Architecture and Polychrome Monumental Sculpture in Mesoamerica* (ed. E. Hill Boone), Harvard University Press, Washington.

Skira, P. (1989) *Still Life: A History*, Rizzoli, New York.

Tyler, C. (2010) Darkness and depth in early Renaissance painting, in *IS&T/SPIE Electronic Imaging Proceedings of SPIE 7527, Human Vision and Electronic Imaging*, vol. XV, 75270V.

Wallert, A., Hermens, E., and Peek, M. (1995) *Historical Painting Techniques, Materials, and Studio Practice*, Getty Conservation Institute, University of Leiden, The Netherlands.

Westland, S., Ripamonti, C., and Cheung, V. (2012) *Computational Colour Science Using MATLAB*, 2nd edn, John Wiley & Sons, Ltd., Chichester.

3

The Present: Materials, Making, Capturing and Measuring

3.1 Introduction: Universal Knowledge

In 1755, Denis Diderot (1685–1759) predicted that people in the future would not spend their time reading but rather 'devote themselves to investigation which will be new, or which they will believe to be new'. He further suggested that those unable to come up with their own ideas, 'will be busy night and day leafing through these books, taking out of them the fragments they consider worthy of being collected and preserved' (Yeo, 1996, p. 1).

The gathering, storing and retrieving of knowledge, as demonstrated by the Encyclopaedias of the eighteenth century, was crucial to gaining insights of practical knowledge, the skill of the craftsperson and furthering the knowledge of the philosopher. Over a century earlier, Francis Bacon (1561–1626) also reflected on the idea that knowledge needed to be practical and relate mankind to the world. The two-volume *Cyclopaedia* by Ephram Chambers (1680–1740), *An Universal Dictionary of Arts and Sciences*, was published in England in 1728 (http://www.xn--cyclopdia-l3a.eu/) and was a first attempt in the eighteenth century to summarise arts and scientific advances. In 1750, Denis Diderot was employed to undertake an overhaul of the Chambers' Cyclopaedia and, along with contributions from over one hundred writers, including François-Marie Arouet (better known as Voltaire, 1694–1778), Jean-Jacques Rousseau (1712–1778) and mathematician Jean-Baptiste le Rond d'Alembert (1717-1783). Their quest was to bring a wider body of knowledge together and to assemble the sciences, arts and manufacture into a dictionary of universal knowledge.

As an Enlightenment enterprise of the eighteenth century, the *Encyclopédie* fuelled a culture of intellectual debate in the so-called Enlightenment salons that brought together discursive activity from different social and cultural backgrounds. The impact of the *Encyclopédie* on subscribers and society was that it did not just provide knowledge but *new ideas*, which were perceived to potentially challenge, subvert and destabilise the economic balance, the class system and religious control of the time. For example, Diderot's desire to explain the *secrets* of the mechanical arts, and to show all the stages of manufacture, according to the wealthy guilds, could potentially reveal and undermine their precious knowledge. In fear of reprisal, many of his collaborators withdrew from the project and Diderot was

2.5D Printing: Bridging the Gap Between 2D and 3D Applications, First Edition. Carinna Parraman and Maria V. Ortiz Segovia.

Companion website: www.wiley.com/go/bridging2d3d

forced to work in secrecy. Eventually the 28 volume *Encyclopédie, ou Dictionnaire Raisonné des Sciences, des Arts et des Metiers* (http://xn--encyclopdie-ibb.eu/index.php) was published in 1772.

As a recorder and journalist, Diderot was not an expert, but an obsessive collector and conveyor of information. He provided curious insights into seventeenth and eighteenth century science and technology and the crossover with the mechanical arts. Diderot was certainly aware that people would not become experts by reading his book, but his methods of gathering knowledge was a way to demonstrate and describe the sum of human practical knowledge in a way that was commonplace to the reader. As suggested by Richard Yeo, writing in the twentieth century, the term 'common place' was understood differently in the eighteenth century. The Encyclopaedia could be considered as a 'store house of material' or as a 'public service', where readers could use the book to 'prompt their memory' for experts or as a 'single point of reference' for non-experts (Yeo, 1996, p.167).

How might Diderot's collection of information about practice on *des sciences, des arts et des metiers* in the eighteenth century relate to twenty-first century making? Through the *Encyclopédie*, Diderot attempted to bring together the commonality and development of skill and talent through making, problem solving by iteration and intuitive knowledge (Sennet, 2009). Echoing the Baconian ideal, the purpose of knowledge was to understand the practical. Diderot sought to find skilled people who could demonstrate their work and, based on his observations, attempted to systematise the experience of making through descriptions and detailed illustrations of the tools and instruments (Pannabecker, 1994). He chose to demonstrate these skills and reflect on their practice. His endeavours highlight the concerns of the twenty-first century of the need to maintain skills, to ensure a greater understanding of the crossovers between knowledge and materials, and possibly how these skills and insights are necessary for new ways of manufacturing and engaging with digital technologies. Diderot's observations in 1755 also reflected a typical structure of society: skilled makers actively engaged in problem solving and innovating, those engaged in working and making, and lastly observers, philosophers and consumers. His obsession with attempting to gather different knowledge highlights the differences between tacit knowledge – knowledge that has been gained over years is intuitively applied and cannot be easily expressed – and by explanation – similar to show and tell or demonstration – and from a practitioner perspective – endeavours by scientists, makers, designers, who employ leaps of faith based on tacit knowledge, who think and make, work across disciplines, work collaboratively and break so-called boundaries.

3.2 The Relationship of Digital Technologies, Knowledge of Materials and Skills

Reflecting on a book on 2.5D printing, what is the relevance and relationship of printing to digital technologies, knowledge of materials and skills? The primary answer is in the need for continued development of skills, the need for diversity of technologies and a search for a range of smarter materials to support it. These include inks, pigments, tools, old materials, new materials and materials in waiting. These materials also need to incorporate the complete design workflow – from cradle to cradle, as explored in Chapter 4. It is imperative that skills and innovation are continued to be supported by governments, research councils, and through collaboration between academia and industry that can drive towards novel and smarter products.

We may also ask if there is a threatened loss of skill in traditional practice and established knowledge, and how may new practices be achieved through innovation and new ways of thinking and working? On the one hand, we could suggest that, rather than inventing new materials that simulated existing stuff, what if we took the materials we already have and worked out how to print them? This answer seems very simplistic and obvious, but throughout history we have found that new insights are gained through a constant reappraisal of existing methods and materials to find new solutions for new applications. There is also a wealth of new and emerging materials that are transforming our lives (Brownell,

2010). From a combination of different ingredients of the new and the old, these are providing inspiration and presenting new possibilities for new applications.

In the twenty-first century, a question that is repeated often is 'have we lost touch with the things we make?' (Sennet, 2009) Is the designer more reliant on a screen to draw things, the physical making removed and, therefore lost the *secrets* of the mechanical arts? The idea of 'tacit knowledge' – a priori knowledge and expertise – that we bring to making objects evolves with experience and practice. As explored by the Hungarian–British critical thinker Michael Polanyi in the late 1950s, 'Great industries ... have realized in these days that they were carrying on their activities in the manner of an art without any clear knowledge of the constituent detailed operations' (Polanyi, 1958). Is there a continued requirement for understanding the crossovers between the disciplines? One the one hand, current manufacturing methods are driven by computer-aided software that is not necessarily empathetic to the material. On the other hand, we are overwhelmed with the potential of new materials and products that can transform our lives.

Over the last century there has been significant study into the evolution of materials. In all fields of making – from craft to engineering – there is a complex symbiotic interrelationship with materials and making: the idea of how an object may look, perform, and the problem solving required to reach that end product using different tools and materials. As Canadian philosopher Marshal McLuhan observed in the 1960s, as a new technology is introduced, each will impact human patterns of behaviour relating to the medium, which, in turn, will also influence the design process by amplifying or accelerating existing processes (McLuhan, 1964). And may also result in the redundancy of an existing process. Ben Shneiderman, in his book *Leonardo's Laptop* (Shneiderman, 2002), describes the shift in technology over the last 20 years, 'The old computing was about what computers could do; the new computing is about what users can do.' This careful line of enquiry highlights the more recent problems and split in communication and practice between the arts and sciences. This split is echoed in the 1959 Rede Lecture. C.P. Snow described how the arts and sciences were divided into 'two cultures' (Snow, 1959). Even in the twenty-first century, the arts and sciences' engagement with colour is generally regarded as being 'in splendid isolation' whereas the relevance of combining the sciences and the arts is often called into question (Love, 2011), the relevance of which continues to be debated (Wilson, Hawkins and Sim, 2014). However, as exemplified by fifteenth century polymath Leonardo da Vinci, demonstrating his knowledge of science and technology through his drawings and art, his assumption was that ideas could be explored through experience and the empirical process of *trial and error*. According to Venturi, one of Leonardo's best students, 'The rational intellect (*l'espirit géometrique*) guided him (Leonardo) in everything, whether in the art of analysing an object, in the stricture of his discourse, or in the care always to generalize his ideas' (Turner, 1995, p. 86).

Of course, depending on whether one is a beginner or an expert, there may well be a difference between expectations and outcome, and this will change over time as the learner's skills develop, and by reducing that gap between frustration and accomplishment, and from competency to mastery.

Current additive manufacturing methods to build an object employs a layering approach, which involves lasers, glue or heat to bind many layers. This might seem to be a highly inefficient and time-consuming way of printing or making, but, on the other hand, it is also a useful tool for exploiting and investigating new materials, and exploring other methods for testing machine and material compatibility. Moreover, the introduction of entry level, build-it-yourself and cheap desktop fabricators means that these machines can be exploited in unexpected ways and can achieve surprising results. For example, because of their low price they can be hacked or modified, and without the cost of invalidating a warranty, alternative materials can be used, and machines can be tested or broken.

Managing expectations, as demonstrated at trade expos, exhibitions, the many design expos across the world, or according to the typical Hype Cycle (https://www.gartner.com) of emerging technologies, according to Amara's Law, 'we tend to overestimate the effect of a technology in the short run and underestimate the effect in the long run' (http://www.pcmag.com/encyclopedia/term/45949/laws). Reflecting on exhibitions that consider the relationship of digital technologies and craft skills, many thought-provoking examples were demonstrated at the Victoria and Albert Museum (V&A) and Crafts

Council exhibitions: *The Power of Making* (2011) and *What is Luxury* (2015). At the 2015 *Surface Design* panel discussion, chaired by Eleanor Young, Executive Editor of *RIBA Journal*, the panel explored the question 'Breaking the mould: are modern techniques bringing back the idea of craftsmanship into modern design?' At the 2011 V&A and Crafts Council exhibition, *The Power of Making* curated by Daniel Charney (2011), the objective was to showcase contemporary craft practitioners, to demonstrate the diversity of craft making and applaud the crossovers between the hand-made, digital fabrication and mechanical technologies. In the catalogue of the same name, the authors describe how craft skills are very much alive, despite the ongoing negative and conflicting counterarguments, 'since the industrial revolution there has been a mourning for the loss of manual labour' (Miller, 2011, p.16) and 'we live in a world that constantly tempts us to be superficial and transient' (Miller, 2011, p. 22). Richard Sennet, in his book *The Craftsman*, echoes the malcontents, but also considers the idea of mechanisation in a different way and proposes how 'might we in our very comparative imperfection learn something positive about being human?' (Sennet, 2009, p.81). He compares the craftsman's desire to make by hand, incorporating variety and flaws, versus the anxiety of 'the rigorous perfection of the machine' (Sennet, 2009, p. 84). He describes two different types of machines, that of the replicant - that mimics human functions in order to best serve our needs (pacemakers, artificial limbs), or as a robot – a machine that is more of an extension of our physical self but larger and stronger, that works faster but never tires (printing presses, paper mills, computers). As Sennet explains, the replicant shows us as we are, the robot as we might be' (Sennet, 2009, pp. 84–85).

Although digital output has increased over the last decade, we have also experienced a reappraisal of analogue processes and renewal of so-called heritage industries. By heritage, we shall also include processes that have been in use over the last 50 years and are now in decline. Whilst most users today will capture, design and print an image employing a digital workflow, there are many analogue print processes that have been adapted for digital output. This may well be due to the qualities of the fine art print remain unsurpassed, particularly where the technique characterises the appearance of the image: colour, texture, gloss, opacity and translucency, versatility, longevity and material provenance. According to Stephen Knott in his book *Amateur Craft History and Theory*, we are entering a third industrial revolution (Knott, 2015). Although amateur craft is more aligned to technique, the need for traditional craft skills and analogue processes are still in demand, and are being reinvestigated and renewed (see Figure 3.1). According to Mark Miodownik, materials scientist, cofounder of the Materials Library at University College London and author of *Stuff Matters*, traditional materials such as wood, so-called *Wood 2.0*, and

Figure 3.1 Examples of an intricately carved wood panel in Hangzhou, China.

with the help of digital technologies, are being given a new lease of life (Miodownik, 2015). The Internet and You Tube also bridges the gap as a didactic digital aid and as a medium to learn and share skills. Demonstration-by-QuickTime movies are able to overcome language barriers – from dry-stone walling to mastering a new tool in Illustrator. New networks are forged via micro-industries, which are not constrained to a region or a country but can be accessed on a global level.

Our emotional responses to manufactured objects are inextricably bound to the types of materials we use, the tools and how both are combined. Wood represents endurance, sustainability and tradition, whereas plastic gives rise to anxieties about cheapness, recycling and landfill. However, new materials continue to be a source of fascination. Is it authentic? What can be made with a new material? How does it feel? How might it transform our life? How can this be measured? The incorporation of different materials can fulfil both an emotional and functional role. The scientific discipline dedicated to measuring our sensual understanding of images, materials, tastes and textures is psychophysics. For example, we may ask how two printed colours may compare under a particular lighting condition or which wood grain is more pleasing. Natural materials may age and develop a patina over time. We could argue that a simple wooden chopping board that is smoothed and oiled is more pleasurable to chop food upon and present food, compared to a plastic board. Over time, the wood takes on different qualities, cuts and dents that belies its use and location. Furthermore, different woods from around the world convey subtle nuances of colour and grain that, when combined, smoothed and polished, can present an extraordinary and visually rich appearance. The different woods used to create the Kew Mural (see Figure 3.2) were collected from trees that had been damaged or destroyed by a storm in 1987. The different woods included: Walnut, Larch, Lime, Oak, Yew, Ash, Hickory, Honey Locust and Elm and demonstrate a variety of colours and textures.

The term 'truth to materials', originally associated with William Morris and The Arts and Crafts of the nineteenth century, and later borrowed from the twentieth century *Brutalist* style of architecture, practitioners considered the importance of using the most appropriate materials to construct buildings and create objects. For example, a wooden table constructed from low-grade industrial chipboard that is varnished to bring out the complexity of the grain, may appear more pleasing than a printed wood

Figure 3.2 Carved by Robert H. Games, the Kew Mural incorporates different woods from Kew Botanical Gardens, London.

Figure 3.3 *Herringbones*, by Yael Mer and Shay Alkalay, Raw Edges, as shown at the Salone del Mobile, Milan, 2016 (courtesy: Raw Edges) (http://www.raw-edges.com/#/herringbones-1).

effect veneer. Likewise, as demonstrated in Figure 3.3 by Yael Mer and Shay Alkalay of Raw Edges Studio V (http://www.raw-edges.com), by simply dipping wood into coloured dye at 45 degrees, the colours expand along the grain of the wood to create subtle coloured patterns, which is simple yet elegant. Although generated from waste materials, a panel made from chopped-up bank notes or recycled plastic bottles, may also be visually surprising. Figure 3.4 shows a range of panels made by Smile Plastics, who have been inspiring designers and architects since the early 1990s (http://www.smile-plastics.com).

There are many innovative ways to combine traditional materials with novel methods of manufacture or to work with composites to change physical properties. There is also a wealth of inspirational books that combine traditional and CAD methods of manufacture. In many cases, none are new, but

Figure 3.4 Different materials used to create colourful panels (courtesy: Smile Plastics).

tend to be more of a reinvestigation that may re-evaluate how digital technologies or contemporary manufacturing processes can transform old materials to create novel surface qualities and textures. As an example, Marcel Wanders' *Crochet Table*, and Matthias Wisniewski's *Deco Chair* use epoxy resins and fabrication methods that are traditionally used to make fibreglass boats or patch holes in cars, has now been transformed into an elegant and aesthetically pleasing material (see Figure 3.6). Fibreglass embedded in resin is usually concealed under layers of paint, but is now being used in combination with flat crocheted cotton material. As demonstrated by Dutch designer Wanders' *Crochet Table*, this delicate piece of textile in its original state has no stability, but when saturated in epoxy resin and moulded to a shape can provide a strong robust structure (see Figure 3.7). The *Deco Chair*, created by German designer Matthias Wisniewski, is fabricated from carbon fibre and epoxy resin. It combines both craft sensibilities of crochet alongside modern casting materials. Wisniewski's chair resembles old cast iron garden chairs of the nineteenth century. It is interesting to note that in the original cast iron furniture, although made of heavy cast iron, the nineteenth century craftsman attempted to capture the natural flora and fauna, often reflecting the gardens and hot-house plants which they were placed near. Figure 3.5 shows a cast iron bench that is placed in a garden setting surrounded by ferns (see also Case Study 1: Nature Printing in the Nineteenth Century).

Over the last few years, commercial efforts have concentrated on the supply and demand of hardware, software and materials, and research into overcoming barriers to methods of printing and printing materials. Whilst 3D printers can make unique objects that traditional manufacturing methods are unable to make, these objects are, more often than not, fabricated using unappealing materials. This issue was also considered by the owner of ideas-based TED lectures, Chris Anderson, 'Laser cutters, 3D printers and CNC mills are cool but like all desk top manufacturing machines they are limited in what they can make both in materials and complexity'. His alternative is what he called a 'full-on Universal Fabricator', which is akin to a 'Star Trek Replicator, it's a machine that can make almost anything on command' (Anderson, 2013, p. 217). The emergence of the 3D print industry has demonstrated the need for equipment and materials to be developed in combination so that new applications can be developed. One of the most important aspects of the emergence of industrial printing is the

Figure 3.5 Cast iron bench with fern motif along the back arms and legs, Berrington Hall, Herefordshire.

Figure 3.6 Matthias Wisniewski, *Deco Chair*, carbon fibre and epoxy resin (courtesy: Matthias Wisniewski) (http://mateodesign.jimdo.com/art-design).

Figure 3.7 Marcel Wanders, *Crochet Table* (2001), cotton, epoxy (30 cm × 30 cm × 30 cm). Produced and distributed by Moooi (courtesy: Marcel Wanders) (https://www.marcelwanders.com).

demand for materials and it is often highly specialist markets (dental, medical, aerospace and engineering) that are driving this supply.

Until the advent of 3D printing, three-dimensional fabrication and manufacturing had consisted of many different forming, machining and construction methods, for example casting, forging and injection moulding (using metal, plaster, ceramic, rubber), fabricating and weaving (using clay, fabrics, fibres), welding, joining, bending and laminating (using wood, plastics, fibres) and machining and cutting (using wood, metal, plastics). This is certainly not an exhaustive list, but conveys the complex and range of manufacturing processes. In order to match material to construction method, the engineer craftsman or designer requires considerable materials expertise and knowledge of how to manipulate materials according to the material's properties and how the product will be used. As described throughout these chapters, this knowledge is diverse, nuanced and has often been passed down over centuries by artists, studios, apprenticeships and across cultures via trade routes and as travelling journeymen. From an evolutional perspective of materials and the advancement of civilisations through the ages – Stone, Copper, Bronze, Iron – methods have developed and been exploited according to availability of materials levels of innovation, know-how and tools.

Of course, there is a difference between the craftsperson with a tacit material understanding of how to make objects versus a material scientist who is able to characterise materials according to their composition (atoms). It was not until the ability to analyse materials on a micro or atomic scale that technologists have been able to characterise, group and understand how different materials behave, or how materials could be combined to produce new materials for different applications. This discipline did not really emerge until the 1950s when the term 'materials science engineering' (MSE) began to be used by the Department of Metallurgy in Northwestern University, Illinois, USA. According to Cahn, there are three conditions for understanding and control of materials: composition (atoms), its microstructure (how the atoms are arranged) and performance (how they behave, e.g. thermodynamics and kinetics) (Cahn, 2001). In the field of materials science, it is vitally important that materials are understood from a cross-disciplinary perspective, from science as well as engineering. Many historic discoveries of materials have been made by those who have applied their knowledge from one discipline to another or realised new opportunities and applications (Ferguson, 2007). The past has shown how some were discovered by accident. For example, the first synthetic dye, *Mauveine*, was discovered in 1856 by pharmacist William Henry Perkin (1838–1907) (http://www.rsc.org/Chemsoc/Activities/Perkin/2006/minisite_perkin_mauveine_non_flash.html). The hydrocarbon aniline, an oily unwanted by-product of coal, was discovered to produce an intense purple colour when dissolved in alcohol. The impact of this resulted in a new range of synthetic colours including fuchsine, aldehyde green and synthetic indigo.

Recent research and explorations have resulted in yet more of an explosion of new ideas as well as finding new uses for existing materials, or smart materials that demonstrate multiples uses. The variety of materials that can be used are increasing and are set to grow quickly over the next few years, including conductive, bio- and nanomaterials, composites, ceramics, polymers, gels, pigments, pastes and liquids. Recent studies into the development of 'super materials', which are classified as 'in-waiting', suggest that laboratories around the world are inventing ingenious materials, for example memory glass, where terabytes of information can potentially be stored in 3D memory crystal glass (a similar ingenious material) and 'self-healing polymers', which have a potential application for renewing holes or cracks in structures such as buildings or bridges, or could even have the power to renew themselves. These super materials, like others 'in-waiting', are not yet robust enough for the marketplace. For example, the aerogel, which was developed in the 1930s by Samuel Kistler (1900–1975) and described as the world's lightest solid, remained on hold until developments into copper nanofoams by Brown University in Providence, Rhode Island, could utilise the concept via an inexpensive method for removing environmental pollutants (Cartwright *et al.*, 2014). Aerogel is neither a liquid nor solid, but water that is trapped under surface tension within a mesh-like solid. Kistler went on to develop a silica aerogel, which at 99.8% air is the lightest solid in the world.

Likewise, the very common carbon material of graphite – the oldest material that makes black marks on paper – has taken on renewed interest through the invention of Graphene. Graphene is a 2D material and belongs to an extraordinary class of materials that are only one atom in thickness. Graphene, because it is only on atom thick, is highly translucent, but is also well known for its high strength, stiffness and conductivity. It may well have the potential to replace silicone chips or introduce touch sensitivity to screens, smart textiles and buildings (Miodownik, 2015).

These sorts of materials demonstrate the potential for materials that can be printed. Some are already under way, such as hydroxyapatite – a phosphate mineral that makes up teeth and bones, bio-glass, and bone and tissue-loving materials for dental or facial reconstruction, or biomaterials for tissue and organ regrowth. Now we have the opportunity to make a bridge to study how these can be effectively applied.

3.3 Methods to Capture and Measure Texture

Capturing texture and optical properties of surfaces has been a challenge over the last couple of decades. Today, we find many methods to do one or the other, but it is still difficult to find something that would exactly fit a workflow to reproduce the appearance of a complex surface. A typical situation is to capture geometrical aspects of surfaces with enough resolution, but this may well be to the detriment of accurate colour and reflection properties, or vice versa. The size of capturing systems, their field of view, their availability, their functionality, their capacities, their price and even the lack of standardization between manufacturers have posed a hindrance to the development of a complete solution that could address several applications at once.

On the other hand, methods for measuring texture have undergone significant development compared with methods for capture. Metrology institutes, as exampled by collaboration EURAMET (https://www.euramet.org) in Europe (Laboratoire National de Métrologie et d'Essais, Physikalisch-Technische Bundesanstalt, National Physical Laboratory) and worldwide (Bureau International des Poids and Mesures) (http://www.bipm.org/en/about-us/) work on providing metrological references and accreditation that have been certified or approved by different organisations. Goniochromatic measurements, reflection properties and surface roughness, among others, can be performed with high precision, but the market remains limited to traditional applications such as paints for the automobile and decoration industries. Their focus is precision rather than applications. Therefore, these measuring devices are specifically designed for a single purpose and usually end up measuring more (in the sense of accuracy) than what is needed for a reproduction pipeline. The trade-off with measuring devices is usually related to the size of the samples that can be measured: the more precise they are, the smaller the sample they measure. Besides quality control, the main task of measuring devices in a reproduction workflow is to measure specific surface parameters of tiny samples to replicate them afterwards across larger areas. However, measuring devices are not yet capable of capturing entire objects.

To put it in perspective, the current status of capturing and measuring for a surface reproduction pipeline can be compared to early scanners that would only work for one printing technology (inkjet, electrophotography) at a time or for a single parameter at a time (colour or black and white, images or text). Some devices that work well in acquiring complex geometries do not provide any optical information (in some cases even colour is not accessible); others work well in gathering reflection properties (bidirectional reflectance distribution function – BRDF), but do not provide geometrical information; and others provide a mix of both but at very low resolution or as an approximation. There is an increasing need to consolidate the features and specifications that would fit a good range of reproduction scenarios in a complete capturing system.

3.3.1 Commercial and Specialist Scanners

Commercially available scanners provide a more conservative function than those developed for Education and Research, but have the advantage of being available for general use. Their prices range from between a few thousand to hundreds of thousands of dollars depending on their resolution, portability and specification. They are commonly based on classical principles of interferometry, laser scanning, profilometry, stereophotography, goniophotometry and multifocal microscopy. Often scanners and 3D printers and supporting software are purchased as a package. These are useful as a starting point: by scanning an existing object that can then be manipulated using the software, rather than building an object from scratch and learning the software that is necessary. High-resolution scanners include the 3DPIXA stereo camera combining line scanning technology with fast stereo algorithms, where height data are calculated from two images of the same object (https://www.chromasens.de/en). The Cruse (https://crusescanner.com) museum and large format scanners designed for artworks up to 200 cm × 300 cm and 3D photo-station are capable of multilevel scanning of objects up to 25 cm^3 The REFLET 180 is a goniospectrophotometer for measuring scattering from diffusing samples or specular near-planar surfaces and the GON 360 is a goniometer for angle-dependent transmission and reflection measurements of a surface of up to 10 mm diameter. Some of the more popular 3D scanners include iSense, MakerBot Digitizer Desktop 3D Scanner and Fuel 3D (http://www.toptenreviews.com/computers/scanners/best-3d-scanners). MakerBot Digitizer and Fuel3D SCANIFY, iSense DAVID SLS-2 are packaged as low-end scanners for fast acquisition of small objects.

The following list outlines the range of technical methods used to capture and measure, texture and draw from cultural heritage research projects as examples of applications (with thanks to Xavier Aure for an explanation of the 3D and 2.5D methods of scanning).

Interferometry/structured light scanning. A known image pattern is projected on to the surface of the object or surface and the deformation of the captured signal is interpreted from the topographic changes in the surface. A typical structured light system consists of a projector and one or two camera sensors. The signals can range from simple black and white patterns to grey tones or more complex patterns (Akca, 2011); Bertani *et al.*, 1995; Guidi *et al.*, 2004; Bunsch and Sitnik, 2010; Karaszewski *et al.*, 2014; Zaman *et al.*, 2014). Some systems can also capture colour information; performance depends on the projector (frequency of the signal, sharpness) and the sensor capturing the deformed signal (camera). There are many external factors that can impact the acquisition quality of the scan, for example, occlusion, gloss, highly reflective surfaces, transparency, as well as effects of ambient light.

Laser triangulation scanning. Similar to interferometry, a laser scanning beam is projected on to the surface of the object and the deformation of the beam is read as proportional deformations of the surface. The colour of the laser beam dictates the colour of the surface that can be captured accurately, for instance, red lasers would perform very well with reddish surfaces but not with greenish ones. Two or more colours of laser could be used to avoid this issue. The distances in between the object, the laser and the sensor determine the size of the capturing area and the resolution. If only one laser is used, the system will only provide information in one dimension at once (http://www.factum-arte.com/pag/5/Laser-scanning) . It is also possible to capture colour by analysing the reflected light from RGB lasers (Hess and Robson, 2013; Blais *et al.*, 2005). Laser triangulation systems have generally very good accuracy. One of the disadvantages is the relatively short depth of view that makes scanning big objects more complicated. The depth of field can also be quite small, which can cause problems while scanning objects with variable depth and require stitching of different range maps to obtain complete models. Another important issue is the difficulty in capturing reflective, dark and transparent surfaces. The National Research Council of Canada developed a colour laser scanner for cultural heritage objects and as early as 1995 undertook the first laser scanning exploration of paintings. The first painting scanned with their system was *Auvers, Street Descending* by the French artist Jean-Baptiste-Camille Corot (1796–1875). The painting was heavily varnished and therefore appeared relatively flat. However, the images captured by the system revealed a rich brush stroke surface relief, masked by the layers of varnish. The scanner provided a depth resolution of 10 μm (0.01 mm) in the z direction and 50 μm (0.05 mm) in the x and y directions. Other studies of paintings by the same group included the scanning of a Renaissance painting, three works by Renoir and the creation of a virtual version of the

Mona Lisa (Blais *et al.*, 2007). The NRCC colour laser scanner has also been licenced to Arius 3D, a commercial company specialising in laser-based optical scanning systems. Laser scanning technology specific to paintings has been developed at the Factum Arte Foundation (http://www.factumfoundation.org). The Lucida Laser scanner is a two-camera, one-laser scanner with bidirectional recording. Their projects have involved the scanning of *Wedding at Cana* (1563) by Paolo Veronese (1528–1588) (http://www.factum-arte.com/pag/38/A-facsimile-of-the-Wedding-at-Cana-by-Paolo-Veronese) and *The Assassination of Saint Peter Martyr* (1507) by Giovanni Bellini (1459–1516) (https://www.factumfoundation.org/pag/630/Scanning-a-painting-by-Bellini-br-at-i-The-National-Gallery-i-London), among many others.

Photogrammetry. This is an image-based modelling technique that is used to obtain three-dimensional models from standard two-dimensional images. Models are produced by photographing the object from multiple angles and processing the resulting overlapping stereoscopic images through automatic image correlation algorithms. These are generally available from a variety of commercial and open-source software. Photogrammetry derives the 3D coordinates of points from the overlapping images. The resulting point cloud can then generate a polygonal three-dimensional mesh. This technique can produce very dense point clouds, although the algorithms used to create the polygonal structures work with smoothing constraints, which can cause loss of information and detail (Remondino and El-Hakim, 2006). An advantage of photogrammetry, compared to most laser scanning systems, is that it can simultaneously capture surface geometry and colour. This method is dependent on the quality and accuracy of the photographs, that is higher image resolutions produce denser models. As a general rule, photographs must have uniform exposure, high contrast, short distance between the camera and the subject. Furthermore, the object must fill the frame and have at least a 60% overlap between each pair of images. In order to determine the scale of the subject, one or more objects of known dimensions need to be visible in the image. However, a wide number of studies exist that use photogrammetry to survey and model areas that do not require high surface accuracy, such as archaeological sites, monuments and buildings (Remondino and Rizzi, 2010; Kersten and Lindstaed, 2012). It has also been used to record small three-dimensional objects that are relatively easy to capture from all directions (Barstanti, Nicoli and Guide, 2013; Barreau *et al.*, 2014; Samaan, Héno and Pierrot-Deseilligny, 2013; Gallo, Muzuppa and Bruno, 2014). Some studies have used photogrammetry in combination with other imaging techniques such as RTI or 3D scanning to study rock art (http://culturalheritageimaging.org/What_We_Do/Fields/rock_art/ [accessed May 2015]) and objects (Zeng *et al.*, 2012; Andrade *et al.*, 2012). Research on the accurate recording of painted surfaces is far less extensive. Photogrammetry is a mature technique and has been used as early as 1987 (Cather, 1991) in the Brancacci Chapel in Florence, Italy, to capture and quantify deformations on mural paintings and the documentation of pre-Hispanic wall paintings in Mexico (Lucet, 2013). The capture and processing of accurate surface details from canvas or panel paintings demonstrate some problems due to specularity and dark areas. There are studies that have used photogrammetry to quantify the overall surface deformation of paintings (D'Amalio and Brutto, 2009; Abate *et al.*, 2014). The best applications of photogrammetry in paintings are in combination with other recording techniques such as multispectral imaging and laser scanning (Mohen, Menu and Mottin, 2006; Pires *et al.*, 2007).

Reflectance transformation imaging. RTI is a computational photography method that captures surface texture and colour of an object and provides virtual three-dimensional information using traditional photographic techniques. The camera and object are both fixed and RTI images are obtained by photographing the object many times, using illumination points that are set at many different known angles. The software processes the images by calculating and stitching all the pixels according to the different lighting angles. The interface enables the user to interactively examine the surface by virtually illuminating the surface of the object from any direction. The RTI images produced are similar to two-dimensional photographs, but reflectance and colour information is contained within every pixel, so when the images are virtually illuminated from different angles they reveal micro details of the object's surface topography as if one was illuminating the real object. Through the application of a variety of filters it is possible to emphasize the surface relief to uncover details that are impossible to record by other imaging techniques. Most RTI methods produce similar results (Earl, 2011; Barbosa, Sobral and Proença, 2007), which include lighting domes with a fixed camera and a light array, lighting arcs of various sizes, a highlight method with a fixed camera, a hand-held light source and reflective spheres. The first two methods allow for faster capturing of the images, but usually present a limit to the size of the object that can be captured. The highlight method offers more flexibility in terms of the object size, but requires more time and preparation. In general, RTI offers certain advantages over other 3D scanning techniques. The cost of the equipment necessary to obtain satisfactory RTI images is much lower, the accuracy and sample density can be far better than many 3D scanners and the size range of objects and materials that can be captured, including highly specular reflective material, such as gold, performs well over most 3D devices. The RTI method, originally known as Polynomial Texture Mapping (PTM), was first introduced by Tom Malzbender from Hewlett Packard Research Laboratories in 2001, as an application for rendering three-dimensional models using digital photographs. From the beginning, this technique was considered to have significant potential for documenting the surfaces of cultural heritage objects. For example, research on PTM and RTI applications has included the study of cuneiform texts (Malzbender *et al.*, 2000; Hameeuw and Willems, 2011; Willems, 2011), scrolls (Zuckerman, 2004), numismatics (Mudge *et al.*, 2005; Kotoula and Kyranoudi, 2014), rock art (Mudge *et al.*, 2006), archaeological objects (Earl, Martinez and Malzbender, 2010; Macdonald, 2011), Byzantine art (Zányi *et al.*, 2007) and paintings (Manfredi *et al.*, 2013, 2014; Padfield, Saunders and Malzbender, 2005; Tamayo, Andres and Pons, 2013). Recent studies on RTI applications demonstrate the possibilities of combining RTI images with other imaging techniques. The integration of data from multispectral imaging with RTI has proven successful for enhancing the ancient texts in the Dead Sea scrolls (*Integrating Spectral and Reflectance Transformation Imaging Technologies for the Digitization of Manuscripts and Other Cultural Artifacts*, http://palimpsest.stmarytx.edu/integrating/index.html [accessed: March 2015]; Caine and Magen, 2011) and to emphasize texture and brush stroke in a replica from a wall painting decoration scene of the Tomb of Menna in Egypt (http://acrg.soton.ac.uk/blog/2796/). These particular applications could prove valuable in the study and presentation of paintings.

Goniospectrophotometry. This is an optical scatter measurement of the angular distribution of light. Different options can be used as the source of light and capturing device, but the main idea is the same for every combination of options. The surface is illuminated by the light source and the reflected light is measured at different angles as a function of wavelength. No geometry information is obtained by this method.

Multifocal microscopy. This is a form of light microscopy that images different focal planes of a surface by forming a depth image to represent the topography of the surface of the object. It works on small samples of near planar surfaces.

Profilometry (contact type device). Similar to the laser scanner, this is also a one-dimensional measuring system. The system is similar to a phonograph where a needle-like or stylus measures small surface variations through vertical displacements as a function of position. The radius of a round tip at the bottom end of the stylus determines the resolution on the vertical position and the horizontal resolution is controlled by the scan speed. The result is an analogue signal that represents the surface topography as a function of the lateral position. The measurements are very sensitive to noise as the tip radii could be in the nanometer range. The preferred option is to have near-planar surfaces for measuring.

3.3.2 Scanning for Cultural Heritage

In the specific case of painted surfaces, there are only a few methods that will meet the high requirements of accuracy needed for the capture, analysis and reproduction of painted surfaces. Within a wide range of sensors and due to the fragile nature of paintings, only non-contact optical sensors can be considered. These are classified into passive and active methods. In active methods, a light beam or pattern is projected over the surface and captured by an adequate sensor, which measures the location of the image in space at the illuminated point. Amongst these systems, structured light and laser triangulation are the most important. In passive methods, the light reflecting from the object and the illumination of the scene are computed to derive the shape information from two-dimensional photographs. More recently, other pseudo three-dimensional imaging techniques such as Polynomial Texture Mapping and Reflectance Transformation Imaging have proven useful to document surface relief features. Their ability to communicate both colour and shape efficiently make them excellent tools to study and document paintings (Case Study 5: Capturing Texture of Paintings for Museum and Heritage). As a cheaper entry point for consumers or small conservation studios, for example, the DAVID SLS-2 combines structured light technology through a video projector that is mounted on to a sliding rail and a camera to capture the surface details of a painting.

Some scientific consortiums composed of academic and industrial partners are working on a range of solutions specific to artefacts in cultural heritage. Two projects funded by the European Commission XD Reflect and Prism (https://www.xdreflect.eu/ and https://cordis.europa.eu/result/rcn/36788_en .html) explored two different approaches to the subject: the first is more related to metrology and industrial applications and the other is closer to human perception of the physical properties of objects. Other laboratories and universities, such as Rochester Institute of Technology (USA), Fraunhofer (Europe) (see Figure 3.8), Institute for Research in Computer Science and Automation (France), Getty Research Institute (USA) (see Figure 3.9) and National Gallery London (UK), are already working with scanners that have been specifically developed for dedicated research projects and cultural heritage applications.

For instance, CultLab3D at the Fraunhofer Institute has a pipeline capable of digitizing geometry, texture and optical material properties of cultural heritage artefacts weighing up to 50 kg and dimensions of up to 60 cm height and 60 cm diameter, with a geometrical resolution that falls in the micrometre range. The cost, portability and technical performance regarding translucency and metallic objects cannot be accurately estimated. The same group is working on a robot for larger artefacts and an aerial digitizing solution for excavation sites (http://www.cultlab3d.de). Their main goal is archiving, which is not necessarily linked to image or object reproduction.

The Scientific Department at the National Gallery London have been testing 3D scanning technology to record and measure the surface of paintings. They have been using the Lucida laser scanner designed by Manuel Franquelo and custom built by Factum Arte (Figure 3.10). The Lucida scanner outputs highly detailed and accurate 3D information as 2D greyscale depth-map files, which can then

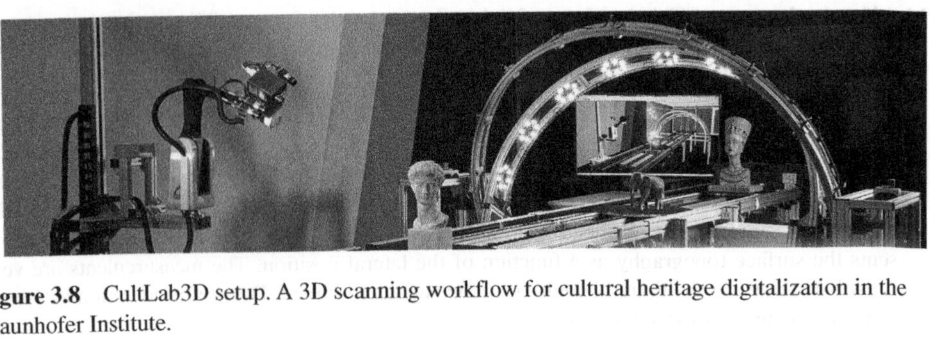

Figure 3.8 CultLab3D setup. A 3D scanning workflow for cultural heritage digitalization in the Fraunhofer Institute.

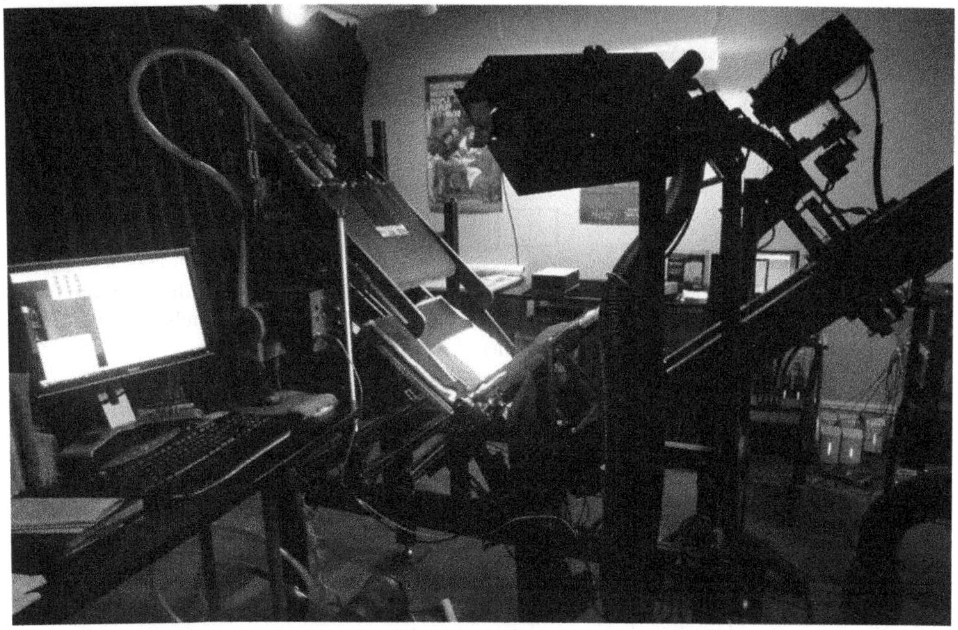

Figure 3.9 Getty Research Institute scanner (courtesy: The Walters Art Museum, Baltimore).

Figure 3.10 Lucida 3D Scanner, designed by the artist Manuel Franquelo and built by Factum Arte, has been specifically to record the surface of paintings and low relief objects (courtesy: Alicia Guirao).

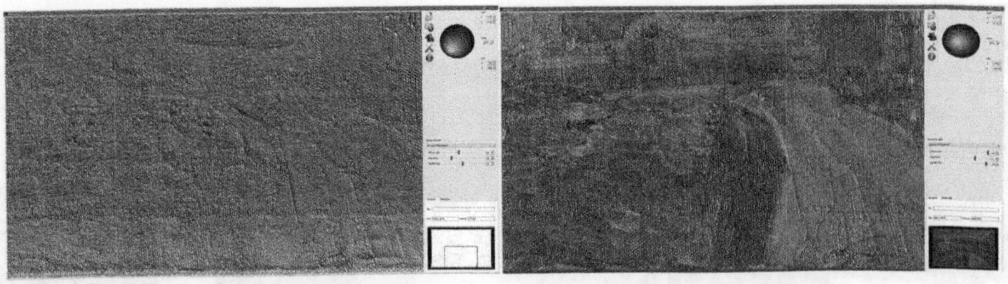

Figure 3.11 Greyscale RTI generated from a number of shaded renders lit from different directions (left). Colour RTI generated after registration and substitution of the colour channels from the shaded renders with the colour image of the painting (right).

Figure 3.12 Showing the render of the 3D model of the painting and (right) detail of the 3D model of the painting.

be used to generate 2D shaded renders of the textured surface. This information can be used for a variety of applications, including conservation and monitoring of paintings, virtual ways for looking at the surface of a painting, such as using RTI, and accurate 3D data for printing reproduction.

However, methods for capturing, visualisation and printing of textured images remain problematic. Greyscale rendered images can only display surface texture as static images that cannot convey texture information as a full interactive 3D model. Furthermore, these greyscale images do not contain colour information and the level of detail of the surface texture requires a vast amount of computing power to visualise and interact with the final processed 3D model. Researcher Xavier Aure, working collaboratively with the Centre for Fine Print Research and National Gallery London, has been studying different workflow scanning methods to combine 2D colour images and 3D data captured at the National Gallery with RTI images to assist in a more efficient and accessible visualisation of the surface texture of paintings (see Figures 3.11 and 3.12) (Case Study 5: Capturing Texture of Paintings for Museum and Heritage). An example of the complete rendered painting can be accessed online via Sketchfab (https://sketchfab.com/models/4737e8b449ba4d4eb10d43b4ca2c652e). As the painting is rotated from left to right, the influence of raking light on the surface of the painting is visible (see also Chapter 2, Section 2.4).

3.3.3 In Pursuit of Standards: Metric versus Quality

Since it is essential for any industrial process to be able to communicate nominal characteristics of the parameters present in their measuring instruments, the International Organization for Standardization (ISO) has provided a set of standards that attempt to assist this aspect. For instance, ISO 12781 and ISO 25178 present a reference for texture parameter measurements as a primary method for evaluating functional properties of material such as friction and wear. However, the standards have fallen short in referencing the breath of measuring methods and therefore its adoption has been delayed.

Even if the technological barriers regarding physical measurements of surfaces are lifted, the questions that remain are: how are these results related to human perception, at what degree are they relevant and, finally, what is their influence on the perception of what is pleasing? Even though different studies, through experiments with computer simulations on a screen, have indicated a link between perceived roughness and perceived glossiness (Ho, Landy and Malony, 2008; Giesel and Gegenfurtner, 2010; Fleming, Dror and Adelson, 2003; Malony and Brainard, 2010), these still lack a verification with real samples (Baar, Brettel and Segovia, 2016); with the addition of more variables such as colour, there is room for improvement.

When considering material reproduction workflows, we need to consider not only measuring common physical characteristics but also the methods of quantifying, (de)correlating, measuring and/or capturing more intuitive characteristics. When considering different measuring or capturing devices, users are often limited by balancing application and price. A realistic benchmark does not exist to evaluate distinct technologies and hardware specifications but productivity alongside maturity will bring along improvements in the field.

3.4 Methods to Represent the Appearance of Texture

The diverse applications and interests in material surfaces and the corresponding scope of measuring and capturing devices have ironically hindered the development of a uniform reproduction pipeline. On the whole, in the academic and the commercial sectors, the use of proprietary software, methods, modelling tools and formats to communicate measurements is widely spread across platforms and operating systems.

The ASTM F42 committee has been working to define a cross-platform AM file format for 3D printing (released in 2011). The need for a universal format has been long overdue, possibly due to larger industrial organisations being less willing to share a common easily exchangeable format. The measuring and the capturing representations are not necessarily suited for reproduction as the devices for capturing were not originally conceived as pipeline devices but as independent ones. The capabilities of the chain are limited by incomplete file formats, such as STL, that provide information about a reduced set of material properties. The colour and reflection features of a surface material are usually avoided on purpose because the goals that have been prioritised until now are synthetic rendering on screen or shape reproduction by fast prototyping processes. Not only is the information incomplete but also the existing information is not readily usable due to the lack of communication protocols across a material reproduction scenario.

In a typical 2D printing environment we can describe pictures as bits, bytes and pixels, but in material reproduction there is not a clear ‘atomic’ unit or particle in which a 3D surface could be subdivided. Possibly, the amount of variables, and therefore the combination of them all, has prevented the definition of a minimal unit. Similar to a pixel, such a unit must be loaded with enough spatial and reflection information to be something on its own. In studies of texture we see a tendency to look for a primitive unit that would allow repetition, rendering, synthesisation and archiving. For instance, a texel, a texture element or texture pixel, is the fundamental unit of texture space used in computer graphics. In the field of visual perception there are textons, which refer to fundamental microstructures of images considered as the atoms of preattentive human visual perception (Julesz, 1991), a concept that remains quite blurry with few reliable mathematical models. More recently, we have realised the 3D textons, which describe textures composed of ridges, grooves, spots or stripes, or combinations thereof. Associated with each (3D) texton there is an appearance vector that characterizes the local irradiance distribution (Leung and Malik, 2001).

The effects of masking, shadowing, specularity and mutual illumination cause the texture to appear drastically different according to the lighting and viewing directions, which increases the problem of texture characterisation and reproduction. There is a strong relationship between surface and subsurface in objects. Usually we are more aware of the surface or the skin of an object and we consider less about what is below the initial layer. It is important to know that the perception of the surface is a result of both layers, either because the outer layer is somewhat translucent and allows us to see the interior layer or because the inner layer has a certain shape that affects the geometry of the opaque exterior layer, and therefore its appearance.

In printing, the subsurface could be the media on which the skin layer is deposited or it could be made of an additional material or paint added to the print process. We usually refer to the appearance given by the combination of surface and subsurface layers as the physical texture. There also exists the idea of visual texture where the appearance is only given by a pattern drawn on the surface layer but is not present on the subsurface. To clarify, we refer to texture here in the sense of physical texture, the combination of layers with different properties.

On the other hand, in computer graphics and graphic design the use of texture is more related to visual texture and patterns created due to contrast and colour changes. In general, the geometry of the 3D object is created on a mesh-based model composed of triangles (other geometries are available) sharing information about placement in the tridimensional space with respect to a reference point. The final skin of colours and visual texture is usually represented by a plane that is mapped to the points or vertices and facets on the mesh. Therefore, the final appearance of the surface texture is given by the combination of the shape of the object, the visual texture mapped to the object and the viewing geometry. Thus, the topography of the texture in this case is fully absorbed by the shape of the object in the grid as the visual texture is limited to provide data related to colours and flat patterns. If the plane describing the texture is no longer a plane but something on relief (a 3D component added) the final surface texture would be the result of the shape of the object (the mesh shapes), the physical texture mapped to it and the viewing geometry. As long as we are conscious of the difference between the two, either path could achieve the same appearance. Although similar appearances may be achieved, there may be some textural appearances that can only be accomplished through the use of the full description of the physical texture as the mesh model may miss some key details associated to our perception.

The tendency is to divide the description of a physical texture into geometrical and spectral parameters, but it is still not clear whether this is perceptually relevant. Some texture classifications also take us in this direction. For instance, we might hear someone saying that a given texture is regular, uniform, structured, periodic and stochastic with respect to reflectance, scale and shape. Academic departments working in the field of visual textures have taken the lead in texture analysis by building parametric and statistical models of textures for synthesisation and analysis. Based on these data, a range of useful databases have been developed, including VisTex by the Media Lab, which is part of the Massachusetts Institute of Technology Media Lab (http://vismod.media.mit.edu/vismod/imagery/VisionTexture/vistex.html), MERL BRDF (http://www.merl.com/brdf) by the MERL Lab (Mitsubishi Electric Research Laboratories) and CUReT (http://www.robots.ox.ac.uk/~vgg/research/texclass/setup.html) by the Visual Geometry Group in the University of Oxford. More elaborate databases including segmented real images and combinations of reflection and statistical data have emerged recently, such as OpenSurfaces (http://opensurfaces.cs.cornell.edu) from Cornell University (Bell *et al.*, 2013). More vivid sets are proposed by scientists involved in appearance capture studies such as the Flickr Material Database at Massachusetts Institute of Technology, the TSV-BRDF (http://www1.cs.columbia.edu/CAVE/projects/time_var/images/database.gif) by the CAVE Lab at Columbia University and Holly Rushmeier's test set from Yale University (https://sites.google.com/site/mam2014samples/home). To our knowledge, just a few physical texture databases (spatial and reflective information of a material) are available, for example, PhoTex by the Texturelab (http://www.macs.hw.ac.uk/texturelab/resources/databases/photex/) at Heriot Watt University, UK. Examples are given in Figures 3.13 to 3.16.

Researchers at Columbia University and Utrecht University have collaborated in an extensive investigation of the visual appearance of real-world surfaces. (Sponsored in part by REALISE of the

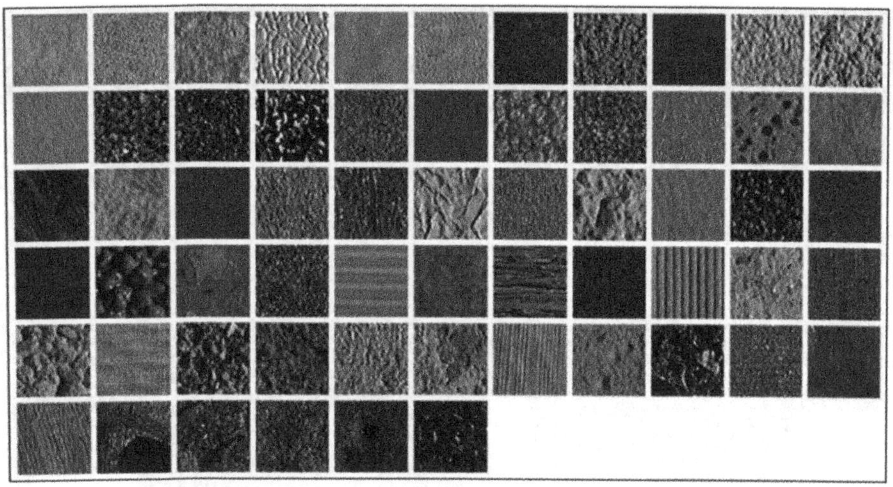

Figure 3.13 CUReT database sample. There are 61 materials that we commonly see in our environment; each has been imaged under 205 different viewing geometries and illumination conditions. Its limitations include the lack of scale change and limited in-plane rotation during capturing and the inconsistent intraclass variation in the selection (Computer Vision Laboratory, Department of Computer Science, Columbia University).

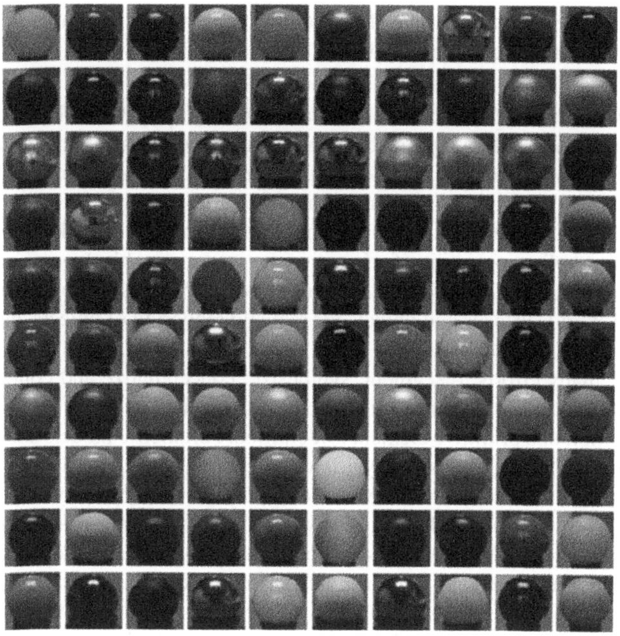

Figure 3.14 MERL-BRDF database samples. It contains material reflectance properties of 100 materials. Each reflectance function is stored as a densely measured Bidirectional Reflectance Distribution Function (BRDF). Only isotropic materials were considered (courtesy: Mitsubishi Electric Research Laboratories, Inc.).

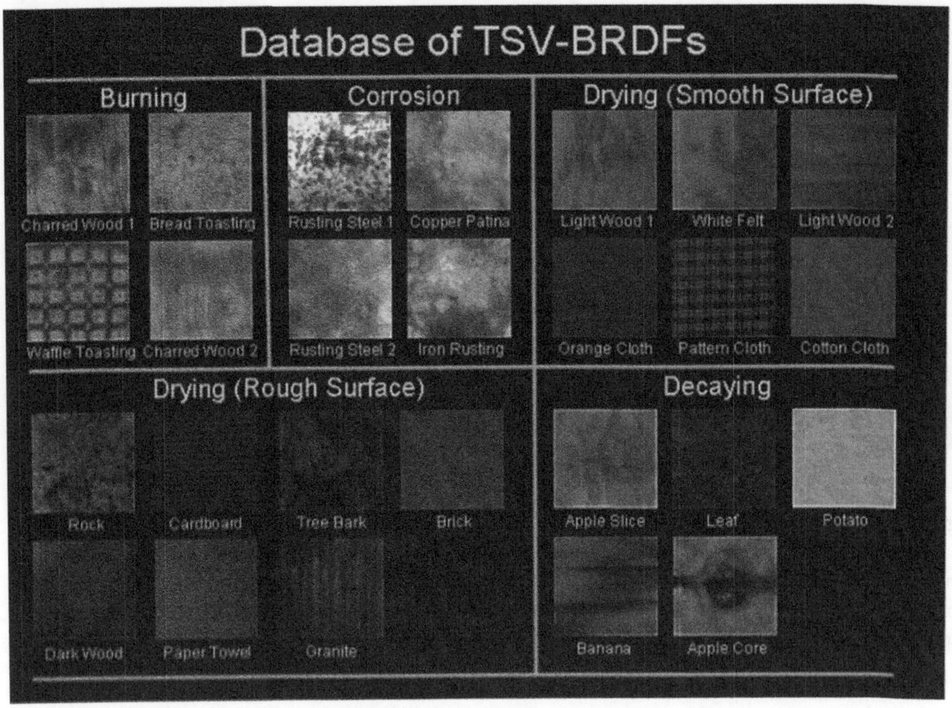

Figure 3.15 TSV-BRDF database sample. The database is composed of time-varying takes of different materials, both natural and man-made.

European Commission, the National Science Foundation and by DARPA/ONR under the MURI Grant No. N00014-95-1-0601 has resulted in three databases: http://www1.cs.columbia.edu/CAVE//exclude/curet/.index.html; http://www.robots.ox.ac.uk/~vgg/research/texclass/setup.html.) Different research groups have used this resource to test methods of analysis of texture classification, for example, The Visual Geometry Group, University of Oxford (http://www.robots.ox.ac.uk/~vgg/research/texclass/setup.html). From an application point of view (i.e. computer graphics) organisations such as the Swedish-based company Quixel (http://quixel.se) are systematically capturing high-definition images or *megascans* of real-world surfaces and objects for realistic rendering of objects, buildings and scenes (https://megascans.se).

Regardless of the many efforts to segment, synthesize, conceptualize and collect surface textures, the problems of communication, classification and contextualization of such textures will not be easily solved until the definition of a primitive element is identified. Parametric models and statistical methods will likely extend visual texture models to physical texture models that would be more arduous to handle given the storage size, computational time and comprehensible complexity they will present.

Beyond academia, how useful are these studies on visual texture or how applicable are these databases for real-world applications? Image-based, texture databases are an increasing requirement for rendering virtual environments and applications, such as interior design, fashion and architectural models, to film and animation (Case Study 12: Painting Machines). In the twenty-first century CGI has become so ubiquitous that unless promoted as a new breakthrough in the news, special effects often go unnoticed. As a commercial application, the content of an Ikea's catalogue is almost entirely computer graphically generated. In 2006, Ikea's first venture into CG was the complete rendering of a *Bertil* pinewood chair, and, at this time, 25% of the products in the Ikea catalogue were CG re-creations. Today more than 75% of the single product images are CG. As explained by Martin Enthed of Ikea

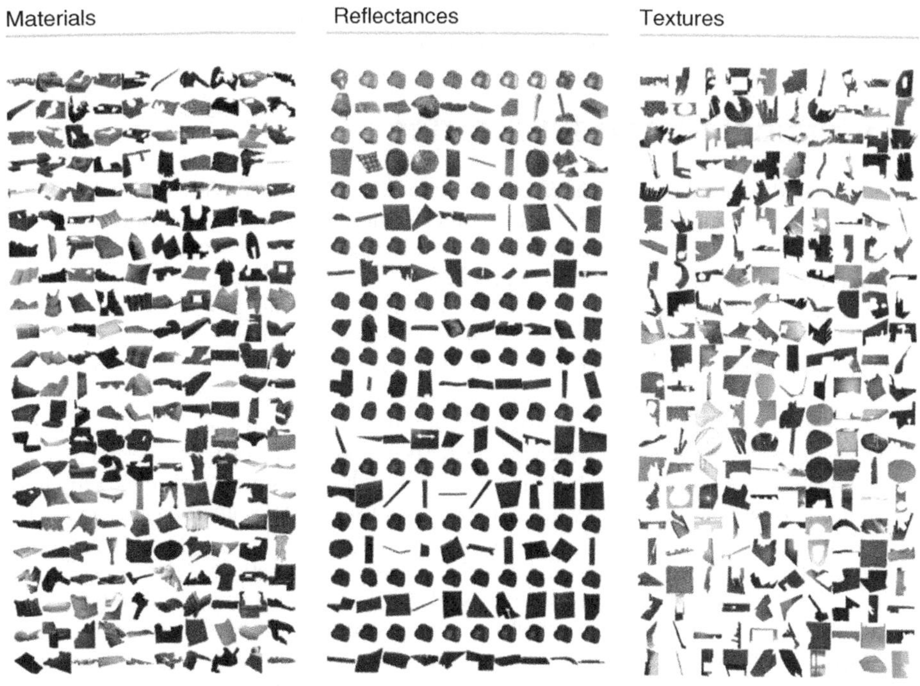

Figure 3.16 OpenSurfaces database sample. Database composed of textures segmented from real images and annotated with several surface properties (material, texture and contextual information). The database is limited to consumer products surfaces (such as furniture and clothes); thus, no natural scenes are part of it.

Communications AB, the objective for creating a CG library of stock items was to economically and effectively manage the different tastes and requirements of different countries. For example, kitchen design and content differ from country to country, and components such as white goods are expensive to ship; therefore, as kitchens are the most expensive and complex environments it is cheaper to photograph individual items and arrange these in a virtual environment. Using rendering programs such as 3D Studio Max and V-Ray, Ikea have created a library of 25 000 3D models, which can be shown separately or combined to create a scene. Each model is created at high resolution so they can be as multifunctional as possible, so that each model may be used in catalogues or printed on a much larger scale as backdrops for walls in showrooms.

Other examples include the DIY paint market. With increased interest in DIY makeover programmes, there is an imperative for paint manufacturers to ensure the colour in the paint sample book is the same as the paint in the tin and that, returning to our example of a yellow paint that is applied to a textured wall, it also matches the expectations of the DIY user when applied to different surfaces (see Chapter 1, Figures 1.5 to 1.7). In order to inspire the home decorator, paint companies are developing more opportunities for the home decorator to use their hand-held devices through the use of apps to make colour selection choices in their homes. This may be useful to gain a contextual understanding of the appearance of the room when painted. The apps suggest two approaches: the first enables the user to photograph their room and then select from paint swatches from the company's existing colour paint range, which is then used to virtually *render* the wall. The app includes a tool that allows the user to crop or mask selections or to incorporate other paint colours. The second approach assists the decorator to create a custom colour palette. The user takes a photograph of their room and the app generates a colour palette based on the paint company's existing colour products. These apps are an entertaining way to virtually decorate your room without too much effort (albeit the real room has yet to be painted).

3.5 Physical Material Libraries

As described, there is a wealth of materials and innovative ways of classification. An important requirement for any designer, whether they are architects, interior designers, or product designers, is to be able to specify materials and reproduce textures accurately for their client. Furthermore, the designer needs to convey all the qualities of a product, so the client can gain insights of the look and feel of a rendered scene or object prior to production. Computer rendering programmes, such as Rhino and AutoCAD, ZBrush and TurboCAD, are now used for rendering a range of objects and environments, from prototypes to kitchens, and the extent to whether these rendered images are convincing for the client may make the difference between securing a deal between one supplier or another. Online sites such as 3D Warehouse offer different textures that can be used to render objects and buildings with the appropriate materials, of which some have been captured using scanning and photogrammetry techniques, as previously described, and therefore leading to more convincing and authentic looking products. However, increasingly there are needs for users and designers to gain access to real textures and materials.

One impact of the rise of new textures and materials has been the variety of expos, books and magazines, online sites, databases and libraries, whereby students, designers, engineers and architects can tap into and to understand how a material might behave or understand more about their functions and applications. Cross-disciplinary studies have included *Transmaterial: A Catalogue of Materials That Redefine our Physical Environment*, which was published both as an online resource and a book. Events such as *The Surface Design Show* at London's Business Design Centre included a dedicated area named *Future Thinking* that showcased emerging trends in materials, colours and surfaces. Likewise *Design Boom* (www.designboom.com), an online magazine for designers, includes a product library and many articles on materials. Universities are introducing students to new materials as part of their curriculum and establishing materials libraries in fabrication centres. There are a number of different transfer printing processes that either require immersion in water (www.hydro-graphics.co.uk) or application of moisture (www.Shockline.it) to release the printed image from the carrier film or paper. The Fresco Shockline transfer is an interesting decal for decorating internal and external walls. *Rematerialise*, an early initiative by Jakki Dehn (https://research.kingston.ac.uk/rematerialise/html_and_flash/searchwelcome.htm) was established at Kingston University, UK, in 1994 as a materials library and online resource that provided information on sustainable products. Funded in 2003 by the Arts and Humanities Research Council, the *Rematerialise* Library now houses over 1200 materials, which demonstrates more sustainable resources or materials that come from renewable resources. The Institute of Making, University College London (http://www.instituteofmaking.org.uk/materials-library), provides a hands-on materials library for the 'curious and material minded'. UCL have also created a downloadable app, which lists materials according to different categories that range from their chemical composition to 'wonderstuff'. From the context of architectural and interior applications, the Materials Council (http://www.materialscouncil.com) was established by Ian Hunter and Brad Turner, former materials researchers at Foster + Partners architects and founding partners of Architonic. In order to address designers' needs to gain access and material experience, the aim of the council was to promote 'a greater appreciation of materials that allow the development of innovative solutions that would not have emerged through other research processes'. The Materials Council works with architects and designers and manufacturers, and in recognition of their demographic, the council uses terms such as 'taster' or 'curated', which are more analogous to terms used in a restaurant or an exhibition. Similar, but commercially centred. organisations include The Material Lab (see Figure 3.17), which was created and established in 2006 by a Stoke on Trent company Johnson Tiles. Johnson Tiles are an innovative ceramic tile company that has been in production since 1901. Always on the search for material inspiration and

Figure 3.17 Material Lab Studio, London (courtesy: Tim Ainsworth).

product specification, The Material Lab provides ideas and specifications alongside a visual resource of new materials and new trends for the materially motivated researchers, designers and architects. Whilst the large central unit provides inspiration, the smaller samples on shelves along the walls of the Lab are arranged in colour order and for designers to select sample swatches to take away (see Figure 3.18).

Figure 3.18 A library of samples arranged according to colour at the Material Lab, London.

There are also many worldwide materials libraries dedicated to material and texture inspiration. Of course, these libraries need to be accessed by researchers and for the materials to be handled, for example: https://www.materialbiblioteket.se in Stockholm; https://www.materio.com in Paris, Belgium, Prague, Shanghai; https://www.es.materfad.com in Barcelona; https://www.matrec.com in Ancona; and product stores such as Modulator in Berlin (https://www.modulor.de), who also offer sample boxes containing small swatches of 199 different products. Their suggestion being that 'neither the best description of the materials nor excellent photographs can compare to actually holding and seeing them in your hands'. As an online accessible site, Matweb in the United States (http://www.matweb.com) is a searchable online database of engineering materials and is also a technical resource containing data sheets.

3.6 Methods for 2.5D Printing

In the previous chapter we considered the ways artists interpreted the appearance of materials and methods for the representation of textures through painting and computer graphics. In this chapter we have explored how designers have incorporated unconventional or recycled materials to create new products. We have explored methods and hardware for physically capturing, measuring and modelling surfaces, as well as the emergence of physical and virtual libraries and resources that can assist the designer. The following section will reflect on historical and emerging printing technologies that have responded to the demands of printing and packaging production.

From its earliest beginnings printing was concerned with the reproduction of writings. Printing was also a way of exploring, proofing and the distribution of ideas, images and products. Printing blocks, tools and materials could be considered as multidimensional but, due to more commercially expedient planographic processes, such as offset lithography and (now) inkjet, the reproduction was usually in two dimensions. The printing press and studio has evolved from the fifteenth century workshop involving many people performing different specialist tasks to a twenty-first century digital print machine in which all the mechanics are hidden and can be operated by one person. Of course, this is a simplistic view and, as we are aware, digital printing is not always straightforward and rapid prototyping is not necessarily rapid.

More recently we have started to explore how the print industry will function in the twenty-first century. As an increasing component of the Internet of Things, as the industry becomes more networked, integrated and functional, what extra benefits may it provide in terms of added value and enhancing knowledge about our products? Furthermore, as printed surfaces that incorporate different dimensions, embedded features and textures is a new and emerging field, whether our experience is somehow extendable or adaptable to multidimensions is a difficult question, but we certainly hope that the answer is positive.

If we consider relief printing as a bridge between 2D and 3D printing, it is possible to seek a way of appropriating from both digital and words and merge some of the boundaries that result in annoying obstacles to overcome higher dimensional challenges. The current set of features on which current 2.5D printing methods have concentrated their efforts is well founded on the applications they are initially aimed at, namely, fine art reproduction, decoration for packaging and reproduction of different material appearances. One can think of relief printing as an extension of 2D printing, even though a direct mapping of concepts in between the two domains is not evident or easily achievable, but it does give the opportunity to 'recycle' some of the work designed for 2D printing. Conversely, one could also think of relief printing as a simplified version of 3D printing, even if many aspects of surfaces with a relief are not something that has yet been sufficiently achieved using 3D printing technologies. The knowledge acquired regarding printing materials, techniques and macrogeometry could be applied or adapted to cases with a reduced dimension.

In general, the process of printing can be divided into:

- Groundwork (what): problem solving, ideas, sketches, drawings, text, designs, pictures, templates.
- Printing materials (with what): inks, powders, plastics, gels, paints, metals, organic products (cells), elementary materials, wood, acrylics, paper, concrete, fabrics, wool, glass.

- Printing device (how): nozzles (piezo, thermal), extruders, brushes, presses, woodblocks, matrices, stencils, carvers, embossers, pens, stencils, stamps, moulds.
- Finishing (now what): coating, laminating, cleaning, cutting, heating, material removal, polishing.
- Bridging the gap (know-how): tacit knowledge, training, iteration, creativity, skills in process and materials, file formats, standards, software, printing parameters, profiles.

The groundwork takes care of the preliminaries of printing, the 'getting ready' for printing things. In preparation for digital printing, knowledge of the printing device, print profiles, file format and print parameters are necessary to achieve the task, which is often the hidden aspect of the technology. To achieve a well-balanced digital print, the technical know-how is analogous to the mastery of creating a coloured woodblock print, where the artisan needs to understand all the parameters of the process, from the original drawing, transcribing the image in reverse on the block, colour separation and registration of each block, the pressure required to print the blocks and paper weight and surface.

Next, we have the printing material. Here we are talking about the finished product and the stuff we look at and handle. Are these appealing? Desirable? Functional? It could simply be a combination of paper and ink or toner as in regular office printers or more complex things such as ceramic for teeth and bone replacement. Once we know the 'what' and the 'with what' to print we need to answer the 'how'. Printheads can come in different shapes and configurations that are either analogue or digital based. Woodblock printing and painting are good examples of analogue production and desktop printer and CNC cutters and routers are examples of digital production. Finally, the last elements are, as expected, the finishers. They are meant to protect or enhance the product, add something that is missing during printing (such as gloss) or make the final adjustments to the print for it to be ready.

Although printing is generally considered as a planographic process, there are many traditional methods from the past to add a 3D element to the surface that include metal dies or inks to create embossed (raised) or debossed (recessed) relief images and designs in paper and other materials. Those working in *two-plus* dimensions could therefore consider an appraisal of analogue methods. 2.5D printing technologies may be considered and presented as a combination of the following methods:

- The combination of traditional tools and analogue materials that may be reappraised using digital technologies but using traditional printing methods.
- A reappraisal of traditional materials to explore how traditional tools and computer=aided methods can be applied for new fabrication methods (see, for example, Figure 3.19).
- The measurement of a surface to reproduce the appearance of texture and colour as a printed relief surface and the development of generic materials that are designed for reproduction for printed relief surface.
- The development of new materials for the physical deposition of functional materials for specific applications using computer-aided methods or traditional printing methods.
- An exploration of the flexibility in the direction of printing; for example, a printhead that can move in any direction and can be located in any point.
- Development of more energy efficient processes and recyclable products.

As in any revolution, time is important as an effector of change; few printers may take the challenge but many will benefit at the end. This bridge will have many practical applications and sensory impacts if this can be suitably addressed (Case Study 3: 2.5D Printed Tactile Books and Artworks).

3.6.1 Analogue versus Digital

There are two worlds of printing and in many instances over the last few decades, the digital and analogue, which have evolved symbiotically, and many methods are now considered as a hybrid between digital and analogue. For example, images may be hand drawn, then digitalised and digital halftone

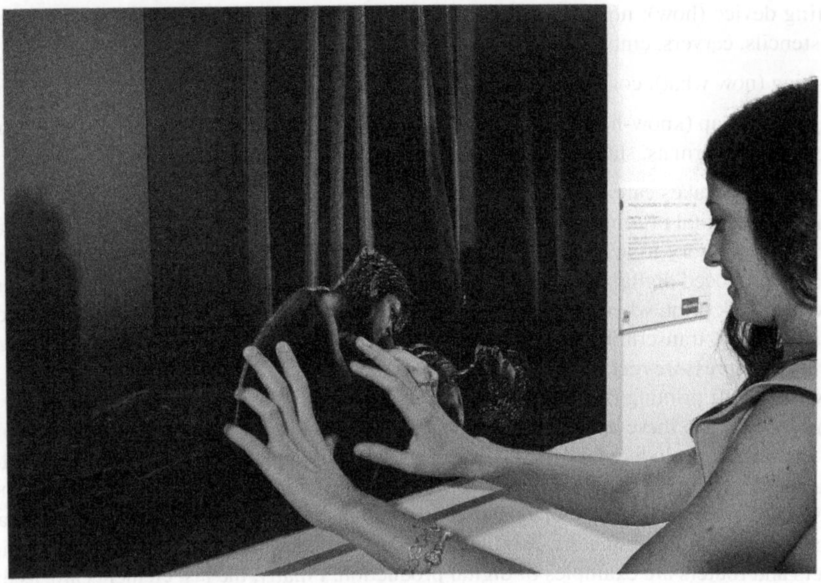

Figure 3.19 Texture reinterpretation for tactile experiences using the Didú technique as developed by Estudios Durero (Courtesy: Estudios Durero).

separations are output for screens and plates as part of the printing workflow. Although digital printing processes are perceived to be increasingly preferred for printing applications, there are many analogue methods that demonstrate unique surface and tonal qualities that cannot be fulfilled by digital technologies. Analogue methods still provide alternative opportunities for prototyping, specialist printing applications, low-volume production, and increase opportunities for many materials that are not available in digital printing.

Analogue printing is traditionally used to create 2.5D and textured surfaces, or as 2D decals that can be applied to textured, complex or curved surfaces. There are a number of different transfer printing processes that either require immersion in water (www.hydro-graphics.co.uk) or application of moisture (www.Shockline.it) to release the printed image from the carrier film or paper. The Fresco Shockline transfer is an interesting decal for decorating internal and external walls. Whilst this is essentially a 2D printed transfer and does not add texture to the wall or the object, it is possible to enhance or change the qualities of the object, for example, to transform an injection moulded plastic to look like Kevlar or wood grain.

This has been particularly useful for the screenprinting industry in the production of applied areas such as biomedical sensors, solar cells and flexible electronics, where sharp lines and high-quality resolution are necessary for electronic circuits. Due to its longevity in the market, there are a range of inks specially developed to obtain surface effects for sign and display (glosses and varnishes), decorative effects for packaging (powder additives such as glitters, metallics), textured surfaces for clothing such as expanding Puff Ink, which is an additive to Plastisol inks to create a 3D soft texture, and hard layers (UV curable inks) for raised or embossed surfaces as used for Braille. Screenprinting is highly useful to deposit thick films, pastes, viscous fluids and larger particulates such as ceramic or metallic pastes. It is increasingly used for specialist areas such as printed circuit boards, radio frequency identification chips (RFID) and biomedical sensors (Case Study 14: Analogue Printing Methods).

Heavier gauge films are used to deposit a thicker film of ink, which are more suitable for braille, tactile prints and electronic circuits. Thick film circuits are widely used in the automotive industry where a greater degree of reliability is required. This is enhanced by the use of a thicker stencil. In order to achieve thick deposits of ink a larger mesh size is necessary, typically between 156 and 60 tpi. A thicker clear stencil is also important to ensure an easy transfer of ink, smooth surface finish and no mesh marks.

Another example is pad printing, which still occupies a niche but significant sector of the printing market. Pad printing is an intaglio process that is used to print on to complex, compound angles and textured surfaces. It is commonly used to print on to objects that are already fabricated, that may be delicate and require decoration or graphical information (numbers, logos), for example a key-pad on a telephone or coins and medals (Case Study 4: Coins and Medals). Another example is a golf ball, which has dimples and is therefore difficult to print cleanly and precisely without missing information. The soft silicone pad is able to conform to the surfaces and print detail.

Pad or Tampo printing is a three-stage printing process and is similar to offset printing, transfers a layer of ink from a flat plate to a soft silicone pad and then on to the object. In more detail, a gravure or engraved plate is inked and wiped clean, the soft silicone pad is pushed into the plate and picks up the ink. The ink adheres to the surface of the silicone pad. The pad is positioned directly above the substrate, is pressed down and wraps around the object while stretching very little, but with enough pressure to transfer the ink from pad to substrate. The silicone has a very low energy so the ink is released easily. The shape of the pad will differ depending on the shape and size of the object. There are two main shapes: round pads and rectangular pads. Shapes can also be customised to fit more complex shapes.

The pad hardness (i.e. durometer) is measured on a Shore scale (according to the system 00 Si) and ranges from 39 Shore 00 Si (very soft) to 70 Shore 00 Si (very hard). The hardest is suitable for printing precise printed images and specialist applications such as watch faces. The ink layer is much thinner than screenprinting inks, typically less than 1 μm (0.001 mm) and therefore for opaque inks, for example printing white lettering, it is necessary to print more than one layer (Thompson, 2007). Figure 3.20 shows examples of different inks for a range of products.

In 2015, in terms of value, digital print accounted for 13.9% of all print and printed packaging, but just 2.5% of total world volume. Although the anticipated trend towards digital will increase, in 2020

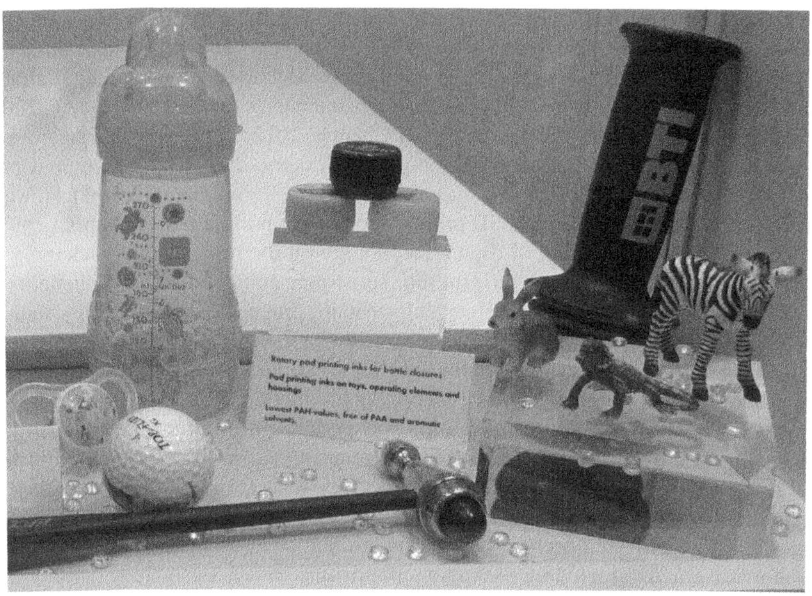

Figure 3.20 Examples of different inks for a range of products as demonstrated by Marabu inks at DRUPA (http://www.marabu-inks.co.uk/products/product-overview/pad-printing-inks.html).

the value of digital printing is estimated to increase to 17.4% and 3.4% in volume, it still demonstrates a huge gap between digital and analogue printing. The increase in digital is increasing mainly through the variable data, direct mail, short runs and print on demand (http://www.piworld.com/article/three-key-trends-future-digital-versus-analog-printing. These figures demonstrate there is still a significant market for traditional printing, which should not be underestimated.

Digital relief printing technologies are beginning to demonstrate an interesting potential for the reproduction of textured surfaces that, so far, has been technically impossible or too expensive to produce using traditional dies unless a product is mass-produced in thousands. An example is an incubator project between the Design Academy in Eindhoven and Canon Océ who experimented on the production of a range of decorative panels using 2.5D technologies, patterns and colours (https://www.designacademy.nl/Research/StrategicCreativity/Latest/tabid/2889/articleType/ArticleView/articleId/2213/Introducing-25D-Printing.aspx). As computer-aided technologies, such as laser cutting, print-on-demand printing and bespoke or personalised printing, on to a wide range of media is becoming increasingly possible, there will be significant overlaps between the analogue and digital methods in the future.

3.6.2 Materials for Printing

The colour print has evolved through technological innovation in materials, pigments, tools, presses and methods. This incremental process of refining, comparison and redefining has evolved through a balance between quality and commercial expediency, mechanisation and speed. Current standards in inkjet technology have developed out of nineteenth century photomechanical processes and methods of halftoning. In the late twentieth century electrostatic and inkjet technologies enabling colour, text and image to print in one pass have shaped how we make and print images. As the refinement process of inkjet has attained higher levels of accuracy, it is now able to emulate all the marks, colours and qualities of both photographic and traditional print techniques.

However, the print quality of both traditional and digital printing is dependent on a range of factors: the gamut of the colourants, the texture and colour and absorbency of the substrate, light scattering properties, colour management, halftoning methods and viewing conditions. Since its introduction to the market in the 1980s, inkjet colour printing has advanced significantly: from poor-quality fugitive inks and inaccurate colour management methods to high-quality, accurate multicolorant printing devices and high-speed presses. This subject has been explored extensively including digital imaging (Sharma and Bala, 2002; Kries, 2015), high dynamic range imaging (McCann and Rizzi, 2002), computational colour science (Westland, Ripamonti and Cheung, 2012) and colour management (Green, 2010a, 2010b). Now that the current state-of-the-art colorimetry, digital halftoning and printing is mature, innovation in the field has begun to concentrate on technologies that direct inkjet towards more novel industrial applications, for example 3D printed ceramics or printed circuit boards.

In order to understand the impact of digital technology on the printing and packaging and production industry, the emerging trends in printing, substrates and products, and insights into the future is DRUPA – a valuable place to start (https://www.drupa.com). Held every four years in Dusseldorf, it is the largest printing and finishing expo in Europe. This four-year gap between events is a useful timeframe to gain an understanding the changes in the market. In 2016, DRUPA expo hosted 1828 stands, exhibiting across 158,237 square metres of space, and a total of 260,165 visitors from 183 different countries. Based on our experience of visiting the expo in the 1990s, there has been an observable shift from traditional screenprinting for large format printing; fast-forward to 2016 where now high-speed digital web-fed, super-wide format inkjet and flatbed UV curing dominate the market. As demonstrated in 2012 at DRUPA, interest in the development of the more permanent UV curable and decorative inks has resulted in an expansion into the application of inkjet into alternative fields, including packaging, publishing, textile design and outdoor display. It has also highlighted a greater concentration on the product - how the technology could be used and its application as an end product; for example print on demand, direct mail and photo-books. At DRUPA 2016, the emerging trends as demonstrated in 2012 have largely been consolidated into high-end product and packaging, whereby branding and personalisation, authentication and security, tagging, tracking and data using the Internet alongside speed and precision printing are the keywords and market drivers (see Figures 3.21 to 3.25).

Through advances in printhead technology, new decorative inks are increasingly available and ubiquitous for the inkjet market. Whilst UV cure inks have primarily been used for lithography, screenprint and flexography for over a decade, it has been largely due to improvements in the chemistry of inks making it possible for curing inks to be inkjetable that has made the difference for direct digital manufacturing, 2.5D and 3D printing and rapid prototyping. Other significant factors have been developments in piezo inkjet head technology, the ability to print larger pigment sizes and at higher speeds.

Figure 3.21 An example of the library of textures that can be printed using the Mimaki UJF printer (courtesy: Mimaki).

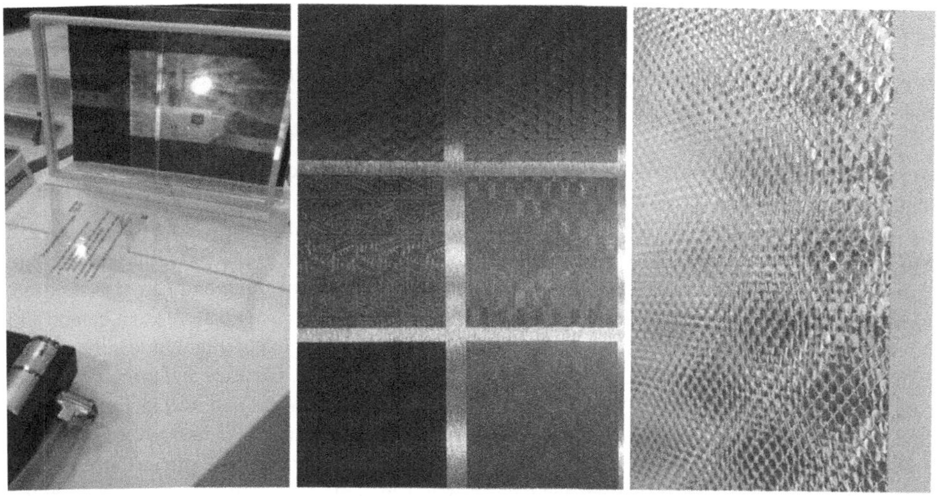

Figure 3.22 As seen at DRUPA 2016, showing different examples of clear ink: transparent colour screen (Fujifilm), printing on to foil (Mimaki) and lenticular printing on to acrylic (SwissQPrint).

Figure 3.23 LDP wood effect textured prints (courtesy: LDP) (https://www.lintor.com).

Figure 3.24 Examples of direct to small object printing using the Mimaki UJF printer (courtesy: Mimaki).

Other factors include the opportunity to incorporate decorative inks such as metallic, gloss, white, fluorescent and security printing inks that are only visible under UV light.

Alongside the ability to print a range of decorative inks, curable inks can be used to build up a surface to create texture, and which is the most significant potential impact for 2.5D printing. These inks can be used to create embossed, raised or decorative surfaces for packaging and can also be used to mimic different materials and textures. Figure 3.21 demonstrates some 2.5D patterns that can be added to the surface of images or objects. To create a printable file, an image and a pattern can be designed using dedicated software tools. The software utilises dedicated texture libraries and printer commands that are specific to the printer RIP. The design is created as a series of layers that may include the image, cut and crease paths, gloss and white varnish. The RIP interface will enable the user to control the quality and speed of the print, the height of the relief, the gloss or matt finish. Some printers are capable of printing running multiple passes that lay down clear, white and colour, so that three groups of printhead channels can simultaneously print in any combination.

Figure 3.25 Example of specialist direct to object printing using a Kebab module on the Mimaki UJF printer (courtesy: Mimaki).

To create a surface texture using UV inks, the surface topography is built upwards through an incremental process of applying and fusing a thin layer of material on top of another. The texture can only be achieved without an overhang (as can be achieved in 3D printing) but with the benefit of the extra dimension, which contains surface features and elevations that contribute to the appearance of a textured surface or low relief, which can be useful for creating high-quality prototypes for the packaging and pre-press industry.

Clear inks can be used as a single gloss, as a varnish to enhance areas or to create a raised relief surface by overprinting layers of clear ink. Using preset texture libraries, gloss can be used to print on to metallic materials to create special effects (see Figure 3.21). Clear inks can either be printed as a matte varnish or a glossy varnish, which can be determined through the RIP software. A glossy varnish finish can be achieved through a partial UV exposure (one UV light) or complete exposure (two UV lights) to create a matte finish. The two-lamp UV-curing system means that clear ink can either be cured matte or it can be cured glossy (Baar, Brittel and Sergovia, 2016). At DRUPA 2016, HP Experimental and Fujifilm demonstrated a clear ink that can be used to create translucent filters for projectors and lenticular screens. Likewise, SwissQPrint include large format lenticular printing patterns that can be used for window displays (Case Study 11: Microstructural Texture).

White is also used to create a textured base layer on to which an image is applied. For example, Lintor (http://www.lintor.com) have produced a range of textured wood effect prints. This could be considered as the most interesting and innovative challenge yet to the traditional print market and is increasingly more adaptable and practicable for the design and packaging market. As demonstrated at DRUPA 2016, the majority of wideformat manufacturers have invested heavily in the development of different inks and multiplatform methods for commercial and bespoke 2.5D printing (HP, Scitex, Durst, Fuji, Xiekon, SwissQPrint, Canon, Océ).

Regardless of the advantages that UV curable inks present, due to their tendency to crack their applications are limited to rigid or near-rigid surfaces. When several layers of fully polymerised UV ink are printed as in relief printing, the probability of cracking is even higher. Nevertheless, at DRUPA 2016, new ink developments were presented by Fujifilm and Mimaki that demonstrated flexible UV curable inks. Fujifilm introduced a hybrid ink that is a blend of solvent-based (dried through heat) and UV inks

-- promising the best of both worlds -- a finish that is flexible enough to resist cracking on to leather, for example, or delaminating during thermoforming processes (flat prints that are given a 3D form thanks to heat), and Mimaki have introduced invisible, thermoformable and stretchable inks.

Decorative metallic inks are a more recent introduction (Roland, Scodix, Fuji). Typically a silver base layer is printed, with the addition of CMYK spot colours as a second layer, to create a wide range of colours that create a metallic appearance. These inks are currently suspended in a solvent-based carrier and are printed by a conventional inkjet process. Fujifilm announced at DRUPA the development of coloured metallic inks that can be printed in a single pass. Likewise, Scodix's Metallic, Inkjet-Braille and Rainbow products promise a new dimension for greeting cards and packaging.

In relation to inks as a product for traditional plate printing (offset litho, flexography), Merck and Co. have developed a range of metallic lustres for different printing applications, which offers new design possibilities, especially for packaging and promotional materials. The iridescent and embossing effects are achieved by printing a background image or colour and then by printing a pearlescent or metallic lustre pigment on top to create a range of lustre and decorative effects (http://www.merck-performance-materials.com/en/printing/printing.html). Whilst there is no physical tactile surface, the Virtual Embossing 3D (VE3D) effect printing process from Merck creates the illusion of three dimensions. The print motif appears raised, but the surface remains completely smooth.

Instead of printing several ink layers to build relief surfaces, a relief surface can be directly generated from the inks. Similar to inks available for screenprinting and textile applications, HP Experimental presented an expanding ink that is initially printed flat and can obtain a thickness of around 1 mm. An alternative approach developed by Dimense for wallpaper applications (Case Study 2: Wallpaper Design), their proprietary wallpaper contains a carrier film (similar to a puff vinyl additive), which reacts thermally as part of the printing process to create a relief surface.

In contrast to the UV and eco solvent-based inks, the HP Latex ink is a water-based curing system, which dries instantly and is scratch resistant. It incorporates a second optimizer printhead, which reduces banding and can print on to a range of uncoated materials, including flimsy textiles as well as heavyweight canvas. Dye sublimation inks offer a range of opportunities for short-run applications such as t-shirts and branding of sports wear. For example, RolandDG's wideformat Textart XT640 images are printed back to front on to a carrier-paper, which is then applied using heat on to a range of materials including polyesters. The types of colours include fluorescent yellow, fluorescent pink, cyan, magenta, yellow, black, orange and violet. Epson's Sure Color SC-F2000 t-shirt printing presents a direct-to-garment printer that is then fixed through heat, which includes CMYK + white ink for printing on coloured backgrounds. Epson also offer a transfer paper than can be used for printing on to rigid surfaces. For superwide print applications, there are an increasing number of direct to soft signage printers, such as the five-metre wide EFI™ VUTEk® FabriVU 520, which uses water-based CMYK dispersed dye inks and is specially designed for wall or theatre drapes, or the EFI Quantum 5, which uses LED UV curing technology and can print on to rigid and flexible materials.

HP's ElectroInks and Landa NanoInks print technology present a challenge to the traditional offset litho printers. The small size of the particles ensures that the printed translucent film of ink does not impact the underlying appearance or quality of the substrate. HP's ElectroInks are specially designed for the HP Indigo printer range. These inks are supplied in paste form, can print up to 16 000 resolution continuous tones to the edge of the paper, with no banding or streaking, and can print on to many different substrates, including metallicised and clear film. The 12 000 digital press can print pink, white and colours and paper can be changed on the fly. In competition to the HP Indigo, the Landa Nanographic uses dry toner electrophotographic NanoInks. As the name of the ink implies, these inks are tiny particles that are applied dry to the substrate, thus reducing any need for post drying.

Inkjet printing including 3D printing is primarily limited as a layerwise technology. It is also limited by the distance between the printhead and the substrate; therefore, printing on to textured materials or complex surfaces requires the printhead to be detached from the print bed and be driven robotically using a multiaxis approach. Thus alternative ways have been sought to address this shortfall, for example by incorporating a *Z* height to ensure correct height of printhead to object. This is useful for small object branding and personalisation or specific printing applications on to already made objects, on rigid and flexible materials. For printing on to cylindrical objects such as candles, cosmetics bottles and wine bottles, a kebab is used that is driven by rotary rollers. These smaller printers enable on-demand short-run personalisation of products (see Figure 3.20). Another recent example is a company that uses facial recognition software to custom print customer's faces and heads on to LEGO® figures (http://www.telegraph.co.uk/men/the-filter/get-your-face-3d-printed-as-a-lego-minifigure; http://funky3dfaces.com).

Direct Write technologies (DW) is a term used to describe a family of application methods -- deposition, continuous jetting, extrusion -- on to a two- or three-dimensional surface. The term can also be used to describe laser etching and engraving, but in this context we will consider the use of tools to deposit inks, paints and pigments. The most common and ubiquitous DW technology is inkjet printing, which deposits dots to create a thin layer of ink. The important rheological factors of fluid materials for DW printing to consider are: flow through the deposition tool, whether they retain their shape once deposited, can mix or fuse to the substrate and neighbouring layers, and behave in a predictable and controllable way.

Other methods of deposition include: a nozzle that extrudes fluid materials (that may range in viscosity from ceramic inks to a screenprinting ink), a brush or nib that can deposit more delicate and fluid-like materials, an aerosol that can spray inks or ink-like materials as a mist, and a jet that uses an array of printheads to print single drops of ink or as a continuous stream. Inks and materials are required to be viscoelastic, which means they need to flow freely under shear through a nozzle, but retain shape after the shear stress is released. Extrusion-based methods enable more robust materials to be deposited. These range from materials that are already liquid, or solids and powders that are heated to become liquid and will return to solid form following extrusion. Methods and materials for deposition and extrusion could be compared to icing a cake. The material properties of an extrusion system are important for their ability to flow. The quality of the paste material needs to be consistent, semi viscous and, by using a constant force, can be extruded consistently and evenly through a nozzle. Therefore, the required rheological quality is that the a paste that will cure or dry in the same shape as it has been extruded, and does not spread or deform. Depending on the computer algorithm for creating the vectors, the extruded path (commonly referred to as roads) is either one continuous line, for example one continuous upwardly spiralling line (analogous to building a coil pot), or a series of short lines that are overprinted to form a mesh structure.

The building material is loaded into a container, where it is maintained at a constant state. The material may be liquid (ceramic, chocolate, concrete) or if dry (as powder, pellets or filament), may be subjected to heat from coils around the extrusion chamber to change into a liquid. The materials are either fed by gravity, whereby dry powder drops to the bottom, or is fed on a reel (filament) under vacuum by force, for example a *Moineau* screw pump or peristaltic pump. A pump is able to deliver precise amounts for a range of viscosities ranging from 0.025 cubic centimetres (cc, or cm^3) to 0.1 cc. Once the material is extruded, the material may be cured or left to evaporate, or cool to solidify. The important consideration, as with all additive layer methods, is to ensure the materials have bonded successfully to the previous layer to create a solid shape.

Since the rise of the smaller desktop printers, there has been much interest in rapid fabrication and digital technologies over the last few years, and the idea of a home 3D printer next to the computer has become an achievable goal. For the home user, websites such as Materialise (https://i.materialise.com) and Shapeways (https://www.shapeways.com) offer customers free apps to customise existing files, upload designs and sell their products through the online store. There are a multitude of 3D printing bureaus that offer a range of services from prototypes, scale models, to useable products. The types of materials that can be used are rapidly increasing and present the most interesting aspect of the technology and include: thermoplastics, ABS or PLA filaments, UV-sensitive epoxy resin, starch powder and metal powders, including stainless steel, titanium alloy, nickel alloy and cobalt chrome.

3.6.3 Printing Functional Materials

There are an increasing number of laboratories exploring the crossovers between functional materials for specialised applications. As described in Case Study 11: Microstructural Texture, there are a range of printing applications including: security, dental, medical and optical markets. The following examples demonstrate functional materials and how these are beginning to impact different markets.

Printed electronic pastes and inks present a significant growth area for textiles, biosensors and touch screens (http://www.printedelectronicsnow.com/issues/2015-03-01/view_features/conductive-inks-drive-growth-in-flexible-and-printed-electronics). The important requirement for electronic inks is conductivity, reliability and adhesion to different surfaces. Traditionally applied through screenprinting, ink manufacturers such as Sun Chemical have developed inkjettable nano-silver inks. Likewise, Dupont's stretchable conductive inks can be printed on to different types of flexible materials (http://www.dupont.com/products-and-services/electronic-electrical-materials/printed-electronics/products/inkjet-silver-conductor-inks.html). Voxel, which began as a start-up company in 2015, are exploring the potential of flexible silver inks for printed electronics and printed antennas (see Figure 3.26).

Engineers from MIT and Singapore University of Technology and Design (SUTD) are using microstereolithography to create shape-memory polymers that can predictably morph in response to temperature. These could have an impact on a range of applications, for example soft actuators that turn solar panels to face the sun or micro drug capsules that open upon early signs of infection (http://news.mit.edu/2016/3-d-printed-structures-remember-shapes-drug-delivery-solar-panel-0826.

Led by Hod Lipson, the team at Creative Machines Lab at Columbia University, are working on developing a 3D food printer that can print edible foodstuff (https://www.youtube.com/watch?time_continue=72&v=VUvDCAG4P64). The food printer holds eight slots for frozen food cartridges. It is straightforward to print with a single ingredient, but according to Lipson, the challenge is to work with different ingredients together, and then to cook it as well. The impact of this technology could be a reduction of allergic reaction to foods (http://www.smithsonianmag.com/innovation/3d-print-your-own-breakfast-180960048/#v2SlG7AhK2fePRpZ.99; http://engineering.columbia.edu/news/hod-lipson-3d-food-printing).

Revolutionary approaches are being undertaken by companies in the field of regenerative medicine and bone tissue engineering. Organovo or 3Dynamic Systems Ltd (3DS) are studying ways to reproduce living human tissue. 3DS, a spinout company of Swansea University (https://www.swansea.ac.uk/press-office/news-archive/2014/university-developed3dtissuebioprintingtechnologynowcommerciallyavailable.php), are exploring bioprinting to deposit different tissues that have the potential to create transplantable complex

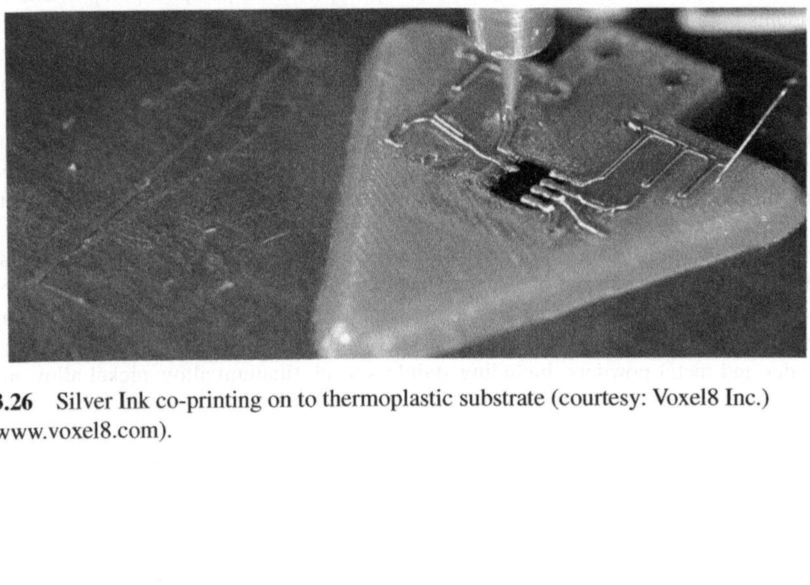

Figure 3.26 Silver Ink co-printing on to thermoplastic substrate (courtesy: Voxel8 Inc.) (https://www.voxel8.com).

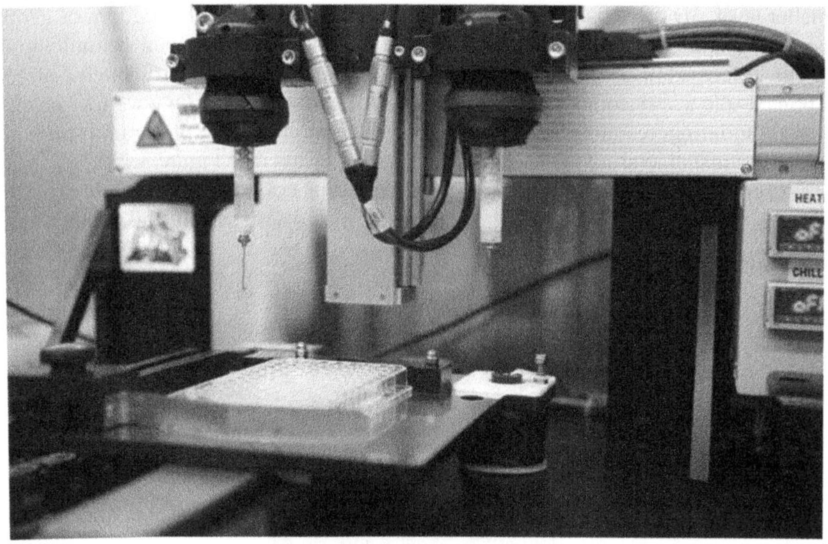

Figure 3.27 Organovo's bioprinter Novagen (courtesy: Organovo) (http://organovo.com/science-technology/bioprinted-human-tissue).

tissues on demand. Likewise, Organovo, based in San Diego, California, are working to create tiny building blocks made of living human cells that mimic the specific function of the native tissue (Figure 3.27).

In collaboration with Architects Foster and Partners, building contractors Skanska, and Loughborough University, new methods for building construction are being explored on a grand scale using a 3D concrete printer (https://www.dezeen.com/2014/11/25/foster-partners-skanska-worlds-first-commercial-concrete-3d-printing-robot/). Other building projects include the 3D Print Canal House (http://3dprintcanalhouse.com) in Amsterdam that has been printed with Macromelt, a printing material made of 80% vegetable oil, and even a 3D printed castle (http://www.totalkustom.com/3d-castle-completed.html.

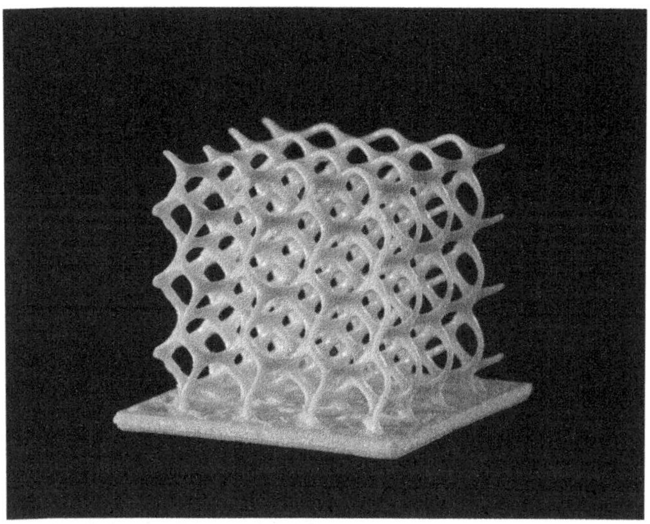

Figure 3.28 Ceramic piece made by Lithoz (courtesy: Lithoz) (http://www.lithoz.com/en),

On a smaller level, 3D printed ceramics present exceptional mechanical (freedom of geometry and material) properties and are a significant growth area for a range of functional ceramic applications, from fuel cells (McDowall, 2016) and filters (http://www.fch.europa.eu/project/cost-effective-and-flexible-3d-printed-sofc-stacks-commercial-applications) to tableware (http://www1.uwe.ac.uk/research/researchimpact/3dprintingofceramics.aspx). Lithoz specialises in a range of ceramic-based materials using LED additive manufacturing methods to print ceramics parts (see Figure 3.28).

Although not considered as a functional product but interesting nevertheless, is a highly specialist study of ancient Egyptian faience, which is a self-glazing ceramic material. David Huson, Stephen Hoskins and

Figure 3.29 David Huson and Peter Walters, 3D printed ceramic tableware, Centre for Fine Print Research, University of the West of England, Bristol (courtesy: David Huson).

Figures 3.30 David Huson, Kate Nash, Peter Walters and Stephen Hoskins, 3D printed Egyptian faience, Centre for Fine Print Research, University of the West of England, Bristol (courtesy: David Huson).

the team at the Centre for Fine Print Research, University of the West of England, Bristol, have developed a printable faience ceramic material (Huson, 2013; Huson and Vaughan, 2015). During the drying process, the soluble salts in the material migrate to the surface forming an alkali rich coating; during the firing these alkali salts react with the silica in the body material, resulting in a finished object that is glazed and vitrified at the same time (Figure 3.29). Traditional glazing methods require the ceramic object to be fired, then dipped in a coloured glaze and then fired again. This describes the efflorescence faience glazing process; the other faience glazing process, cementation glazing, is a diffusion process where the silica-rich body is packed into a *saggar* surrounded by a glazing powder. The image of three 3D printed faience hedgehogs in Figure 3.30, modelled from an Egyptian figure (1975–1640 BC) at the Ashmolean Museum, Oxford, shows the extraordinary range of surface and texture, and also the unpredictability of the process.

References

Abate, D., Menna, F., Remondino, F., and Gattari, M.G. (2014) 3D painting documentation: evaluation of conservation conditions with 3D imaging and ranging techniques. *ISPRS-International Archives of the Photogrammetry, Remote Sensing and Spatial Information Sciences*, 1, 1–8.

Akca, D. (2011) 3D modeling of cultural heritage objects with a structured light system. *Mediterranean Arhaeology and Archaeometry*, 12, 139–152.

Anderson, C. (2013) *Makers: The New Industrial Revolution*, Random House, London.

Andrade, B.T., Bellon, O.R.P., Silva, L., and Vrubel, A. (2012). Digital preservation of Brazilian indigenous artworks: Generating high quality textures for 3D models. *Journal of Cultural Heritage*, 13, 28–39.

Baar, T., Brettel, H., and Segovia, M.O. (2016) Relating optical and geometric surface characteristics for gloss management in printing applications. *Electronic Imaging*, 2016, 1–14.

Barbosa, J., Sobral, J.L., and Proença, A.J. (2007) Imaging techniques to simplify the PTM generation of a bas-relief, in *VAST'07: Proceedings of the 8th International Symposium on Virtual Reality, Archaeology and Cultural Heritage*, 2007.

Barreau, J.B., Nicolas, T., Bruniaux, G. *et al.* (2014) Photogrammetry based study of ceramics fragments. *International Journal of Heritage in the Digital Era*, 3, 643–656.

Barsanti, S.G., Micoli, L.L., and Guidi, G. (2013) Quick textured mesh generation for massive 3D digitization of museum artifacts. *Digital Heritage International Congress (DigitalHeritage)*.

Bell, S., Upchurch, P., Snavely, N., and Bala, K. (2013). OPENSURFACES: A richly annotated catalog of surface appearance. *ACM Transactions on Graphics (SIGGRAPH 2013)*, 32.

Bertani, D., Cetica, M., Melozzi, M., and Pezzati, L. (1995) High-resolution optical topography applied to ancient painting diagnostics. *Optical Engineering*, 34, 1219–1225.

Blais, F., Taylor, J., Cournoyer, L. *et al.* (2005) Ultra-high resolution imaging at 50 μm using a portable XYZ-RGB color laser scanner, in *Research Gate*.

Blais, F., Taylor, J., Cournoyer, L. *et al.* (2007) Ultra high-resolution 3D laser color imaging of paintings: the *Mona Lisa* by Leonardo da Vinci, in *7th International Conference on Lasers in the Conservation of Artworks* , 17-–21 September 2007, Madrid, Spain.

Bunsch, E. and Sitnik, R. (2010) Documentation instead of visualization – applications of 3D scanning in works of art analysis, in *Electronic Imaging*, IS&T/SPIE International Society for Optics and Photonics, San Jose.

Cahn, R.W. (2001) *The Coming of Materials Science*, Elsevier Science Ltd, London.

Caine, M. and Magen, M. (2011) *Pixels and Parchment: The Application of RTI and Infrared Imaging to the Dead Sea Scrolls*, EVA, July 2011, London.

Cartwright, J., Adee, S., Webb, R. *et al.* (2014) Wonderstuff, seven materials that will revolutionise your world. *New Scientist*, 36–41.

Cather, S. (ed.) (1991) *The Conservation of Wall Paintings*, 13–16 July 1987, Courtauld Institute of Art and the Getty Conservation Institute, London.

Charneey, D. (ed.) (2011) *Power of Making, The Importance of Being Skilled*, V&A and The Crafts Council, London.

D'Amelio, S. and Brutto, M.L. (2009) Close range photogrammetry for measurement of paintings surface deformations. *International Archives of Photogrammetry, Remote Sensing and Spatial Information Sciences*, 38, 6.

Earl, G.E.A. (2011) Reflectance transformation imaging systems for ancient documentary artefacts. *Electronic Visualisation and the Arts*, EVA, 2011, London.

Earl, G., Martinez, K., and Malzbender, T. (2010) Archaeological applications of polynomial texture mapping: Analysis, conservation and representation. *Journal of Archaeological Science*, 37, 2040–2050.

Ferguson, C. (2007) *Historical Introduction to the Development of Material Science and Engineering as a Teaching Discipline*, UK Centre for Materials Education, Liverpool. Available at: http://www.materials.ac.uk/pub/materials-history-intro.pdf.

Fleming, R.W., Dror, R.O., and Adelson, E.H. (2003) Real-world illumination and the perception of surface reflectance properties. *Journal of Vision*, 3, 347–368.

Gallo, A., Muzzopappa, M., and Bruno, F. (2014) 3D reconstruction of small sized objects from a sequence of multi-focused images. *Journal of Cultural Heritage*, 15, 173–182.

Giesel, M. and Gegenfurtner, K.R. (2010) Color appearance of real objects varying in material, hue, and shape. *Journal of Vision*, 10, 10.

Green, P. (ed.) (2010a) *Color Management: Understanding and Using ICC Profiles*, John Wiley & Sons, Ltd, Chichester.

Green, P. (2010b) *Common Color Management Workflows and Rendering Intent Usage*, in (ed. London College of Communication, UK), John Wiley & Sons, Ltd, Chichester.

Guidi, G., Atzeni, C., Seracini, C., and Lazzari, S. (2004) Painting survey by 3D optical scanning: The case of 'Adoration of the Magi' by Leonardo da Vinci. *Studies in Conservation*, 49, 1–12.

Hameeuw, H. and Willems, G. (2011) New visualization techniques for cuneiform texts and sealings. *Akkadica*, 132, 163–178.

Hess, M. and Robson, S. (2013) Re-engineering Watt: A case study and best practice recommendations for 3D colour laser scans and 3D printing in museum artefact documentation. Available at: http://discovery.ucl.ac.uk/1411525/1/23-Lacona-IX-Hess.pdf.

Ho, Y.-X., Landy, M.S., and Malony, L.T. (2008) Conjoint measurement of gloss and surface texture. *Psychological Science*, 19, 196–204.

Huson, D. (2013) 3D printing of self-glazing ceramic materials: An investigation inspired by ancient Egyptian technology, in *NIP29 Digital Fabrication 2013 International Conference on Digital Printing Technologies*, Society for Imaging Science and Technology, Seattle, USA.

Huson, D. and Vaughan, K. (2015) 3D printed self-glazing ceramics: Process and materials development, in *NIP31 Digital Fabrication 2015 International Conference on Digital Printing Technologies*, 2015, Society for Imaging Science and Technology, pp. 416–420.

Julesz, B. (1991) Textons, the elements of texture perception and their interactions. *Nature*, 91–97.

Karaszewski, M., Lech, K., Bunch, E., and Sitnik, R. (2014) In the pursuit of perfect 3D digitization of surfaces of paintings: Geometry and color optimization. *Digital Heritage. Progress in Cultural Heritage: Documentation, Preservation, and Protection*, pp. 25–34.

Kersten, T. and Lindstaedt, M. (2012) Potential of automatic 3D object reconstruction from multiple images for applications in architecture, cultural heritage and archaeology. *International Journal of Heritage in the Digital Era*, 1, 399–420.

Knott, S. (2015) *Amateur Craft History and Theory*, Bloomsbury, London.

Kotoula, E. and Kyranoudi, M. (2014) Study of Ancient Greek and Roman coins using reflectance transformation imaging. *E-Conservation Magazine*, 25, 74–88.

Kriss, M. (ed.) (2015) *Handbook of Digital Imaging*, John Wiley & Sons, Ltd, Chichester.

Leung, T. and Malik, J. (2001) Representing and recognising the visual appearance of materials using three-dimensional textons. *International Journal of Computer Vision*, 43, 29–44.

Love, T. (2011) Science and the arts, in *Litrefs Articles*. [Online]. Available from: http://litrefsarticles.blogspot.co.uk/2011/01/science-and-arts.html.

Lucet, G. (2013) D survey of pre-Hispanic wall painting with high resolution photogrammetry. *ISPRS Annals of Photogrammetry, Remote Sensing and Spatial Information Sciences*, 1, 191–196.

Macdonald, L.W. (2011). Visualising an Egyptian artefact in 3D: Comparing RTI with laser scanning. *Electronic Visualisation and the Arts*, 155–162.

Malony, L.T. and Brainard, D.H. (2010) Color and material perception: achievements and challenges. *Journal of Vision*, 10, 19.

Malzbender, T., Gelb, D., Wolters, H., and Zukerman, B. (2000) *Enhancement of Shape Perception by Surface Reflectance Transformation*, Hewlett-Packard Laboratories, Palo Alto, California.

Manfredi, M., Williams, G., Kronkright, D. *et al.* (2013) Measuring changes in cultural heritage objects with reflectance transformation imaging. *Digital Heritage International Congress*, 1, 189–192.

Manfredi, M., Bearman, G., Williamson, G. *et al.* (2014) A new quantitative method for the non-invasive documentation of morphological damage in paintings using RTI surface normals. *Sensors*, 14, 12271–12284.

McCann, J.J. and Rizzi, A. (2011) *The Art and Science of HDR Imaging*, Chichester, John Wiley & Sons Inc.

McDowall, J. (2016) Glastonbury festival harnesses the power of pee. [Online]. *Resource*, 17 June 2016, [Online]: http://resource.co/article/glastonbury-festival-harnesses-power-pee-11181.

McLuhan, M. (1964) *Understanding Media, The Extensions of Man*, Routledge, London.

Miller, D. (2011) The power of making, in *Power of Making, The Importance of Being Skilled* (ed. D. Charney), V&A and The Crafts Council, London.

Miodownik, M. (2015) Wood 2.0 reaches for the skies as digital techniques give it a new lease of life. *Guardian*, 10 June 2015. Available at: https://www.theguardian.com/technology/2015/jun/10/plywood-reaches-for-the-skies-wood.

Mohen, J.P., Menu, M., and Mottin, B. (2006) *Mona Lisa: Inside the Painting*, Harry N. Abrams, New York.

Mudge, M., Voutaz, J.P., Schroer, C., and Lum, M. (2005) Reflection transformation imaging and virtual representations of coins from the Hospice of the Grand St. Bernard. in *Proceedings of 6th International Symposium on Virtual Reality, Archaeology and Cultural Heritage (VAST2005)* (eds M. Mudge, N. Ryan, and R. Scopigno), The Eurographics Association, pp. 29–39.

Mudge, M., Malzbender, T., Schroer, C., and Lum, M. (2006) New reflection transformation imaging methods for rock art and multiple-viewpoint display, in *VAST, 2006*, The Eurographics Association, pp. 195–202.

Padfield, J., Saunders, D., and Malzbender, T. (2005) Polynomial texture mapping: a new tool for examining the surface of paintings, in *ICOM 14th Triennial Meeting*.

Pannabecker, J.R. (1994) Diderot, the mechanical arts, and the Encyclopédie: In search of the heritage of technology education. *Journal of Technology Education*, 6.

Pires, H., Marques, P., Henriques, F., and Oliveira, R. (2007) Integrating laser scanning, multispectral imagery and GIS in C&R documentation practices: A first approach using two XVIth century wood paintings from Convento de Cristo in Tomar in *XXI CIPA Symposium*, 2007, Athens, Greece.

Polanyi, M. (1958) *Personal Knowledge: Towards a Post-Critical Philosophy*, Routledge & Kegan Paul Ltd, London.

Remondino, F. and El-Hakim, S. (2006) Image-based 3D modelling: A review. *The Photogrammetric Record*, 21, 269–291.

Remondino, F. and Rizza, A. (2010) Reality-based 3D documentation of natural and cultural heritage sites – techniques, problems, and examples. *Applied Geomatics*, 2, 85–100.

Samaan, M., Héno, R., and Pierrot-Deseilligny, M. (2013) Close-range photogrammetric tools for small 3D archeological objects. *ISPRS-International Archives of the Photogrammetry, Remote Sensing and Spatial Information Sciences*, 1, 549–553.

Sennet, R. (2009) *The Craftsman*, London, Penguin.

Sharma, G. and Bala, R. (eds) (2002) *Digital Color Imaging Handbook*, CRC Press, Boca Raton, Florida..

Shneiderman, B. (2002) *Leonardo's Laptop: Human Needs and the New Computing Technologies*, The MIT Press, Cambridge, MA.

Snow, C.P. (1959) *The Two Cultures*, Cambridge University Press, London.

Tamayo, S.N., Andres, J.V., and Pons, J.O. (2013) Applications of reflectance transformation imaging for documentation and surface analysis in conservation. *International Journal of Conservation Science*.

Thomson, R. (2007) *Manufacturing Processes for Design Professionals*, Thames and Hudson, London.

Turner, R. (1995) *Inventing Leonardo, The Anatomy of a Legend*, Papermac, London.

Westland, S., Ripamonti, C., and Cheung, V. (2012) *Computational Colour Science Using MATLAB*, 2nd edn, Johnn Wiley & Sons, Ltd, Chichester.

Willems, G., Verbiest, F., Moreau, W. *et al.* (2005) Easy and cost-effective cuneiform digitizing, in *6th International Symposium on Virtual Reality, Archaeology and Cultural Heritage (VAST 2005)* (eds M. Mudge, N. Ryan, and R. Scopigno), Eurographics Association, pp. 73–80.

Wilson, B., Hawkins, B., and Sim, S. (eds) (2014) *Art, Science, and Cultural Understanding*, Common Ground, Illinois.

Yeo, R. (1996) Ephraim Chambers's Cyclopædia (1728) and the tradition of commonplaces. *Journal of the History of Ideas*, 57, 157–175.

Zaman, T., Jonker, P., Lenseigne, B., and Dik, J. (2014) Simultaneous capture of the color and topography of paintings using fringe encoded stereo vision. *Heritage Science*, 2, 1–10.

Zányi, E., Schroer, C., Mudge, M., and Chalmers, A. (2007) Lighting and Byzantine glass tesserae, in *EVA London Conference*, 11–13 July 2007, London.

Zheng, J., Haiyan, G., Yi, W., and Mingwei, S. (2012) 3D reconstruction of small-sized cultural relics based on laser scanning and close-range photogrammetry. *International Journal of Digital Content Technology and Its Applications*, 6, 196–205.

Zuckerman, B. (2004) Every dot and tiddle: a consideration of the limitations of computer imaging for the study of Dead Sea scrolls, in *Double Takes: Thinking and Rethinking Issues of Modern Judaism in Ancient Contexts. Studies of the Shoah* (eds Z. Garber and B. Zuckerman), University Press of America.

Bibliography and Further Reading

Adam, A.J., Planken, P.C., Meloni, S., and Dik, J. (2009) Terahertz imaging of hidden paint layers on canvas. *Opt. Express*, 17, 3407–3416.

Aita, H., Oh, W., Kubo, K. *et al.* (2008) Light-induced bone cement–philic titanium surface. *Journal of Materials Science*, 43, 1552–1558.

Akca, D., Gruen, A., Breuckmann, B., and Lahanier, C. (2007) High definition 3D-scanning of arts objects and paintings, in *Optical 3-D Measurement Techniques VIII* (eds A. Gruen and H. Kahmen), Zurich, Switzerland, pp. 50–58.

Ashbaugh, J.C., Berns, R.S., Darling, B.A., and Taplin, L.A. (2009) Artist material BRDF database for computer graphics rendering, in *Proceedings of the IS&T/SID 17th CIC*, pp. 62–68.

Avery, M.P., Klein, S., Richardson, R. *et al.* (2015) Through a glass clearly: the challenge of glass 3D-printing, in *Proceedings of the IS&T NIP31 International Conference on Digital Printing Technologies and Digital Fabrication*, pp. 280–284.

Baar, T. and Ortiz Segovia, M.V. (2013) Colour and texture appearance modelling of 2.5D prints, in *12th Congress of the International Colour Association (AIC)* (eds L. Macdonald, S. Westland and S. Wuerger), Newcastle upon Tyne, UK.

Baar, T., Brettel, H., and Segovia, M.V.O. (2013) A survey of 3D image quality metrics for relief print evaluation. *Colour and Visual Computing Symposium (CVCS)*, pp. 1–6.

Baar, T., Samadzadegan, S., Brettel, H., Urban, P., and Segovia, M.V.O. (2014) Printing gloss effects in a 2.5D system, in *IS&T/SPIE Electronic Imaging, Proceedings of SPIE, Measuring, Modeling, and Reproducing Material Appearance*, vol. 9018, pp. 90180M-8

Baar, T., Shahpaski, M., and Segovia, M.V.O. (2014) Image ghosting reduction in lenticular relief prints, in *IS&T/SPIE Electronic Imaging, Proceedings of SPIE, Measuring, Modeling, and Reproducing Material Appearance*, 90180N

Baar, T., Samadzadegan, S., Urban, P., and Ortiz Segovia, M.V. (2016) Interrelation between gloss and texture perception of 2.5D-printed surfaces, in *IS&T/SPIE Electronic Imaging, Proceedings of SPIE, Measuring, Modeling, and Reproducing Material Appearance*, pp. 1–6.

Ball, P. (1999) *Made to Measure: New Materials for the 21st Century*, University Press, Princeton.

Beland, M.C. and Bennett, J.M. (2000) Effect of local micro-roughness on the gloss uniformity of printed paper surfaces. *Applied Optics*, 39, 2719–2726.

Benn, C. and Morgan, L. (2005) *Breakdown Printing: New Dimensions for Texture and Colour*, Oxted, Committed to Cloth.

Benoit, A., Le Callet, P., Campisi, P., and Cousseau, R. (2008) Quality assessment of stereoscopic images. *EURASIP Journal on Image and Video Processing*.

Bensalma, R. and Larabi, M.C. (2013) A perceptual metric for stereoscopic image quality assessment based on the binocular energy. *Multidimensional Systems and Signal Processing*, 24, 281–316.

Berns, R.S., Chen, T., Wyble, D.R., and Chen, L. (2012) Practical total appearance imaging of paintings, in *IS&T Archiving Conference*, Copenhagen.

Berns, R.S. and De la Rie, E.R. (2003) The effect of the refractive index of a varnish on the appearance of oil paintings. *Studies in Conservation*, 48, 251–262.

Bibelriether, F., Jillek, W., Pfänder, N., Rudolph, F., Schmitt, E., Yusop, N.M., and Lesyuk, R. (2010) Print the printed circuit board – inkjet printing of electronic devices, in *Proceedings of IS&T NIP26: International Conference on Digital Printing Technologies and Digital Fabrication*, Austin, Texas, pp.715–719.

Blais, F., Taylor, J., Cournoyer, L. *et al.* (2005) Ultra-high resolution imaging at 50 μm using a portable *xyz*-rgb color laser scanner, in *International Workshop on Recording, Modeling and Visualization of Cultural Heritage*, 22–27 May 2005, Ascona, Switzerland, NRC 48099.

Bose, S., Vahabzadeh, S., and Bandyopadhyay, A. (2013) Bone tissue engineering using 3D printing. *Materials Today*, 16, 496–504.

Briones, V., Aguilera, J.M., and Brown, C. (2006) Effect of surface topography on color and gloss of chocolate samples. *Journal of Food Engineering*, 77, 776–783.

Brownell, B.E. (2005) *Transmaterial A Catalog of Materials that Redefine our Physical Environment*, Princeton Architectural Press, New York.

Brownell, B.E. (2008) *Transmaterial 2: A Catalog of Materials that Redefine our Physical Environment*, Princeton Architectural Press, New York.

Bucklow, S. (1997) The description of Craquelure patterns. *Studies in Conservation*, 42, 12.

Bunsch, E., Sitnik, R., and Michonski, J. (2011) Art documentation quality in function of 3D scanning resolution and precision, in *Proceedings of IS&T/SPIE Electronic Imaging*, San Francisco, 78690D.

Campisi, P., Neri, A., and Scarano, G. (2000) Reduced complexity modelling and reproduction of colored textures. *IEEE Transactions on Image Processing*, 9, 510–518.

Cather, S. (1987) The conservation of wall paintings, in *Courtauld Institute of Art and the Getty Conservation Institute*, Getty Publications, London.

Cootes, T.F., Edwards, G.J., and Taylor, C.J. (2001) Active appearance models. *IEEE Transactions on Pattern Analysis and Machine Intelligence*, 23, 681–685.

Dalal, E.N. and Natale-Hoffman, K.M. (1999) The effect of gloss on color. *Color Research and Application*, 24 (5), 369–376.

Dik, J., Janssens, K., Van Der Snickt, G. *et al.* (2008) Visualization of a lost painting by Vincent van Gogh using synchrotron radiation based X-ray fluorescence elemental mapping. *Anal. Chem.*, 80.

Duffy, S.M. (2013) Multi-light imaging for heritage applications. [Online]: http://www.english-heritage.org.uk/publications/multi-light-imaging-heritage-applications/Multi-light_Imaging_FINAL_low-res.pdf.

Emrith, K., Chantler, M.J., Green, P.R., *et al.* (2010) Measuring perceived differences in surface texture due to changes in higher order statistics. *Journal of the Optical Society of America*, 27, 1232–1244.

Eriksson, Y. (1998) *Tactile Pictures: Pictorial Representations for the Blind, 1784–1940*, Acta Universitatis Gothoburgensis.

Fontana, R. (2003) Integrating 2D and 3D data for diagnostics of panel paintings, in: Salimbeni, R., (ed) *Optical Metrology for Arts and Multimedia* (ed. R. Salimbeni), SPIE, pp. 88–98.

Forés, A., Pattanaik, S.N., Bosch, C., and Pueyo, X. (2009) BRDF Lab: a general system for designing BRDFs, in *XIX Congreso Espa nol de Informática Gráfica (CEIG'09)*, San Sebastián, Spain.

Gallo, A., Muzzupappa, M., and Bruno, F. (2014) 3D reconstruction of small sized objects from a sequence of multi-focused images. *Journal of Cultural Heritage*, 15, 173–182.

Ged, G., Obein, G., Silvestri, Z., Rohellec, J., and Viénot, F. (2010) Recognizing real materials from their glossy appearance. *Journal of Vision*, 10, 18.

Ghosh, A., Heidrich, W., Achutha, S., and O'Toole, M. (2010) A basis illumination approach to BRDF measurement. *International Journal of Computing Vision*, 90, 183–197.

Häkkinen, J., Kawai, T., Takatalo, J. *et al.* (2008) Measuring stereoscopic image quality experience with interpretation based quality methodology, in *Proceedings of SPIE 6808, Image Quality and System Performance V, 68081B-68081B-12-68081B*.

Hanrahan, P. (2001) From geometry to lighting to material models, in *IS&T SPIE Human Vision and Electronic Imaging VI* (eds B.E. Rogowitz and T.N. Pappas), vol. 4299, pp. 13–18.

Harper, C.A. and Petrie, E.M. (2003) *Plastics Materials and Processes: A Concise Encyclopedia*. John Wiley & Sons.

Howes, P. and Laughlin, Z. (2012) *Material Matters, New Materials in Design*, Black Dog Publishers, London.

Huson, D. (2010) The digital fabrication of ceramics by 3D powder printing, in *Proceedings of IS&T NIP26 International Conference on Digital Printing Technologies and Digital Fabrication*, Austin, Texas, pp. 545–548.

Huson, D. (2013) 3D Printing of self-glazing ceramic materials: an investigation inspired by ancient Egyptian technology, in *Proceedings of IS&T NIP29 International Conference on Digital Printing Technologies and Digital Fabrication*, Seattle, pp.14–17.

Huson, D. and Vaughan, K. (2015) 3D printed self-glazing ceramics: Process and materials development, in *Proceedings of IS&T NIP31 International Conference on Digital Printing Technologies and Digital Fabrication*, pp. 416–420.

Hyun, W.J., Secor, E.B., Hersam, M.C. *et al.* (2015) High-resolution patterning of graphene by screen printing with a silicon stencil for highly flexible printed electronics. *Advanced Materials*, 27, 109–115.

Jin, L., Boev, A., Gotchev, A., and Egiazarian, K. (2011) 3D-DCT based perceptual quality assessment of stereo video, in *Image Processing (ICIP) 18th IEEE International Conference*, pp. 2521–2524.

Johnston, L. (2015) *Digital Handmade: Craftsmanship and the New Industrial Revolution*, Thames and Hudson, London.

Karaszewski, M., Adamczyk, M., Sitnik, R. *et al.* (2013) Automated full-3D digitization system for documentation of paintings. *SPIE Optical Metrology*, Munich.

Kehren, K., Urban, P., and Dörsam, E. (2011) Bidirectional reflectance and texture database of printed special effect colors, in *Final Program and Proceedings Color and Imaging Conference, 19th Color and Imaging Conference*, pp. 316-321.

Lan, Y., Dong, Y., Pellacini, F., and Tong, X. (2013) Bi-scale appearance fabrication. *ACM Transactions on Graphics (TOG)*, 32, 145.

Lee, H., Seo, S., and Yoon, K. (2011) Directional texture transfer with edge enhancement. *Computers and Graphics*, 35, 81–91.

Lefteri, C. (2012) *Making It Manufacturing Techniques for Product Design*, Lawrence King, London.

Leloup, F., Hanselaer, P., Versluys, J., and Forment, S. (2007) BRDF and gloss measurements. *CIE Expert Symposium on Visual Appearance*, pp. 28–33.

Leloup, F.B., Obein, G., Pointer, M.R., and Hanselaer, P. (2014) Toward the soft metrology of surface gloss: a review. *Color Res. Appl.*, 39 (6), 559–570.

Leloup, F.B., Hanselaer, P., Pointer, M.R., and Dutré, P. (2012) Integration of multiple cues for visual gloss evaluation, in *Proceedings of Predicting Perceptions: The 3rd International Conference on Appearance*, pp. 52–55.

Liang, H., Cid, M., Cucu, R., *et al.* (2005) En-face optical coherence tomography – a novel application of non-invasive imaging to art conservation. *Optics Express*, 13, 6133–6144.

Lowe, D. (2004) Distinctive image features from scale-invariant keypoints. *International Journal of Computing Vision*, 60.

Mace, D. (2015) Digital printing of packaging – the Tonejet solution, in *Proceedings of IS&T NIP31 International Conference on Digital Printing Technologies and Digital Fabrication*, pp. 404–408.

Magdassi, S. (2010) *The Chemistry of Inkjet Inks*, World Scientific Publishing, London.

Malzbender, T., Gelb, D., and Wolters, H. (2001) Polynomial texture maps, in *28th Annual Conference on Computer Graphics and Interactive Techniques*, Los Angeles, California, pp. 519–528.

Malzbender, T., Samadani, R., Scher, S. *et al.* (2012) Printing reflectance functions. *ACM Transactions on Graphics*, 31 (20), 1–11.

Manchee, D. (2006) *20 Works of Art in the Age of Mechanical Reproduction*. [Online]: http://www.dougmancheeprojects.com.

Martin, T. (2010) *The Truth about Solvent…*. [Online]: https://inkjetprinters.wordpress.com/2010/11/30/truth-about-solvent/.

Matusik, W., Ajdin, B., Gu, J. *et al.* (2009) Printing spatially-varying reflectance. *ACM Trans. Graphics, Proceedings of SIGGRAPH Asia*, 28 (128), 1–9.

Miodownik, M. (2014) *Stuff Matters: The Strange Stories of the Marvellous Materials that Shape Our Man-Made World*, Viking Books, London.

Mudge, M., Voutaz, J.P., Schroer, C., and Lum, M. (2005) Reflection transformation imaging and virtual representations of coins from the Hospice of the Grand St. Bernard, in *The Eurographics*

Association Proceedings of 6th International Symposium on Virtual Reality, Archaeology and Cultural Heritage (VAST2005), pp. 29–39.

Olsson, R. and Sjöström, M. (2007) A depth dependent quality metric for evaluation of coded integral imaging based 3D-images. *Proceedings of 3DTV Conference, IEEE*, pp. 403–406.

Pereira, T. and Rusinkiewicz, S. (2012) Gamut mapping spatially varying reflectance with an improved BRDF similarity metric. *Computer Graphics Forum Proc. Eurographics Symposium on Rendering*, p. 31.

Polzin, C., Spath. S., and Seitz, H. (2013) Characterization and evaluation of a PMMA-based 3D printing process. *Rapid Prototyping Journal*, 19, 37–43.

Rao, A.R. and Lohse, G.L. (1993) Towards a texture naming system: identifying relevant dimensions of texture, *IEEE Transactions on Image Processing*, 1070.

Rao, P., Rosen, M.R., and Berns, R.S. (2005) Performance Evaluation of Profile Maker Professional 5.0 ICC Profiling Software. Technical Report, Rochester Institute of Technology, Rochester, January 2005.

Renninger, L.W. and Malik, J. (2004) When is scene identification just texture recognition? *Vision Research*, 11.

Metal Powder Report (2013) Cyclists take industrial 3D printing for a spin. *Metal Powder Report*, 68, 26–29.

Metal Powder Report (2013) 3D printing of dental restorations. *Metal Powder Report*, 68, 32–33.

Saunders, D. and Cupitt, J. (1995) Elucidating reflectograms by superimposing Infra-red and colour images. *National Gallery Technical Bulletin*, 16, 61–65.

Seeboth, A. and Lötzsch, D. (2014) *Thermochromic and Thermotropic Materials*, Pan Stanford Publishing, Singapore.

Seuntiëns, P.J., Heynderickx, I.E., IJsselsteijn, W.A. *et al.* (2005) Viewing experience and naturalness of 3D images, in *Optics East, Proceedings of 3-D TV, Video, Display IV*, vol. 6016, SPIE, pp. 5–7.

Sharma, A. (2006) Methodology for evaluating the quality of ICC profiles, scanner, monitor, and printer. *Journal of Imaging Science*, 50, 469–480.

Sheets, H.M. (2010) Closer look: A riverscape, from higher ground. *The New York Times*, [Online]: http://www.nytimes.com/interactive/2010/10/03/arts/design/03wayne-grfk.html?ref=design&_r=0 [Last accessed September 2017].

Sher, D. (2014) *Trabecular 3D Printed Titanium Cranial Implant Performed in Argentina*, 3D Printing Industry Limited. [Online]: http://3dprintingindustry.com/2014/06/17/trabecular-3d-printed-titanium-cranial-implant-performed-argentina [Last accessed September 2017].

Shillito, A.M. (2013) *Digital Crafts: Industrial Technologies for Applied Artists and Designer Makers*, A&C Black Visual Arts, London.

Taylor, J., Beraldin, J.A., and Godin, G. (2001) 3D imaging collaboration between the National Research Council of Canada and European Museums and Cultural Organizations, in *Electronic Imaging and the Visual Arts (EVA 2001)*, Montreal, Quebec.

Thompson, R. (2012) *Prototyping and Low-Volume Production*, Thames and Hudson, London.

Tuceryan, M. and Jain, A.K. (1998) Texture analysis, *The Handbook of Pattern Recognition and Computer Vision*, 2, 207–248.

Varma, M. and Zisserman, A. (2004) A statistical approach to texture classification from single images. *International Journal of Computer Vision*, 62 (1–2), 61–81,

Vurro, M., Ling, Y., and Hulbert, A. (2013) Memory color of natural familiar objects: effects of surface texture and 3-D shape. *Journal of Vision*, 13, 20.

Westland, S., Ripamonti, C., and Cheung, V. (2012) *Computational Colour Science Using MATLAB*, 2nd edn, John Wiley & Sons, Ltd, Chichester.

Wills, J., Agarwal, S., Kriegman, D., and Belongie, S. (2009) Toward a perceptual space for gloss. *ACM Transactions on Graphics*, 28 (103) 1–15.

Wyant, J. (1982) Interferometric optical metrology: basic principles and new systems. *Laser Focus*, 18.

Yip, S. (2007) *Handbook of Materials Modelling*, Springer.

Yoon, Y.J., Moon, S., and Hwang, J. (2014) 3D printing as an efficient way for comparative study of biomimetic structures – trabecular bone and honeycomb. *Journal of Mechanical Science and Technology*, 28, 4635–4640.

Zányi, E., Schroer, C., Mudge, M., and Chalmers, A. (2007) Lighting and Byzantine glass tesserae, *Electronic Imaging and the Visual Arts (EVA) London*, 22, 1–8.

Zhang, Z. (2000) A flexible new technique for camera calibration. *Pattern Anal. Mach. Intell. IEEE Transxctions*, 22.

Zou, X., Li, H., Zou, L., *et al.* (2007) Porous tantalum trabecular metal scaffolds in combination with a novel marrow processing technique to replace autograft, in *Tissue Engineering* (ed. J. Fisher), Springer.

4

The Future

'It follows that an art which has fallen into disuse for the period of a generation is altogether lost. There are hundreds of examples of this to which the process of mechanization is continuously adding new ones. These losses are usually irretrievable' (Polanyi, 1958).

4.1 Introduction

How might 2.5D printing impact the future? The primary question looking forward to the future is: *What will the print industry look like and how will it function over the next fifty years?* The chapters and case studies in Chapter 5 explore the richness of our material culture and the need for printing and material diversity. To ensure we can continue to do this, we are reliant on nurturing new research talent through academia and industry, knowledge exchange between academic and industry and collaboration across countries. The sharing and migration of knowledge between countries is vital for cross-fertilization and circulation of ideas, best practice, local knowledge and between different socio and cultural groups.

In commercial terms we see a wealth of opportunity and application through commercial expos, exhibitions, science, design and craft magazines and these in a way are the barometer of the crossovers in a material culture. In education, the crossovers from research to teaching, from doctorate to undergraduate requires teachers and students to be well equipped with a portfolio of basic skills: drawing, understanding of making and materials, alongside good design and communication.

4.1.1 The Future of Education in Art Design and Sciences

In Chapter 3, we explored the historical crossovers between art, science and philosophy; but what is the future of art and science in the twenty-first century? We are beginning to notice that governments are alert to an increasing skills gap and recognise there is an urgent requirement to train the next generation of researchers. Training, skills development and innovation can only come about through education at all levels (Leighton and Mitchell, 2015). The Higher Education Funding Council in England (HEFCE), in recognition of the need for more funding to support key subjects in the sciences, technology, engineering and mathematics (STEM) (http://www.hefce.ac.uk/skills/stem), undertook a boost of funding to schools, but for many companies and industries that need to ensure robust ideas and support a new entrepreneurial generation and environment, there is still room for improvement. Now that schools are under pressure to

2.5D Printing: Bridging the Gap Between 2D and 3D Applications, First Edition. Carinna Parraman and Maria V. Ortiz Segovia.

Companion website: www.wiley.com/go/bridging2d3d

perform in STEM subjects, there has also been growing concern that art and design subjects are seriously overlooked. The Council for Higher Education in Art and Design (CHEAD, 2015) outlined as part of their strategy a need for a greater interdisciplinarity, and in promoting the role of art and design in cross-disciplinary research and teaching (CHEAD, 2016). This has led to significant and positive debates as to what are the priorities for schools, higher education and universities, and of course the impact on students. Do we teach single subjects and do we concentrate efforts in the arts or sciences? Or teach both? One suggestion is that instead of subject-based lessons, emerging wisdom endorses object-based learning (Independent, 2016; Bolkan, 2017; Walsh, 2016). We certainly need to improve the long-term strategy and lean on policy makers. In order to ensure a robust, curious and imaginative minded generation of makers and thinkers, we need to encourage crossovers between art, design and science.

The really important factor here is that design – good design – will continue to play an increasing and more obvious role. It is important to design products, from the creative industries to aerospace, that contribute to enriching our lives. How do we ensure this? Over the last several years, design has re-emerged as a significant topic of interest. Although objects and artefacts have been made since the beginning of human history, there is a growing understanding across many countries, now beginning to recognise the significance and role of *design* in a highly interconnected and competitive global economy. Why it matters is complex because design provides more than economic benefits. It has implications for the cultural and social identity of a country that could also be used as a beacon for the sector by highlighting areas of design excellence, in determining a country or region's brand identity that could contribute to an increase in tourism and cultural identity through design quarters, as an international export or by gaining funding or investment in infrastructure. An example of a study on the impact of design, was Bristol and Bath by Design (2016) a two-year Arts and Humanities Research Council and Design Council funded project to understand the history and value of design in the region (see also: https://www.bristolbathdesign.org).

4.1.2 The Future of Materials and Manufacture

In Chapter 3 we explored the relationship between materials and methods used in the past and, through reappraisal of digital technologies, how traditional materials might be used for new applications. The past has shown how some materials were too advanced to find the right application, or methods could not be resolved because the technology had yet to be developed. Examples include Spencer Silver's attempt in the 1960s to make a strong glue, but resulted in a disappointing 'low tack' adhesive. After many failed attempts to find a marketable application, the Post-It Note proved to be a useful non-slip bookmark. Likewise the synthetic glass-ceramic known as Corning Glass (Gardiner, 2012) was discovered in 1952, but no commercial application was found until Steve Jobs recognised its potential for touch screen phones some thirty years later. Recently discovered through serendipitous exploration by Andre Geim and Kostya Novoselov at Manchester University in 2004, Graphene is the world's first 2D material. Graphene has demonstrated extraordinary properties, for example it is two-hundred times stronger than steel, but is lightweight, flexible, translucent and thermally conductive (https://www.graphene.manchester.ac.uk/explore/the-story-of-graphene), yet due to such a short provenance, its potential and application has really yet to be understood. Due to its short provenance, its potential and application has really yet to be understood. One such application, and of relevance here, is the exploration and development of water-based and inkjet printable 2D material inks (2017) (McManus *et al.*, 2017), which have so far been linked to printable electronics, but could also demonstrate other printing applications. So-called intelligent materials, such as biomaterials and nanomaterials, present a wealth of opportunities and applications, given the right conditions for cross-disciplinary exploration and dissemination (Ternaux, 2006).

On a commercial level, evolving technologies are closely monitored and presented according to novelty, maturity and productivity. The *Hype Cycle*, produced by Gartner.com (https://www.gartner.com) provides branded but useful insights across Gartner's research areas. The curve highlights emerging technology at its peak and where industries have bridged the gap into productivity. Gartner reported in 2012 the emergence of 3D printing at home: 'In this scenario, 3D printing allows consumers to print physical objects, such as toys or housewares, at home, just as they print digital photos today. Combined with 3D scanning, it may be possible to scan certain objects with a smartphone and print a near-duplicate. Analysts predict that 3D printing will take more than five years to mature beyond the niche market" (http://www.gartner.com/newsroom/id/2124315). Five years on, it seems the prediction is mirroring the market. However, the ability to print objects at home may be disadvantaged by insufficient knowledge by the home user as how to make things using CAD software. Furthermore the materials to print objects are still expensive and have yet to be properly developed for durable products. Nevertheless, as demonstrated by online market trends, there is a greater interest in computer-aided technology for laser cutting, engraving, object embellishments and surface printing; these types of innovation are significantly impacting the packaging industry.

Insights into the 3D printing industry are comprehensively covered by trend forecasters, such as Wohlers Associates (http://www.wohlersassociates.com), providing a review and analysis of additive manufacturing and 3D printing worldwide. According to Wohlers, the market overall is growing. Many companies, traditionally associated with the 2D market are now beginning to introduce 3D printing hardware. The additive manufacturing industry has quadrupled in sales over the past five years, reaching $4.1 billion last year, of which $2 billion came from machine and material sales. Two thirds of manufacturing companies are using 3D printing as part of their prototyping workflow (Mearing, 2014). In terms of units shipped, the home and desktop market had outstripped the industrial market; for example, in 2014, the number of desktop 3D printers, costing less than $5000, totalled 139 584, whereas 12 580 industrial printers were shipped. The company reported that although the AM industry grew by 17.4% in 2016, it was down from 25.9% the year before. Much of this decrease in growth came from a decline from the largest manufacturers of AM systems – 3D Systems and Stratasys – which together represent $1.31 billion (21.7%) of the $6.063 billion AM industry. 'If these two companies were excluded from the calculations, the industry would have grown by 24.9%' (Cadalyst, 2017).

Our recent observations at DRUPA 2016 confirmed that, for the moment, conventional printing and digital printing continues to coexist to deliver a different customer experience. There will still be rotary printing presses and traditional printers that will continue to offer alternative solutions in the print industry, but there is an increasing concentration on delivering multiprint platforms that print on to many different materials, can add decorative effects, cuts and finishes, such as gloss, gold, embossing and laser cutting in one pass. With an increase in mobile use there is more emphasis on the ease of transfer from social media platforms, apps and technological applications, methods for storage, file transfer and networks. The print industry is beginning to adjust to different working methods across many different digital platforms, where information is delivered from a range of operating systems, laptops, and hand-held devices such as phones and tablets. In order to ensure that no one market is in waiting or in catch-up mode, the emphasis is on increased productivity through the co-creation and co-development of new materials and hardware alongside the coordination of robotics, cloud communication and integrated infrastructure building of bandwidth capacity in cities. The primary emerging themes for the future could be considered as:

- Circular economy and sustainable manufacturing
- Worldwide print connectivity
- Mass-printing for one
- Security printing and tracking.

4.2 Circular Economy and Sustainable Manufacturing

The print industry is undergoing a profound change. Consumers' expectations and concerns are also growing about sustainability, raw materials, scarcity of resources, and the end of a product's life or how a product is recycled or thrown away. Increasing studies and efforts are being undertaken to understand how to ensure a more circular economy (Webster *et al.*, 2016), how cities and industries can work more flexibly and sustainably, and how to make global manufacturing models more local. However, currently this comes at a cost. In production terms the final cost tends to decrease as more units of the same produce are manufactured; it is simpler to over-order or over-produce because cost models make it cheaper to order in bulk and throw away what we do not need. The other factor here is the issue of overproduction; for example, around a third of textiles do not get even to market. Overproduction may be due to poor quality control, delay in raw materials or shipping so that a product fails to arrive on time. Alternative production solutions are beginning to be introduced through print on demand, bespoke printing, personalised products and developing printer hardware that can perform a range of jobs including, for example, printing as well as creasing and cutting.

Based on the concept of biomimicry and ecological cycles for the growth and sustainability of manufacturing and business (Benyus, 1997), there has emerged a significant body of research studying resilience models, which are being applied to a range of areas from climate change (https://www.rockefellerfoundation.org/our-work/initiatives/100-resilient-cities) to a redistributed manufacturing model for consumer goods (http://www.recode-network.com). The notion of resilience and circular and sustainable economies is also beginning to play an increasing role in our cities. It is estimated that, by 2050, 75% of the population will be living in cities, which means that cities have to learn how to adapt and grow as more pressures are placed upon them. In the publication entitled *The Circular Economy – A Wealth of Flows*, Ken Webster, head of innovation at the Ellen MacArthur Foundation, describes how the linear 'take–make and dispose' economy is a '19th century heritage adrift in the 21st century reality' (Webster, 2016). Technology is conditioned by both the past (old infrastructures and partners and analogue technologies) and current systems (digital connection, open design, fabrication, computer-aided design and modelling) and how the evolution of an economy is driven by irregular bursts of presses and pulses (Schumpeter, 1939), that breaks cycles and old partners to drive change. Otherwise known as creative destruction, by building-in a capacity to evolve, change and adapt, and by developing insights about resilience, an economy can begin to identify how things and organisations are connected and the performance of the whole system. The model of a 'circular' economy, by reusing resources and redistributing manufacturing, brings production back towards a resilient and sustainable economy. The Ellen MacArthur Foundation provides useful resources and case studies and models for a circular economy (http://www.ellenmacarthurfoundation.org). Parallel areas of interest are Open Source Ecology (OSE) as described in Marcin Jakubowski's Ted Lecture entitled *Open Sourced Blue Prints for Civilization* in which he describes the idea of using digital fabrication tools and an 'open source civilization kit' to overcome the cycle of broken equipment and to be able to work out how to repair it (https://www.ted.com/talks/marcin_jakubowski/transcript?language=en). On a similar functional level, the idea behind iFixit (https://ifixit.org) was based on the simple need to take a product apart simply to find out whether it could be repaired, which has led to an online site that provides open-source service manuals. Kyle Weins, founder of iFixit, explains that by preventing access to repair manuals shortens the life of a product, as consumers are unlikely to be able to repair it on their own, are likely to buy a new one and the old one is thrown away (The Great Recovery (2016), Lessons Learned from the Great Recovery: http://www.greatrecovery.org.uk).

Reusing resources and sustainability is not limited to business and manufacturing. There are vast opportunities for people to make things and likewise there is also the aspect of tinkering, once termed as hobbyists. These range from Instructables (http://www.instructables.com),an open source website for making just about anything, and once made to share how to make it, to the Khan Academy (https://www.khanacademy.org), a not for profit education site to teach anyone on a wide range of subjects, to Make Things Do Stuff (http://makethingsdostuff.co.uk), to foster the next generation digital makers. The Tinkering School, founded by Gever Tulley, provides opportunities for children to learn how to make things through workshop activities (http://www.tinkeringschool.com). How Toons (http://www.howtoons.com), as it suggests, combines comic strip with DIY projects and is armed with the 'tools of mass instruction' to assist children to make things at home.

Sophie Thomas and Lucy Chamberlain, working on the The Great Recovery Project (http://www.greatrecovery.org.uk), explains how the stuff that we use and make and throw away on a daily basis may not end up in ground-fill. It is possible, by disassembling objects and by exploring the complexity of materials used for one product,for designers to begin to understand the life of a product, design for a circular economy and reduce unnecessary waste.

The idea of recycling is certainly not new. In 1939, Prime Minister Neville Chamberlain announced the establishment of a department for the reclamation of waste. J.C. Dawes, who oversaw the municipal salvage effort during World War Two, observed how we are a wasteful race and, during wartime, the country had to be mindful of resources because they were so scarce. The recycling effort resulted in the collection of two and a quarter million tons of recyclable materials. In an annual report produced by the Waste Paper Associate, the weight of paper recycled was recorded as a percentage of paper consumed by weight that was collected between 1939 and 1955. Paper was in scarce supply and in 1942 just over 61% of paper was recycled. Simple suggestions such as the reuse of old envelopes meant that as paper became scarcer, then the consumption of paper declined.

Aside from the well-established worldwide Fablabs and online stores that sell 3D printing materials to assist the hobbyist, there are also online 3D printing bureaus (Shapeways, i.Materialise, Sculpteo, Ponoko) that also serve as an online market for those who have designed products. There are also the 3D hubs that connect people who wish to print an object to people who have a machine that is not is use or has spare capacity. They may be businesses or those who have a 3D printer at home. MakeXYZ presents the idea of local production, whereby smaller organisations and individuals are microfactories (https://www.imakr.com; http://www.fastcoexist.com/3051869/google-is-planning-for-a-zero-waste-circular-economy; http://www.fastcodesign.com/; http://www.greatrecovery.org.uk/.

4.3 Worldwide Print Connectivity

Today, we have greater connectivity and more processing power. As Gordon Moore predicted just over 50 years ago, the number of transistors that manufacturers could put on to a microchip would double every 1–2 years. In the 1960s chips contained a few thousand components, whereas chips today hold several billion. This doubling of power and improvement in technology has reflected the rapid transformation of computers and the introduction of smart phones, high-speed Internet and intelligent appliances we use today. Over the last 50 years, manufacturers have attempted to keep in-pace with this law, but is this beneficial for innovation? Has this become more of a curse for the industry? It is now acknowledged that the doubling has started to slow and a different strategy is necessary and is being sought. Until now, the drive towards improving performance of microprocessors has meant efforts have been concentrated into scaling down the elements to fit on to a chip. However, there is a physical constraint to just how small components can be made, therefore requiring more sophisticated methods to print, etch and solder necessary to operate at micro levels (Waldrop, 2016). There is also a notable difference between the business models of large organisations versus SMEs. Moore's Law favours larger businesses, which are able to simultaneously work across a portfolio of products, whereas smaller companies working in incremental developments towards innovation are less able to compete. This shift away from Moore's Law and with access to open-source platforms could potentially facilitate new opportunities in innovation (Huang, 2016).

We are currently experiencing a shift in manufacturing. We have moved from local manufacturing industry, technical and material know-how, to remote production sites, where knowledge and skills, materials and whole manufacturing centres are shipped offshore, and now to a state change where the United States and some European countries are 're-shoring' or 'in-shoring', meaning to return the production that was sent to Asia during the 1980s and 1990s. Economists believe that moving back to local ways of producing will have a significant impact over the next 10 years. Offshore manufacturing has had benefits for the economy, including reduced costs in manufacturing and thus reduced cost in products to the consumer, but it has also led to a loss of local knowledge, industry and production, societal and cultural identity, and where cities that relied on a major industry have experienced a collapse. Furthermore, this economic and manufacturing model consumes resources that are not sustainable and which employ a 'linear' approach, where raw materials are extracted, used and discarded. The environmental and societal impacts are many: goods are shipped long distances, materials are extracted from less-regulated counties, there is a tendency for overproduction and much is left to waste.

Due to the huge cost in purchasing machines, online print service bureaus are steadily on the increase, offering a range of products that uses multichannel business models. Print services have begun to move away from shops and catalogues to deliver a range of products to customers that collaborate with a worldwide network of printers across the web. In this model, the Internet service provider works as the 'front of house', which promotes and sells a range of products, provides inspiration, advice about files and customer service. The network of printers may range from specialist printers making business cards, printed T-shirts, magazines, brochures or posters. One-off or short-run products are ideal for this type of model and can even be turned around and delivered as quickly as within 24 hours. The costs are kept low because small orders that are requested from the different clients are sent as large batches to the printer. The success of the model requires lots of different manufacturers, suppliers and digital presses that can produce printed products and ship them within a short timeframe (24 hours) that makes this model possible. A central service provider operates as an online platform through which the client uploads files, provides advice, assists in ensuring the files are ready to print and then sends jobs to printers. Their emphasis is on monitoring file formats and specifications, quality, delivery time and pricing and testing of new products. Wide format, textile, digital (business cards, folders leaflets), sheet-fed offset (posters, brochures) and premium gifts (pens, umbrellas). Once printed, the product is delivered to the client. If the product needs to be shipped to a different part of the world, then the file can be sent to a printer that is local to the recipient, thus cutting down the transportation costs. Smaller desktop printers, as demonstrated at DRUPA, such as a T-shirt, UV curing for small object marking or printing on to wine bottles means that companies can be modest in size but productive.

According to a recent report (Belson, 2016), on a global basis the average connection speed of countries has increased by 26%. South Korea has invested heavily in their Internet network infrastructure. Listed in the top ten fastest countries, although with a slight decrease, South Korea remains in the lead at 26.1 Mbps. Other countries listed in the top ten range from Norway as the second fastest, at 23.6 Mbps, to the tenth fastest, the Netherlands, at 17.6 Mbps. Therefore, barriers to print worldwide connectivity are the speed of data transfer and without investment in infrastructure the exchange of information, data and productivity could well be restricted.

4.4 Mass Printing for One

In 2012, when Neil Gershenfeld was considering the relationship between digital technology and fabrication of objects, he taught a class on 'how to make (almost) everything' at MIT's Centre for Bits and Atoms. The course was designed to teach students how to use a range of fabrication tools. He described how students attended the course to make use of the facilities in order to *make things*. The crucial question asked was 'What is digital fabrication good for?' His answer was the 'ability to personalise – producing products for a market of one person' (Gershenfeld, 2012).

As the printing industry becomes bigger and more impersonal then there is greater need for personalisation and safeguarding products in transit, protection of intellectual property, copyrighting and ensuring the provenance of products. As described in Chapter 2, there are many organisations dedicated to the accurate reproduction of artworks (Weverbergh, 2012), both legally (commercial reproduction of paintings, posters and postcards as produced by galleries) and illegally (counterfeits and forgeries). This is reflected in the relationship between digital printing, craft and reproduction – for example, 3D printing of ceramics, painting, drawing, letterpress and 2.5D printing, and from packaging to fashion. Whilst it is seemingly removed from traditional origins, there are new opportunities to gain insights into new practices and methods for craft reproduction and an exploration of methods that consider the relationship between analogue and digital processes and how these might co-exist.

These ideas of co-existence and co-evolution are reflected in the group *Print Isn't Dead* (http://www.printisntdeadmag.co/element003) who do not differentiate between traditional and contemporary printing, but embrace the most high-tech digital print innovations of the twenty-first century. As demonstrated in their publication *Element 003*, which provides a detailed description as to how the publication has been made, 'This cover was printed by Pureprint an HP Indigo using 4 hits of white ink on to 290 gsm Fedrigoni Sirio Nero paper and designed with the Patron typeface by Milieu Grotesque.' The words underlined on the cover show the most significant aspects in the production of the book: the print company, the hardware, the paper, the typeface, the designer of the type. What is significant here is their choice of an HP Indigo sheet-fed printer that uses electro-ink 4-colour plus white, which is capable of printing thousands of sheets per minute, can print on to a wide range of paper stock, print spot colours and white.

Established in 1997, the Dutch magazine company *Frame* publish visually rich and inspirational publications by incorporating die-cut, embossed, different papers and surface embellishments (https://store.frameweb.com). They are also an exemplar magazine publisher to have overcome the recent trend of on-screen publishing. Based on the return of magazine shops to the high street, it appears there is a renewed interest in the printed publication and sales of specialist subjects, photobooks, artbooks, mags, comic books and fanzines are on the rise. This may well be due to print-on-demand business models and faster more versatile print technologies that can print on to a variety of substrates, using specialty inks and decorative effects. Digital content and print-on-demand providers are presenting new opportunities for independent book and magazine publishers. Book and magazine print interface solutions such as Blurb, MagCloud, Snapfish, Joomag, Madmagz and Lulu2 (http://www.magcloud.com; https://issuu.com; https://www.joomag.com/en/; https://www.photobox.co.uk; https://www.snapfish.co.uk/2/photobooks; https://madmagz.com, https://www.lulu.com) are online platforms for converting virtual publications into print, now using mature print-on-demand publishing and global distribution networks. Jeremy Leslie – of the Mag Culture Studio – recently interviewed for the Idler (the Idler was established in 1993; it has also returned to the medium of print after a cessation of 3 years), explained that in 2016 he opened a shop dedicated to magazines, stocking 400 different titles from around the world. He says the desire to make print magazines is growing through a combination of creativity and serendipitous discovery of the product (Leslie, 2017, p.81).

Print-on-demand models are now used to exploit personalised direct mail, hypercustomisation and the mass production of objects that are unique and personalised to one person. Traditionally, and to keep costs to a minimum, a mass-produced item was manufactured in the same way to create exactly repeatable results. Customisable printing enables the product to be tailored using different images, codes and security marks that can be personalised to a particular person, location or event. This has been used extensively by the drinks and packaging industry, but what about textiles, or clothing, or shoes? For example, ShiftWear sneakers have included printed side panels made from flexible electronic paper (as used in the Amazon Kindle) and images can be customised and updated using a smartphone app (Budds, 2015). In 2012, Uniqlo teamed with Sharp Electronics and augmented reality developers Holition to produce an in-store *memory mirror* for customers to compare, edit and share outfits. Memory mirrors may well be used for different applications, such as the development of computer software for packaging complex objects. This may involve the use of origami-style nets to ensure good support of a delicate product, or protected during transit, or to reduce excess packaging. For example, Chuck Hoberman's foldable, retractable structures or *Metamaterials* was developed at the Harvard John A. Paulson School of Engineering and Applied Sciences (SEAS) (Burrows, 2017).

4.5 Security Printing

For the future of 2.5D printing, one of the primary trends is security printing and tracking. Watermarking, complex guilloche patterns and magnetic inks have long been used by the banking, legal document and ticket industries, but there is increasing pressure to protect other items, including medicines, consumer packaged goods, luxury goods and artworks, where there are many opportunities for fraud and counterfeiting. Handbags, wallets and clothing present the biggest challenge as a percentage of counterfeit goods, where around 40% are seized at the United States borders and 20% in Europe. Likewise, the percentage of watches and jewellery in the United States is around 30% and just over 20% in Europe. Other items include pharmaceuticals, perfumes and cosmetics, footwear and consumer electronics (Economist, 2015). In an attempt to counteract fraudulent copying, Italian shoe designer Ferragamo has included in the sole of each left shoe, a passive radiofrequency identification (RFID) tag.

Where RFID tags are impossible, 2.5D printing is being explored and developed. The objective is to incorporate as many complex features into a product so that it cannot be easily reproduced and the level of protection correlates with the quality of the brand or level of security. Current methods use both overt (visible) and covert (invisible) marking. Visible may include bar codes, colour QR codes or guilloche patterns that are copy sensitive, raised dots, holographs, magnetic and silver inks, laser marking, extra spot colours and inks that may change colour according to temperature, light or off-gassing. Covert marking may include a clear toner that emits an ultraviolet (UV) light range, red ink that is only visible under infrared and invisible patterns that are embedded into the packaging. Other inks are taggants, which can be visible or invisible and are uniquely microscopically coded materials that are analogous to fingerprints.

Connectivity through the Internet of Things (Burgess, 2017) may well begin to address the problems of tracking counterfeiters and counterfeit goods, estimated to be 5–7% of world trade, and which poses serious challenges for protecting products and brands (see also Section 4.4). Companies are looking to the Internet of Things to track the lifetime of a product alongside methods for innovative packaging to protect and monitor the movement of packaging around the world. It is estimated that within the *Internet of Packaging*, three and a half thousand billion annual packages move around the world each year. Over the next 5 years our everyday (passive) products will be embedded with sensors and actuators. These will become interactive and connected through our (hand-held) devices and other external antennae, which each requires hardware and continuous electronics. In order to track the lifetime of a product, a preserialisation method of printing-embedded codes are used, every single package of which contains a unique code that remains passive until activated. Packages are stored inactive and are only activated when metadata are added. The metadata can provide as much information as required and specific to that product, for example when the product is made, where it is made, its contents and so on. It is also possible to then add further data, for example consignment, country destination, address of the customer, delivery date, drop-off point or installation instructions. Lastly, if it is of a high value or a perishable product that, for example, the product may require monitoring of the temperature of the product, on-going support through maintenance, collection of product or customer feedback. This increasing data collection has begun to give companies in-depth knowledge about their products and customer profile, which can be fed back to the company. This may include stock control, distribution among stores and monitoring of prices; revenue protection, such as protection of brands and prevention of counterfeiting; increased consumer insights by building trust, provenance of product and knowledge about its content; and visibility of distribution, so that customers are able to follow products, feedback to and from customers, enabling companies to obtain deeper insights of the entire life cycle.

There is also an increasing and genuine need for robust methods of authenticating artworks (Chun, 2016) across the art market and likewise, as evidenced by the topic being included in major recent conferences including DRUPA 2016 and professional advisors to the International Art Market Conference 2017 (http://www.paiam.org/upcoming-events/the-paiam-conference-2017.html), where a special session was held on authentication of artworks. A 2016 report from Deloitte stated that: '75% of all stakeholders surveyed agree that authenticity, lack of provenance, forgery, and attribution are the biggest threats to credibility and trust in the art market' (Deloitte, 2016). The UK-based company Tagsmart (http://www.tagsmart.com) works with artists, dealers and galleries in the contemporary art market, having security tagged more than 5000 artworks, and are constantly exploring innovative, more robust ways to confirm provenance, authentication, security and tracking of contemporary artworks.

4.6 Conclusion

This chapter has explored our anxiety status of an overproduction culture and the demands that are made in keeping pace with technology. Customers are increasingly calling for personalised products in small batches and the print industry has to meet these demands. Innovation throughout the industry is concentrating on increasing connectivity between the different industries. We already have remote support services and off-shore call-centres, and as renewable energy becomes more sustainable and access to the Internet is more widely available, we can anticipate more online services from Third World and emerging countries. As the Internet enables a greater connectivity saving and transference of files across the Internet, uploading terabyte-sized files to servers with no delay will also increase and we will be able to keep track of our products as they move between countries and are bought and sold.

References

Belson, D. (2016) *Akami's State of the Internet Q4 2016 Report*, Vol 9, No. 4.

Benyus, J. (1997) *Biomimicry: Innovation Inspired by Nature*, Harper Collins, New York.

Bolkan, J. (2017) STEM/STEAM: science, technology, engineering, math and the arts. *The Journal,* 28 September 2017. [Online]: https://thejournal.com/articles/list/stem.aspx.

Budds, D. (2015) Shiftwear sneakers feature an integrated e-paper screen for infinite customization. *CoDesign*, 1 December 2015, [Online]: http://www.fastcodesign.com/3054038/shiftwear-sneakers-feature-an-integrated-e-paper-screen-for-infinite-customization.

Burgess, M. (2017) *What is the Internet of Things? Wired Magazine, 21 April 2017, [Online]*: https://www.wired.co.uk/article/internet-of-things-what-is-explained-iot.

Burrows, L. (2017) A toolkit for transformable materials. How to design materials with reprogrammable shape and function. *Harvard*, 18 January 2017, [Online]: https://www.seas.harvard.edu/news/2017/01/toolkit-for-transformable-materials.

Cadalyst (2017) Wohlers Report finds slower overall growth, More competition in 3D printing space. *Cadalyst,* 28 April 2017. [Online] https://www.cadalyst.com/hardware/3d-printers/wohlers-report-finds-slower-overall-growth-more-competition-3d-printing-space-3.

CHEAD (2015) Strategy and Policy Outline 2015–2018. *Council for Higher Education in Art and Design*. [Online] https://chead.ac.uk/overview/

CHEAD (2016) Strategy to 2020. *Council for Higher Education in Art and Design*. [Online]: https://chead.ac.uk/overview/

Chun, R (2016) *These Four Technologies May Finally Put an End to Art Forgery, 18 July 2016, [Online]*: https://www.artsy.net/article/artsy-editorial-these-four-technologies-may-finally-put-an-end-to-art-forgery.

Deloitte (2016) *Art and Finance Report 2016*. [Online]: https://www2.deloitte.com/content/dam/Deloitte/global/Documents/Finance/gx-fsi-art-finance-report-2016.pdf.

Economist (2015) Counterfeit.Com *The Economist,* 30 July 2015, [Online]: https://www.economist.com/news/business/21660111-makers-expensive-bags-clothes-and-watches-are-fighting-fakery-courts-battle.

Gardiner, B. (2012) *Glass works: How Corning created the ultrathin, ultrastrong material of the future, in WIRED*, Condé Nast, New York.

Gershenfeld, N. (2012) *How to make almost anything. The digital fabrication rrevolution. Foreign Affairs*, 91, 16.

Huang, A. (2016) The death of Moore's Law will spur innovation. *IEEE Spectrum. [Online]*. Available at: https://spectrum.ieee.org/semiconductors/design/the-death-of-moores-law-will-spur-innovation.

Independent (2016) STEM vs STEAM: How the sciences and arts are coming together to drive innovation. *Independent*, 25 May 2016. [Online]: https://www.independent.co.uk/student/student-life/Studies/stem-vs-steam-how-the-sciences-and-arts-are-coming-together-to-drive-innovation-a7047936.html.

Leighton, S. and Mitchell, P. (eds) (2015) A new STEAM age, *Challenging the STEM agenda in research. The Culture Capital Exchange. [Online]*: https://www.theculturecapitalexchange.co.uk/wp-content/uploads/2016/06/Challenging-the-STEM-agenda-in-research.pdf.

Leslie, J. (2017) *Mags to riches. Idler, May–June* 2017, pp. 75–81.

McManus, D., Vranic, S., Withers, F. *et al.* (2017) Water-based and biocompatible 2D crystal inks for all-inkjet-printed heterostructures. *Nat Nano*, advance online publication.

Parraman, C., Ellen, H., Parry, G. *et al.* (2016) *Bristol and Bath by Design; To understand the economic and cultural value of design in the Bristol and Bath region, Project Report*, University of the West of England.

Polanyi, M. (1958) *Personal Knowledge: Towards a Post-Critical Philosophy*, Routledge & Kegan Paul Ltd, London.

Schumpeter, J.A. (1939) *Business Cycles: A Theoretical, Historical and Statistical Analysis of the Capitalist Process*, McGraw-Hill, New York.

Ternaux, E. (2006) *Material World 2: Innovative Materials for Architecture and Design*, Frame Publishers, Amsterdam.

Waldrop, M.M. (2016) The chips are down for Moore's law. *Nature*, 530.

Webster, K. (ed.) (2016) *The Circular Economy: A Wealth of Flows Isle of Wight*, Ellen MacArthur Foundation.

Webster, K., Johnson, C., Blériot, J., and Gravis, L. (eds) (2016) *A New Dynamic 2: Effective Systems in a Circular Economy*, Ellen MacArthur Foundation, Isle of Wight.Weverbergh, R. (2014) A 3D scan of Van Gogh to create the perfect replica , Whiteboard. Available at: http://www.whiteboardmag.com/3d-scan-van-gogh-perfect-replica-formart

Weverbergh, R. (2014) A 3D scan of Van Gogh to create the perfect replica, Whiteboard. https://www.whiteboardmag.com/3d-scan-van-gogh-perfect-replica-formart/

Bibliography and Further Reading

Aita, H., Oh, W., Kubo, K. *et al.* (2008) Light-induced bone cement-philic titanium surface. *Journal of Materials Science*, 43, 1552–1558.

Braungart, M. and McDonough, W. (2002) *Cradle to Cradle: Remaking the Way We Make Things*, North Point Press, San Francisco.

Cartwright, J., Adee, S., Webb, R. *et al.* (2014) Wonderstuff, seven materials that will revolutionise your world. *New Scientist*, 36–41.

Hyun, W.J., Secor, E.B., Hersam, M.C. *et al.* (2015) High-resolution patterning of graphene by screen printing with a silicon stencil for highly flexible printed electronics. *Advanced Materials*, 27, 109 –115.

Kobayashi, T., Owens, M., Cross, W. *et al.* (2015) Structural color for security printing: patterned robust colloidal crystals, in *Proceedings of IS&T NIP31 International Conference on Digital Printing Technologies and Digital Fabrication*, pp. 395–396.

MacArthur, E. (ed.) (2016) *Effective Systems in a Circular Economy*, Ellen MacArthur Foundation, Isle of Wight.

Melchels, F.P.W., Domingos, M.A.N., Klein, T.J. *et al.* (2012) Additive manufacturing of tissues and organs. *Progress in Polymer Science*, 37, 1079–1104.

Miodownik, M. (2014) *Stuff Matters: The Strange Stories of the Marvellous Materials that Shape Our Man-Made World*, Viking Books, London.

Walsh, B. (2016) What we learn from making. *EdTechUpdate*, 23 June 2016. [Online]: http://www.edtechupdate.com/stem/white-paper/?open-article-id=5241824&article-title=what-we-learn-from-making&blog-domain=digitalpromise.org&blog-title=digital-promise.

Yoon, Y.J., Moon, S., and Hwang, J. (2014) 3D printing as an efficient way for comparative study of biomimetic structures – trabecular bone and honeycomb. *Journal of Mechanical Science and Technology*, 28, 4635–4640.

5

Case Studies

There are many opportunities to contribute to manufacturing by creating visually engaging and well-designed products. Craftspeople, designers and manufacturers have studied the appearance of existing materials, developed methods and sought alternative materials that may mimic their appearance, or they may have simply been inspired by objects and materials to reinvestigate old subjects in new ways. This has often resulted in highly imaginative, convincing and texturally pleasing results. This chapter explores a miscellany of subjects, of which the aim is to: provide a short historical context, contemporary methods that are now being used, visual cross-overs and comparisons and suggestions for further reading. These case studies demonstrate a survey of printing and reproduction methods used in the past and how many have evolved over time. We have explored ideas, traditional processes and materials that may have been modified or recycled in response to emerging technologies and trends.

Case Study 1: Nature Printing in the Nineteenth Century

In the nineteenth century, Victorians were devoted to collecting ferns, to an extent that in 1855, author Charles Kingsley (1819–1875) introduced the term 'Pteridomania' or 'Fern Craze' (Loundon, 2015). There are many thousands of different ferns, each family containing subtly varied genus, and these variations were collected as specimens, pressed, dried and stored in scrapbooks. As they were pressed into the pages of the volumes, the ferns created imprints into the paper, so that, by turning over the pages, the embossing from the proceeding pages are often visible. These were fairly successful in preserving the plants, but over time, pressed green leaves and flowers can lose their colour, become brittle and are easily broken and dislodged.

Visual comparisons can be made with plants that were encapsulated and fossilised in rocks over many millions of years (see Figure 5.1). As the rock is split, an everlasting impression of the plant is embossed on both sides.

Artist and designer William Morris was well-known for his stylised ornate versions of flowers, curling leaves, fern chintz fabrics, prints and wallpapers for G.P. & J. Baker (founded in 1884) and Arthur Sanderson & Sons Ltd. (founded in 1860) (see also Case Study 2: Wallpaper Design). The explorative objective for the late eighteenth and nineteenth century scientist was to gain greater insights

2.5D Printing: Bridging the Gap Between 2D and 3D Applications, First Edition. Carinna Parraman and Maria V. Ortiz Segovia.

Companion website: www.wiley.com/go/bridging2d3d

Figure 5.1 Fern fossil (*Pecopteris miltoni*) from the Carboniferous period 359–299 million years ago. The specimen was found in Bath and NE Somerset.

and understanding of the natural world. Scientific discoveries proliferated alongside industrial revolution, and for the collector, akin to the nineteenth century photographer, the quest was to capture objects and scenes directly from real life. William Henry Fox Talbot's (1800–1877) 'search for a method of printing directly from nature' (Weaver, 1992, p.1) resulted in his exploration into many different photographic experiments. In *The Pencil of Nature* (1844) Fox Talbot (1844) describes how he could faithfully reproduce the leaves of a plant, or any similar thin and delicate object, such as lace, to the same size, 'very perfectly and beautifully impressed or delineated upon it' (see also *The Project Gutenberg EBook*, http://www.gutenberg.org/files/33447/33447-h/33447-h.htm). Talbot explains how he discovered, during the spring of 1834, that he was able to increase the sensitivity of his papers by using much weaker solutions of salt than he had previously tried and, when dry, the paper was washed with nitrate of silver. 'This paper, when exposed to the sunshine, immediately manifested a far greater degree of sensitiveness than I had witnessed before, the whole of its surface turning black uniformly and rapidly: establishing at once and beyond all question the important fact, that a lesser quantity of salt produced a greater effect' (Fox Talbot, 1844, pp. 6–7). Talbot describes how he created a negative image of the object and then, in order to obtain a positive image, he laid a second sheet of sensitive paper on top of the first and exposed it to sunlight in the same way, so that 'when finished, the second paper is found to have received an image of a contrary kind to the first; the ground being white, and the image upon it dark' (Fox Talbot, 1844, pp. 6–7).

The Cyanotype process presented by Sir John Frederick William Herschel (1792–1891) to the Royal Society of London on 16 June 1842, *On the Action of the Rays of the Solar Spectrum on Vegetable Colours and on Some New Photographic Processes*, followed the same contact negative positive method as Talbot, but in Hershel's experiments the paper was coated with the inorganic pigment Prussian blue (hydrated iron hexacyano ferrate complex) to produce a rich deep blue stain to the paper. Photographer Anna Atkins (1799–1871) was quick to realise the potential of Hershel's Cyanotye process. In her illustrated book *Cyanotypes of British and Foreign Flowering Plants and Ferns* (1854) Atkins made 'photograms' by laying flowers and ferns on to paper impregnated with iron salts, which were then exposed to sunlight for several minutes. When washed in water, the area on which the plant had blocked the light remained white, but the area that was exposed came out blue. Whilst neither Talbot nor Herschel's methods achieved a textured surface, these processes laid the foundation for ideas and methods that led to extraordinary printed volumes containing textured facsimiles of plants (Stulik and Kaplan, (2013). Other early methods of nature printing included plants and other flat objects that were inked and pressed on to paper. This destroyed the plant or object and only a limited number of impressions could be achieved; for example, a folio made by Henry Smith in 1857 included inked objects such as snakes.

The method of nature printing that best recreated the intricate detail and textured qualities of the plants was described by Henry Riley Bradbury (1831–1860) at a lecture to the Royal Institution of Great Britain on 11th May 1855, entitled *Nature Printing: Its Origin Objects*. Bradbury had studied at the Imperial Printing Office in Vienna, where he gained knowledge from Alois Auer (1813–1816), who was a botanist, printer and inventor of the process. Auer, working at the Imperial Printing Office in Vienna, was able to faithfully record the hairs, veins and other minute information. The ferns were pressed into soft lead, *gutta-percha* or *gum-plastic* to create moulds from which an electrotype (or galvanoplasty) could be made. The soft impression was coated with graphite and immersed in a solution of copper sulphate and sulphuric acid, through which a low-voltage electrical current was sent (McMillan, 1890). The electrotype created a more durable printing plate and multiples could be generated for the purposes of multiple publication. It is generally believed that the British surgeon, William Montgomerie, introduced gutta-percha in 1843, which is a tough resinous thermo plastic from Palaquium trees (http://www.plastiquarian.com/index.php?id=42). After returning from Vienna early in 1853, Bradbury introduced the technique to his father's printing firm, Bradbury & Evans, in London. In 1853, they took out a British Patent (No. 1164) for what they claimed was a modification of the method. In 1856, Bradbury & Evans of London published a work by Thomas Moore entitled *The Ferns of Great Britain and Ireland*, using colour nature-printed plate engravings (see Figures 5.2 and 5.3). Moore's important work on nature printing was one of the first of the genre to be printed in England. The beautifully executed plates were produced under the direction of Henry Bradbury and are considered to be among the finest pieces of nature printing ever completed. By running one's fingers over these printed artworks, one can detect the raised edges, veins and spines of the plants. His interest in printing incorporated security elements for paper money, a subject he presented at the Royal Institute. His aim was to create intricately illustrated books, including nature-printed books on fungi and trees and a contemporary book on graphic arts in the nineteenth century, but these were never realised (Blunt, 1994).

As developed by Fox Talbot and patented in 1852, one of the many milestones of the nineteenth century (in the context of printing and reproducing images), was the development and the production of photographic images using a printing press. Talbot spent more years researching the photomechanical print than the photographic. Talbot worked on and patented his photomechanical process in 1852 (photographic engraving) and 1858 (photoglyphic engraving). He discovered that by combining a colloid such as gelatine or albumen and potassium bichromate he could sensitise a metal plate, and through the placement of a piece of gauze on the surface of the plate that broke up the surface into a fine grid of lines, and which thus served as the equivalent to a halftone mesh. The sensitised plate with the gauze was exposed to the sun for a few minutes, and he then placed flat objects such as leaves or grasses on to the plate, and again exposed it. Where the sun had fallen on the exposed gelatine, these lines were obliterated, and where the leaf covered the gridded background, these tiny lines remained. After the plate was etched, these crossed lines became tiny reservoirs that captured the ink. He also suggested other methods for segmenting the image such as powdered particles on glass, an aquatint applied to the plate, and a line screen ruled on glass.

When in 1834, Michael Faraday published his laws of electrolysis, correspondence between Fox Talbot and Faraday in 1835, suggests that Talbot was certainly aware of the potential of electrotyping (or galvanoplasty) for platemaking (http://foxtalbot.dmu.ac.uk/letters/letters.html). How Talbot developed his ideas on electroplating is unclear, but there are test plates made by Talbot in the archives at the National Science and Media Museum in Bradford that indicate Talbot was experimenting with this process.

From Talbot's original method, many photomechanical variations quickly evolved, which within twenty-five years, the print industry was able to commercially adapt for high-speed print production.

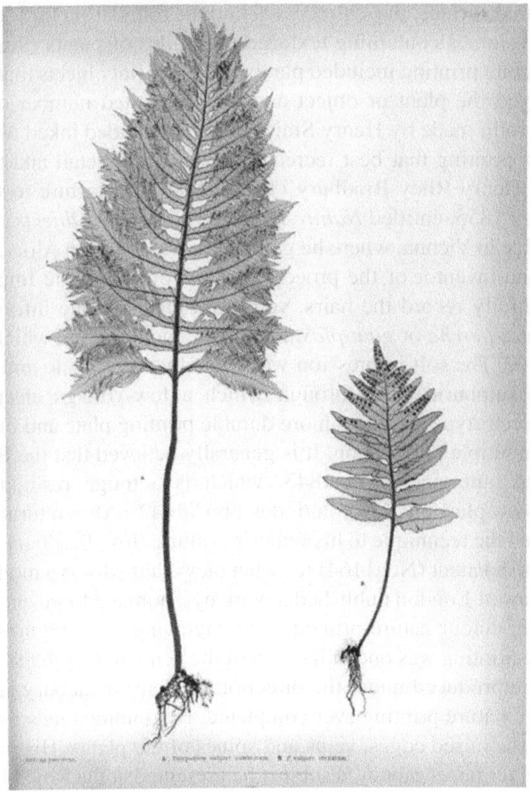

Figure 5.2 The Common Polypody in the book *The Ferns of Great Britain and Ireland*, Thomas Moore, printed and published by Bradbury & Evans of London in 1856 (courtesy: The Peter H. Raven Library, Missouri Botanical Garden).

Working as an artist and scientist, Talbot produced a range of beautiful images by capturing nature in detail, and the process of breaking up tones by means of a screen can be credited to him (Gamble, 1897). Although Talbot's approach did not create an embossed surface, as exampled in the nature prints, other processes involving bichromated gelatine such as the Woodburytype demonstrated the qualities of a physical surface deposition to create a tonal range. The Woodburytype is a beautiful example of relief printing, where the tones in the image were generated through the physical relief of pigments. Walter Woodbury's invention came towards the end of photomechanical printing during the 1860s. The Woodburytype follows the same method as a carbon print (Crawford, 1979, pp. 285–289) and is a photorelief of tinted gelatine that corresponds to the tones of the original image.

Returning to the delicate problem of fragile botanical albums and archives, how might technical advances in the twenty-first century be of use in reconstructing pressed plants and recovering lost information such as the colour of leaves? There has been significant progress in image reproduction for photography, display rendering and printing. In the quest for reproducing different materials and textures either on the screen or with a printer we find many challenges for apparently simple things such as leaves. We might well begin to understand that a leaf – being the primary organ of the plant – is fundamental for capturing light and then by using the chemical chlorophyll to produce energy from sunlight. The obvious difference in appearance between dry or dead leaves (as in a herbarium) and living ones is firstly its greenness of colour, which is due to the presence of chlorophyll. As the leaf stops

Figure 5.3 The Northern Holly Fern in the book *The Ferns of Great Britain and Ireland*, Thomas Moore, printed and published by Bradbury & Evans of London in 1856 (courtesy: The Peter H. Raven Library, Missouri Botanical Garden).

making energy, it changes colour and becomes brown, which is a clear indication of the decomposition and eventually absence of this pigment; then other chemicals present in the leaf can be seen.

Leaves are complex structures and contain a high concentration of light-absorbing substances such as water, cellulose, starch, proteins and chlorophyll. The presence or absence of these substances is responsible for the level of internal scattering of light and therefore determines how light will be reflected and transmitted by a leaf. To understand the optical characteristics of leaves as a function of their biochemical and structural properties, empirical and radiative transfer models have been developed (Jacquemoud and Ustin, 2001). By observing the reflectance and transmittance spectra of dry and fresh leaves we can find the wavelength regions where different components interact with light. In fresh leaves, the visible spectrum is dominated by the absorption of light of photosynthetic particles, whereby the spectra at the near infrared region are mainly affected by structural aspects and the shortwave infrared region sees its absorption predominantly affected by water. On the other hand, in dry leaves, the near infrared region absorption is caused by dry carbon compounds such as cellulose and sugar.

Models to simulate leaf optical properties may range from a leaf that is described as a single layer to more complex descriptions that include cell shape, size and position. Nevertheless, both options require knowing the absorption coefficients of leaf components. A typical radiative transfer model based on the original PROSPECT model (Jacquemoud and Baret, 1990) treat leaves as semi-transparent plates (either compact or non-compact) where the light is reflected, transmitted and absorbed in different proportions (see Figure 5.4), which includes parameters such as the percentage of dry matter, the number of plates (idea of structure), chlorophyll concentration and the water content.

The plate structural modelling does not represent every single kind of leaf and therefore another approach that uses spherical particles instead of plates was introduced by Dawson, Curran and

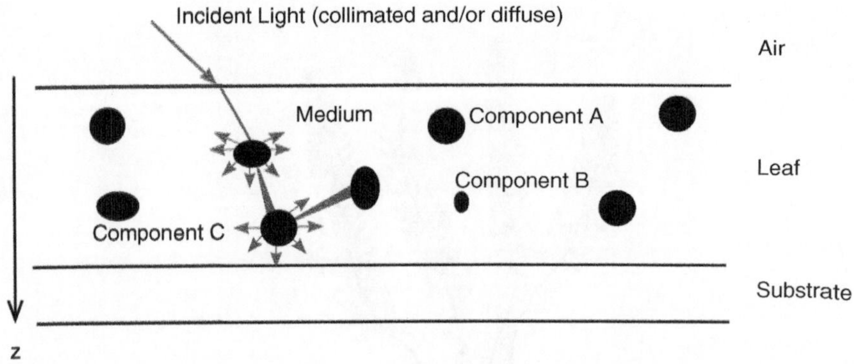

Figure 5.4 Illustration of how light is reflected, transmitted and absorbed in different proportions.

Plummer (1998). This model adds three structural parameters to the previous ones: sphere diameter, space in between spheres and leaf thickness. Another option is the N-flux models, derived from the typical Kubelka-Munk (KM) theory widely used in the printing field for modelling ink layer interaction, which consider a leaf as a slab of absorbing and diffusing material (key scattering). These models provide analytical equations to calculate reflectance and transmittance in terms of the absorption coefficient k, scattering coefficient s and thickness of the slab d, where the radiation field inside the slab consists of fluxes propagating in opposite directions (i.e. KM uses two fluxes) (Figure 5.5).

We can also consider a leaf as a random medium with a spatially varying permittivity and electric fields. The flow of energy in the medium can be modelled by the radiative transfer equation (Ganapol *et al.*, 1998). Radiation is assumed to be lost by absorption and diffusion as it propagates in the medium. Another flavour of this method consists of simulating the radiation transfer as a stochastic process where photons are treated as random variables. Despite of the previous efforts in modelling optical aspects of leaves, only methods based on ray tracing are able to account for the internal leaf structure. Ray tracing models require a complete description and arrangement of individual cells in the medium as well as the optical coefficients of the components of the leaf. These models simulate the propagation (rays) of a few individual photons and deduce their complete behaviour statistically from such rays.

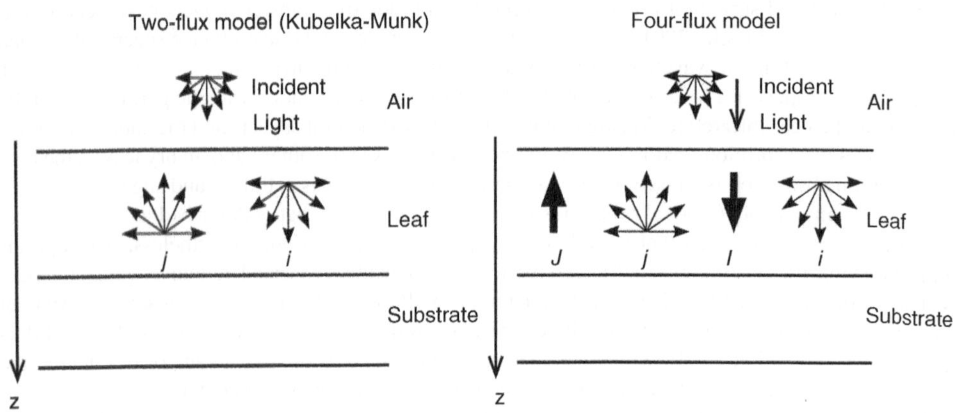

Figure 5.5 Comparison between a two-flux model and a four-flux model.

Applications in computer graphics have contributed to the recent improvements in this area through the use of 3D rendering techniques as well as more sophisticated probabilistic methods. Nevertheless, the interaction between leaf components, leaf structure and leaf optical properties for reproduction processes like printing are yet to be explored and developed. In fact, the inversion of these models to deduce printing parameters has been hindered by the lack of BRDF (bidirectional reflectance distribution function) measurements of leaves and information about their internal structure. The potential to being able to address 2.5D printing solutions for dry or fresh leaves is not only important as a challenging, diverse and dynamic case study but it may also be the answer to theoretical and practical aspects in a more general printing workflow.

It is interesting to note that as this rather small and possibly insignificant investigation into nature printing has evolved, the historical study has highlighted a compact and highly industrious range of discoveries and interest in new materials and processes and a search for beauty through technical innovation.

Case Study 2: Wallpaper Design

Possibly the most complex, numerous, overlooked and enduring product that demonstrates both a range of historical, technological advances and material evolutions is the domestic wallpaper. Wallcovering and wallpaper have for the last five centuries reflected changes in society, domestic tastes, self-expression and interior design. Wallpaper, by the very process of being pasted to the wall implies an ephemeral product. By applying it to the wall, the paper is irrevocably fixed into the space in which it occupies and yet is destroyed at the same time. Furthermore, it can be damaged by humidity, mould, wear and tear, or simply repapered or painted over. Wallpaper could be considered as many things: as an art form or installation, as a substitute for a wallpainting or fresco.

Wallpaper has been used to imitate different materials such as ceramic wall tiles, hessian, damask, woodgrain or marble, or as *trompe l'oeil* to suggest swags, alcoves, wooden panelling or tapestries. In old houses, heavy textiles, tapestries and wall-hangings were originally used to maintain warmth, but as heating was introduced, wallpaper became more decorative and these early wallpapers mimicked the appearance of textiles by including embroidery stitches, patterns, brocade and textures of fabrics (http://www.vam.ac.uk/content/articles/t/textile-influences-on-wallpaper).

The reason for including a case study on wallcovering, is that, like the traditional print, it demonstrates several centuries of provenance, and through print archives and picture books, highlights 2.5D printings as applied design and art form. Still very much in use in the twenty-first century, the papered wall demonstrates the benchmarks and the technology of the past, and, given the advances in print and material technology today, shows what may be further achievable. With advances in technology the printed wallpaper has been revolutionised once again, the industry is enjoying a revival and wallpaper in homes is very much back in fashion (Abrahams, 2013). With the introduction of novel materials, such as embossed blown vinyl, decorative beads, holographic surfaces and metallic inks in the twenty-first century, printing using contemporary flexographic or rotogravure methods has resulted in novel shimmering and interreflective surfaces (see also Case Study 13: Analogue Printing Methods). Examples such as patterns and surfaces designed by Matthew Williamson for Osborne and Little's Eden range demonstrate an exotic design through the use of colour, metals and texture. Furthermore, tiny beads were used to enhance the peacock motif and surface dimensionality (https://www.osborneandlittle.com).

However, it is interesting to note that many contemporary digitally printed surfaces suffer from a loss of surface quality, vibrancy and opacity of colour and require secondary processing; in order to overcome a flat uniform surface, images are printed on to pretextured and decorative papers (see the end of this case study). This rather subjective difference can be experienced by comparing the relationship of paint on paper, for example a wallpaper printed in the late nineteenth century versus a flexo or digital printing of the twenty-first century, which, by using traditional block printing highlights the squash of the thick paint, the raised surfaces, and the ridged edge effect as the block is pulled away from the paper.

Figure 5.6 William Morris (1834–1896), *Acanthus* (1875). Wallpaper block-printed in distemper colours. Given to the V&A by Morris & Co. (reproduced with permission: Victoria and Albert Museum) (http://www.vam.ac.uk/page/w/william-morris).

However, once again there is a balance where economy of time and cost have a significant impact; for example, the *Acanthus* pattern by William Morris (Figure 5.6) incorporated 30 blocks, 15 colours and up to 4 weeks to complete, whereas a digitally printed roll may take only hours to complete.

In Britain, paper that was printed with a repeatable pattern has been used as wall decoration since the sixteenth century. Decorative papers originated as plain paper that was pasted to the wall and then hand-drawn and hand-coloured by artists. Early examples include fragments of wallpaper that were found in the Master's Lodge in Christ's College, Cambridge, and can be dated to first year of Henry VIII's reign in 1509 by Hugo Goes. A reconstruction of a panel fragment made in the V&A (http://collections.vam.ac.uk/item/O176029/wallpaper-goes-hugo) shows a formalised pomegranate design. The pattern resembles textiles and leather wallcovering that were imported from Italy and Spain into this country in the fifteenth century (http://www.vam.ac.uk/content/articles/t/textile-influences-on-wallpaper).

Hand block printing was the most common method and was recorded in the early sixteenth century, but as a mechanically printed product, was not introduced until some Three-hundred years later. A carefully designed wallpaper may well be printed partially using one process and then transferred to a number of other processes, each process carries with it its own characteristics and idiosyncrasies.

In 1712, the British government levied a tax on any patterned, painted or printed wallpaper. The tax that was levied started at a penny a square yard (90 cm^2) and was increased to a shilling in 1809. A method to (legally) evade the tax was to paste plain paper to the wall and use hand stencils. The tax was abolished in 1836. By the late nineteenth century wallpapers were widely used by all classes, in homes and also in public buildings. Traditional wallpaper throughout the eighteenth and nineteenth centuries became increasingly mechanised using rollers and a wide range of technical and material innovations, resulting in the twentieth century when wallcovering became easily affordable (http://www.anstey.uk.com/files/documents/PDFs/Manufacturing/anstey_print_guide.pdf).

Wallpaper Printing Processes

In the sixteenth century, the earliest process of block print used a relief wooden block; traditionally a hard-wearing wood such as cherry or pear was used to prevent warping or cracking (see Figure 5.7).

Figure 5.7 Examples of traditional blocks and block printing from the Anstey Wallpaper Company (courtesy: Anstey Wallpaper Company).

The paper was placed on to a table, which was padded with a wool blanket. Trays of premixed colours contained a pad saturated with paint. The relief block was pressed on to the pad using a lever system and a hook attached to the back of the block was swung, lowered and accurately positioned on to the paper using pins on the side of the blocks as a register. The block was applied as a repeat pattern along the length of the paper. Each layer of colour had to be left to dry for 4–5 hours before the next colour could be applied, and was therefore a time-consuming process.

These strips, drops, rolls or flounces of printed paper, as exampled in the large scale *Acanthus* pattern by William Morris, incorporated 30 blocks and 15 colours. *Acanthus* was printed in two colour combinations – one in shades of green and the other in reddish-brown tones. Wightwick Manor near Wolverhampton is one a few surviving examples of a nineteenth century house built in an arts and crafts style and using arts and crafts decoration (https://www.nationaltrust.org.uk/wightwick-manor-and-gardens). William Morris wallpapers and fabrics were used in 13 of the rooms, of which the *Acanthus* motif was hung in one of the bedrooms. Associated with the British Arts and Crafts Movement and based on natural forms, William Morris began designing wallpapers in the 1860s. By the 1880s Morris wallpapers had become highly fashionable and were printed by Jeffrey & Co. in London, who specialised in the production of high-quality block printed wallpapers. The Victoria and Albert Museum in London (https://collections.vam.ac.uk/item/O74613/acanthus-wallpaper-morris-william/?utm_source=vam.ac.uk&utm_medium=redirect&utm_content=images&utm_campaign=legacy-images-2012) has an extensive archive of his work; his patterns and prints are still popular and in production today, but flexographic rollers now replace the original woodblocks.

In the seventeenth century an added extra dimension was introduced using flocking, which replicates velvet and damask. Flocking was traditionally applied in the same way as block printing as a final layer or a series of layers, and instead of a colour, the printer applied a highly tacky adhesive that held the fibres in place. Tiny flecks of coloured wool were hand-sprinkled on to the paper. Wooden batons were positioned underneath the paper, which were used to agitate the paper and ensure the fibres were evenly spread and adhered to the glue. An interesting nineteenth century example of flock wallpaper is at Tyntesfield House near Bristol (https://www.nationaltrust.org.uk/tyntesfield). The walls of the main drawing room are covered with opulent and expensive red and gold woven damask fabric with a red velvet border, whereas in the hallway to the drawing room, possibly where more traffic and potential wear and tear may be encountered, a gold flock wallpaper, using a similar motif to the drawing room, was used (Figure 5.8). One of the original makers of flock paper, Cole and Sons, in the 1870s has now revived the tradition as a new interpretation and collection of rich colours and patterns (http://www.cole-and-son.com/en/collection-534/).

In the late nineteenth century a range of specialist embossed papers evolved that began as more functional and utility papers. These have withstood the test of fashion and time, and are wallcoverings that offered both practical solutions and beauty. In 1860 French manufacturer Paul Balin (1856–1898) embossed the surface of paper to create realistic imitations of embroidered, lace and silk wall hangings (http://collections.vam.ac.uk/item/O211408/wallpaper-balin-paul). Wallcovering manufacturers such as Anaglypta and Lincrustra produced relief wallpapers and fibrous plaster panels to create the appearance of plasterwork. Lincrusta (http://lincrusta.com) and Anaglypta (http://www.anaglypta.co.uk) are still in production today, and use similar materials and processes. They have successfully reinvented

Figure 5.8 Red and gold woven damask fabric that is applied to the wall (left). Gold flock wallpaper in the hall (right). Tyntesfield House, Bristol, National Trust.

their brands and patterns to appeal to the twenty-first century market. Frederick Walton developed Lincrusta wallpaper in England in 1877. Their objective was to mimic embossed leathers, plasterwork or wood panelling. Comparisons can be drawn to the imitation leather patterns called *Kinkarakawakami* (gold-embossed wallpaper) made in Yokohama in the 1880s and 1890s (http://www.handmadejapan.com/e_/features_/eft008_01.htm). These wallcoverings were designed to imitate the calfskin panels known as Spanish leather. As illustrated in Figure 5.9, the Spanish leather panels demonstrate a highly ornate tooled surface. The method uses rectangular panels of wet leather that are pressed or hammered into a wooden mould; once dry, the panels are then gilded and painted and nailed to the wall.

The Lincrusta wallcovering is made from linseed oil, wood flour and gum rosin that is ground into a soft paste. The paste is applied to a paper backing and a pattern is machine embossed under pressure through steel engraved or iron rollers (Figure 5.10). It can be embossed using intricate designs. In its postprocessed state the rolls of textured paper are off-white in colour and once applied to walls can be decoratively painted to simulate the appearance of leather or more exotic surface finishes through the use of glazes. It was popular to use these papers to simulate a range of effects from traditional white plasterwork to leather panelling or gilding.

It is interesting to note that Walton's father was the inventor of linoleum in 1873 – a hard-wearing flooring that was made of linseed oil, resins and woodpulp, which is spread on to a canvas backing. The Lincrusta was entirely waterproof, could be cleaned and was therefore recommended as a sanitary product for use in public buildings, where it needed to withstand the wear and tear of cleaning that would have destroyed the surface of traditional wallcoverings. Furthermore, over time and as it aged, the material became harder, and it was not unusual to find original Lincrusta still attached to interior walls of nineteenth and early twentieth century Victorian and Edwardian houses. Other versions, such as a wood grain, were used to mimic panelling and were even used to cover doors (Hoskins, 2005, p.157).

Figure 5.9 Spanish leather wallcovering, Dyrham Park, Bath, National Trust.

Figure 5.10 Example of embossed wood grain effect from mid-twentieth century.

Thanks to new investment in the Lincrusta company, the product has been re-launched (Figure 5.11). Essentially using the same raw materials, the company has updated the designs to reflect contemporary tastes, but in parallel to other heritage products, one of the bestselling designs that was launched in 1896 is still in production today.

A notable designer who worked with Lincrusta and Anaglypta was the industrial designer Christopher Dresser (1834–1904), who was one of the most influential ceramic designers of his time and wrote on

Figure 5.11 Contemporary Lincrusta wallcovering (2016). Pigments and glazes are hand applied to Lincrusta wallcoverings to create a dramatic effect (courtesy: Lincrusta).

a range of subjects relating to design and decorative ornament. He also looked to exploit mass-produced products such as Lincrusta and Anaglypta to create highly ornate patterns. His objective was to promote interior design ideas to the late nineteenth century *middling* class consumer, stating that irrespective of their status, they too could afford to live in beautiful surroundings.

Anaglypta – taken from Latin meaning raised cameo or carved in relief – was a commercial evolution and competition to Lincrusta (see Figure 5.12). In 1886, Thomas Palmer, who was an employee of Frederick Walton, developed an embossed paper that was lighter than the Lincrusta. Walton felt there was a conflict of commercial interest so Palmer set up production on his own, and the paper was soon in demand. This robust low-relief wallpaper was manufactured in an enormous range of patterns, and typically supplied as uncoloured, mimicking the appearance of low-relief decorative plasterwork or painted carved wood panelling. However, as it was made of wood or cotton pulp and was hollow backed, it was lightweight and could easily be applied to ceilings.

In the early 1800s Henry and Sealy Fourdrinier developed the Fourdrinier Machine for producing a continuous sheet of paper, and in 1839 Harold Potter, of Potter & Ross calico printers in Lancashire, adapted a calico machine for the purposes of printing on to wallpaper using engraved copper rollers (http://www.coatings.org.uk/Sectors/About_wallpaper.aspx). The impact of mechanisation in the nineteenth century meant increased speed, productivity and variety. The mechanised rotary surface printing, which was introduced in 1839, contained all the elements of block printing using a relief block surface and trays with pads that were saturated with paint, but was able to print more colours and in line. As the cylinders rotated they came into contact with a felt roller that was coated in paint. The felt was constantly charged with paint as it was rolled into the paint trough. As the paper was drawn under the roller it came into contact with the relief surface of the cylinder, and then moved forward to the next cylinder that was charged with the next colour, and so on, until all the colours were printed. Unlike block printing, these colours were printed before the previous colour had dried, and where some wet colours bled into another, resulted in a feathering effect and softening of edges. Other rotary methods such as screen-printing employed cylindrical screens. The squeegee is fixed inside the roller and ink is pumped into the inside of the cylinder. The paper and cylinder move synchronously, and as the squeegee comes into

Figure 5.12 Traditional Anaglypta hollow relief papers that resemble a classic late nineteenth century tiled wall (left) or a brush stroke pattern of the mid twentieth century (right).

contact with the mesh the paint is squeezed through the mesh on to the paper. Rotogravure is a high-quality print process, still used for printing fine detail including postage stamps, fabric and wallpaper. The image is engraved into a steel surface and is often plated with copper cylinder. To print an image, the cylinder is coated with ink and then the surface is wiped clean so that only the engraved cells contain the ink. As the roller rotates and comes into contact with the ink, excess ink is then wiped with a doctor blade, leaving the ink only in the cells. As the paper comes into contact with the roller, pressure is applied and the cells of ink are transferred to the paper. The paper then moves on to the next colour and so on. A more contemporary version of gravure is flexoprinting, which can be printed either as a relief or in intaglio. Flexography, introduced in the 1960s, is a contemporary version of rotary surface printing. Colour separations are exposed on to photopolymer plates, the printing elements are in relief at a constant height and are coated with a layer of ink.

Digital printing in combination with analogue processes has also enabled many more complex patterns and imaginative colour palettes (Figure 5.13), with the addition of decorative elements including microbeads (Figure 5.14), effect foils for optical illusions, gold and silver lustre, and holograms. Fashion designer Matthew Osborne working for Osborne and Little has incorporated these tiny beads to create special embellished effects.

As demonstrated at DRUPA 2016, printer manufacturers have introduced digital-only machines for printing wallpapers. Digital studios are now able to print and fit to wall pattern services, create repeat wallpapers and bespoke wall-sized prints and murals. A vast range of images, colours and effects are possible (www.ohpopsi.com) (Figure 5.15).

In order to compensate for a flat smooth surface, there is also a range of textured papers, for example substrates such as *EasyLife Digital* by Ahrlstrom (https://ahlstrom.episerverhosting.com/en/Products/Building/Wallcover/EasyLife-Digital-Print-/) (Finland) using Xeikon and HP Latex digital machines; BN International (http://printmedia.bnint.com/wallpaper-vinyl) produce *Digifort*, a series of textured vinyl

Figure 5.13 Peacock pattern with beads (W6541-01), designed by Matthew Williamson (courtesy: Osborne & Little) (www.osborneandlittle.com).

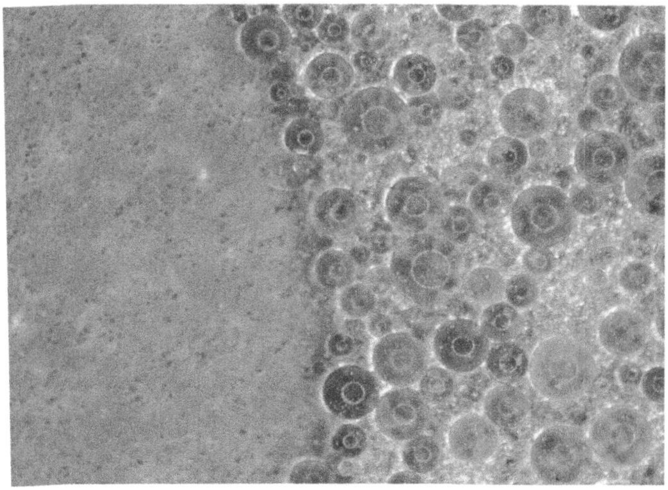

Figure 5.14 Under magnification, different sized glass microbeads are attached to sections of wallpaper to create special effects.

Figure 5.15 Digitally printed wall murals using textures, as exampled here using a weathered wood image (Courtesy: Ohpopsi).

papers that can be printed using (eco) solvent UV or Latex ink; Lahnpaper (http://produkte.lahnpaper.de/en/applications/digital-walldecoration) offer a range of smooth matt and gloss papers such as *Variless*; Mac Tac Deco mural (http://www.mactac.eu/products-indoor-1-29-2034-2463.htm) *DecoIvory* and *DecoMural* and *DecoSatin* are white woven polyester textile-coated media for inkjet printing with a clear acrylic repositionable adhesive; Papergraphics (http://paper-graphics.com/en/products/digital_decor/digimura_ag) offer a range of textures and appearances including *Digimura AG Silver*, which is a brushed silver finish that is suitable for UV printing, *Tactile* (https://paper-graphics.com/en/products/digital_decor/dm2t1300), which contains a lightly raised cross-hatch texture to give the appearance of a woven canvas;

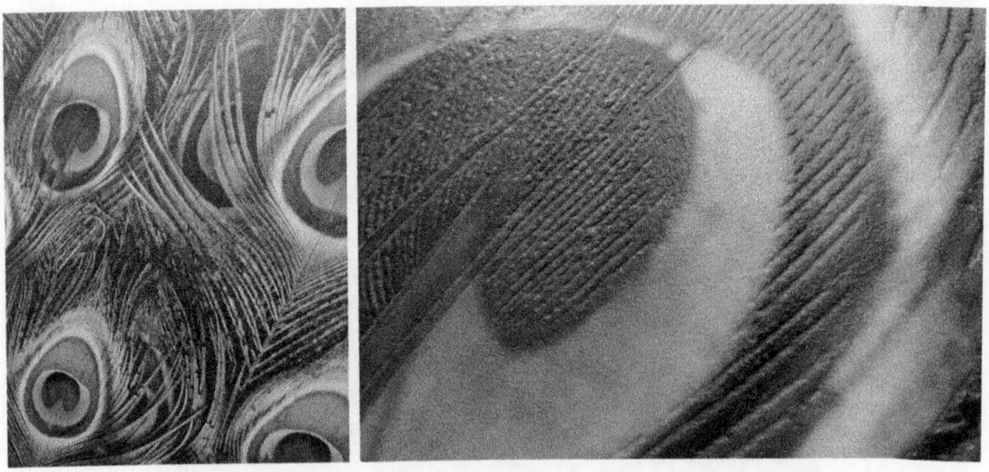

Figure 5.16 Dimensor textured wallpaper by Dimense on show at DRUPA 2016 (courtesy: Dimense) (www.dimense.com).

and Senfa Decoprint (https://www.decoprint.eu/en/products.php), which has a diamond (lustre) coating effect. Landor Phototex (http://www.phototex.co.uk/) is a low-tack self-adhesive paste-free fabric repositionable wallpaper that can be used for water-based as well as solvent inks. The majority of these papers are designed for use with digital printing inks such as Latex, solvent, eco-solvent and UV curables.

The Lithuanian-based company Dimense Dimensor (http://www.veika.lt/) are ink manufacturers and producers of different vinyl wallpapers for inkjet including Decojet, who have developed a cover-coat that incorporates a thermal layer, which, when passed through their printer, will raise the surface to create a texture (Figure 5.16). As part of the workflow process, the designer creates an image in two parts, the CMYK image and a separate layer, in the same way as a special effect layer (spot, varnish, white), which is then exported to the printer.

Over the last decade, as design and interior decoration has experienced significant growth that spans both the high-turnover home DIY to the luxury and high-end market, there has also been a renewed interest in high-quality wallpaper production. Makers of wallpaper still use traditional methods of printing wallpaper including gravure, rotary, flexo, block printing, surface printing and screenprinting, and by incorporating gold and glitter effects and rich colours create shimmering finishes. In recognition of the limits of contemporary inkjet printing and to overcome the appearance of a flat surface, contemporary methods for achieving relief or texture can be broadly divided into pretextured wallpapers and texture or embellishments that are added after or on top of the image, and suggests opportunities for further exploration.

Case Study 3: 2.5D Printed Tactile Books and Artworks

The aim of this case study is to explore and gain insights into how relief-printing technologies can be usefully applied for the benefit of people who are blind or partially sighted. It is hoped that where there is already a significant amount of innovative work already undertaken by charities, there can be mutual benefits for developing texture-driven methods that can lead to an enhanced understanding and appreciation of images and artworks.

Over the last two decades, museums have begun to address how artworks in their collections can be handled. Ancient, fragile and original artefacts cannot be touched, but it is possible to make facsimiles and replicas that can provide visitors to a museum with an improved understanding of cultural objects. This of course may have an impact for any visitor. Blind people are especially restricted by what they can access in museums. Museum audio guides and large print catalogues are the primary methods to find out about pictures, a commentary may highlight various key points of a picture, or may describe an art historical context or aspects of technical interest. Now with new technologies, access to a wider range of visual arts (artworks, picture books and photographs) is increasingly possible, whereby secondhand didactic forms of education are replaced by an enhanced and sensory experience. Novel printing and layering technologies that incorporate a raised or embossed surface have demonstrated new opportunities for printed replicas and tactile surfaces and presents exciting opportunities for incorporating a wide range of different textures. As resolution and print quality has improved and the technology is more accessible, there are now more opportunities to explore texture in artworks in more imaginative ways, especially in relation to tactile printing for blind or visually impaired. Figure 5.17 demonstrates products that can be downloaded for 3D printing; examples here include a patterned surface for dominoes and a texture dice.

The raised dots of the Braille universal language system is the traditional method for touch reading text, which is based on a variation of six dots arranged in two columns of three dots to represent all the letters of the alphabet, punctuation and numbers, of which there are sixty-three possible combinations (see Figure 5.18).

Figure 5.17 Downloadable tactile objects for 3D printing as demonstrated on the Shapeways website (courtesy: Willamette Valley Shapeways) (https://www.shapeways.com/shops/willamettevalley).

Figure 5.18 A typical Braille calendar, showing raised braille dots alongside large font numbers and text.

There are two grades of braille: a direct letter for letter translations system, and a contracted version that includes common words and letter combinations such as 'the', 'not' or 'gg', which can increase reading speed and save space. There are also other braille codes used for a foreign language, accented letters and specialist subjects such as mathematics, science and music (http://www.duxburysystems.com/braille.asp; http://www.omniglot.com/writing/braille.htm#).

On a practical level, the day-to-day identification of products and things is vitally important, and raised dots are printed on to packaging, bottles and cans. A research group at the Department of Clothing, Japan Women's University have developed a method for identifying the colour of clothes. They have acknowledged that blind people, particularly women, have a strong preference for identifying the colour of their clothes (Okudera *et al.*, 2015). The use of colour-coded tags is a simple method that enables blind people to identify the colour of their clothes and to feel confident their clothes are colour co-ordinated (Figure 5.19). The tactile colour identification tag is attached into each item of clothing. The tactile tag comprises ten small raised soft rubber dots that are arranged around a Munsell hue colour circle, each representing one of the primary colours. There are also three raised dots placed vertically in the centre of the circle to represent white, grey and black. In order to identify the representative colour, a bigger hole is punched into the hue circle to indicate the colour (Okudera *et al.*, 2015).

Tactile pictures and tactile picture books are also included in Braille libraries. Traditional approaches use simple lines and patterns to create a distinct and recognisable shape and separate objects to be perceived. Lines are used to describe a simple outline of an object or to define a boundary between two surfaces. Patterns may be used to suggest a change in colour or to fill in a large area or a solid object, or as a specific key (as one would use on maps) to indicate different information (Figure 5.20). Patterns or lines may be also used to suggest a texture, such as the texture of a man's beard.

Books for young sighted children tend to be illustrated using very bold and bright visual images. Artworks as exampled in the work of Eric Carl (1929–), who illustrated some of the most visually iconic children's books, including *The Very Hungry Caterpillar* (1969), which has delighted many young children for decades. In this instance, for the visually impaired, the pictures have been transformed into tactile books, thus assisting children to explore ideas through different textures and scale. As illustrated in Figure 5.21, the top image, as produced by the charity Living Paintings, demonstrates the comparison of scale and how the caterpillar grows over time – when the hungry caterpillar starts eating it is very small, but increases in size as it eats.

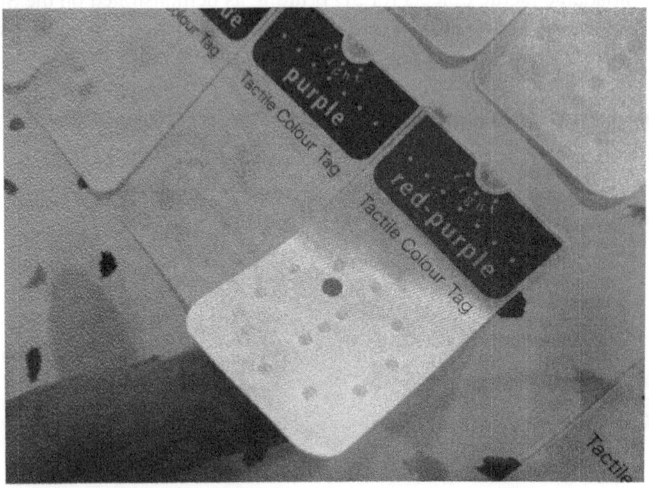

Figure 5.19 Clothes tags, with raised dots for identifying and coordinating the coloured garments (courtesy: Okudera *et al.*).

Figure 5.20 UV curing printing technologies are expanding the opportunities for raised signage (courtesy: Mimaki).

However, what if a child or adult cannot see the pictures in a book? How does the reader convey the meaning of objects or the content of a scene and engage with the reader as an active participant? For children and those who have severe vision impairment the process of learning can take much longer. The eye can differentiate an enormous range of colours and textures, whereas the fingertips can perceive fewer; therefore a picture containing a lot of texture can be confusing. To be effective, a tactile illustration should provide the reader with a tactile experience that, along with the book's words, triggers a connection with the child's own experience of the object in everyday life (Wright and Stratton, 2007).

Touch and feel books and bright simple graphical images assist the very young child in reinforcing the subject. Learning about texture begins around the age of one, and in order to assist children to identify different surfaces – hard soft, rough smooth – a range of tactile materials are used. Yvonne Eriksson of The Swedish Library of Talking Books and Braille explains that a child tends to use their whole hand, by patting the object or surface instead of an exploration using fingertips, which are more sensitive and can find nuances and differences; hence the need for very simple outlines and unambiguous information such as the watermelon image (Figure 5.22).

What makes a picture that is clear and useful to the blind person? The most important aspect in composing a picture is to present unambiguous information by selecting what is the most salient information. Furthermore, an accurate description of the picture that guides the listener to the various points of an image is also highly important, which is then reinforced through touch and encourages the reader to make pictures in their mind's eye. However, in order to understand how the object appears, it is assumed that the user has previous experience of the object. Eriksson explains that it is almost impossible for a person to understand the content of a relief image without some form of description or guidance. Most images are accompanied by text or oral commentary and for those with severe vision impairment or young people learning, a guide is important to help direct their hands and fingers.

A second aspect is consideration of scale, 'for the object to resemble the impression of the true object as a whole, and so it is very seldom one can use the same materials as in the real object, because the scale affects the impression made' (Eriksson, 1998). It is a difficult for a severely visually impaired user to understand the relationship between the texture that they are touching and how the texture may relate to the picture or indeed the real world. Therefore, scale is very important and the way elements in the picture relate to each other, as exampled in Figure 5.21, the designer has included a small caterpillar and a large caterpillar to demonstrate how it has grown. Likewise, the scale of a material needs to correlate to the scale of the image; for example, if fake grass is used to describe grass in a picture, it would feel wrong, because it would be used at the wrong scale; a smaller fibre would be required.

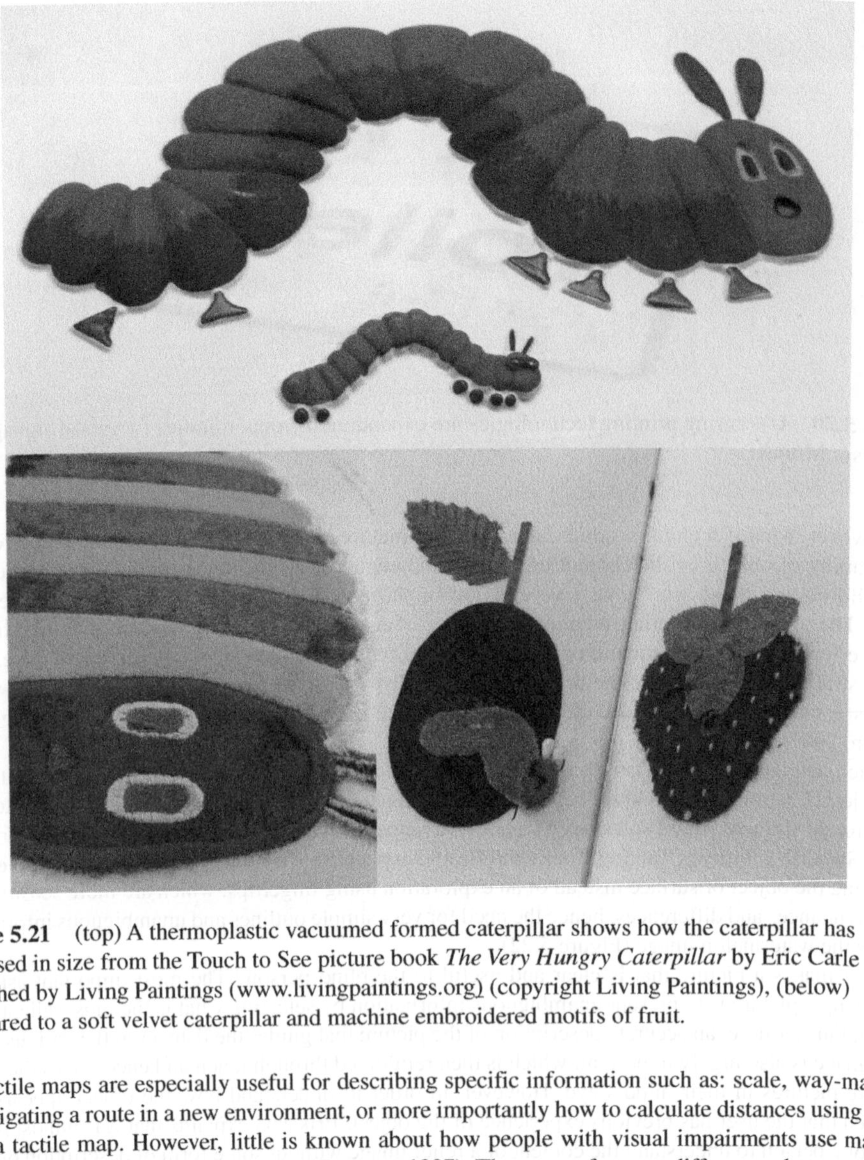

Figure 5.21 (top) A thermoplastic vacuumed formed caterpillar shows how the caterpillar has increased in size from the Touch to See picture book *The Very Hungry Caterpillar* by Eric Carle published by Living Paintings (www.livingpaintings.org) (copyright Living Paintings), (below) compared to a soft velvet caterpillar and machine embroidered motifs of fruit.

Tactile maps are especially useful for describing specific information such as: scale, way-marking or navigating a route in a new environment, or more importantly how to calculate distances using ratios from a tactile map. However, little is known about how people with visual impairments use maps to work out distances (Ungar, Blades and Spencer, 1997). There are of course differences between maps that describe a concentrated environment such as navigating through a building as opposed to finding landmarks on a map (Figure 5.22).

The examples in Figure 5.22 demonstrate a range of ways the designer has considered tactile solutions. All these images were taken during a visit to the Japan Braille Library or the *Nippon Tenji Toshokan* in Tokyo (http://www.nittento.or.jp/en/index.html). Traditional approaches use lines and patterns and non-patterns, which create a distinct and recognisable shape and enable separate objects to identified. The student traces a line or a shape with their finger and by doing so reinforces that shape. This is nicely exampled in the dot pattern of the red watermelon as there is a clear gap around the pips and the edge of the red section. Likewise, the patterns used in the painting are very simple but indicate the wavy lines of the hair and show which are different to the jacket. There is no attempt to give any form to the details of the coat other than a gap to suggest an edge to the collar.

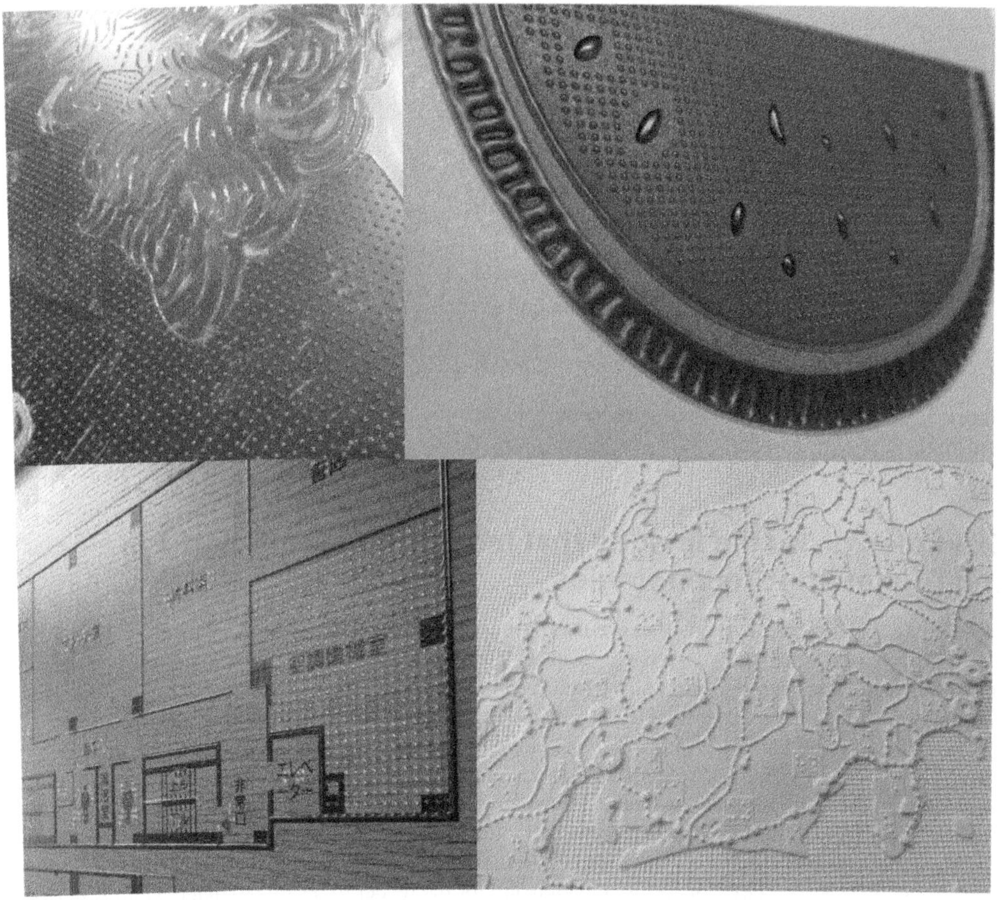

Figure 5.22 Examples of different patterns that are printed and translated as tactile images (courtesy: Japan Braille Library).

The two maps demonstrate very different information. The thermoplastic map (bottom right) is a geographical map of the central region of Japan, which shows the location of cities, towns and mountains. It also shows the network of roads and the demarcation of the different prefectures. Similar to any map, the designer has used different symbols and textured lines to represent different elements. For example, the rings represent large cities, the smaller dots represent towns, the triangles represent mountains, the smooth lines define the outline of the prefectures and the dotted lines represent major road networks. The room map (left) provides a layout of the rooms to scale and includes staircases, lavatories, the names of each room, and highlights a particular room by filling in that space with a dotted pattern, suggesting a quick identification and waymark point for the visitor.

The simplest and most engaging example, called the Tactile Picture Books project (https://tactilepicturebooks.org), was developed by Tom Yeh, professor of computer sciences at University of Colorado Boulder (Figure 5.23). He worked with his students to create tactile books suitable for young children who are learning to cope with their impairment. The simple printed tiles are made from coloured rigid PLA plastics using cheap FDM methods. Similar to a child's picture book, the images are very simple and the images are well defined and raised from the background; a simple text in Braille describes each image.

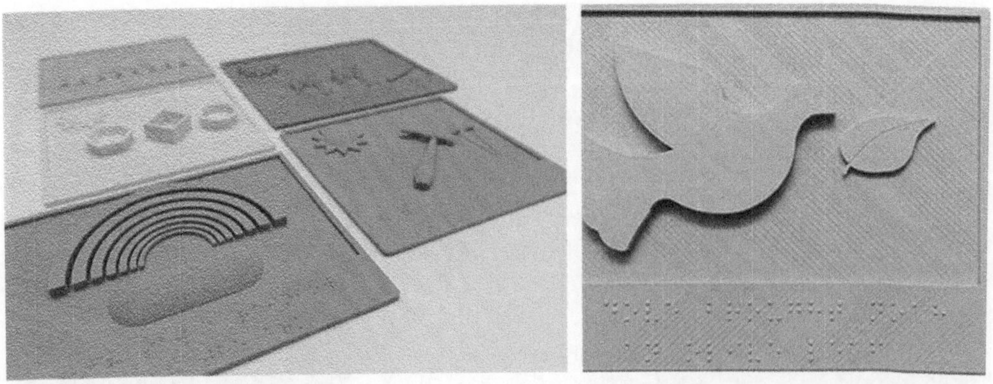

Figure 5.23 Tactile Picture Books' project by Tom Yeh of University of Colorado Boulder (courtesy: Tom Yeh).

Examples of Tactile Images that Incorporate Multiplatform and Digital Print Technologies

In the context of finding methods that are useful museums, how might tactile surfaces be applied to paintings or photographs so that users can visit a gallery, appreciate paintings or gain greater insights into cultural heritage? Furthermore, how might visitors gain more insights and comprehend the relationship between the things they read using tactile images and hear through audio commentary?

The charity Living Paintings (https://www.livingpaintings.org) is a loan-library service providing a variety of *Touch to See* books. An audio guide accompanies each book, so the books can engage the reader as a multimedia experience that is informative, interpretive and didactic. These tactile relief pictures effectively communicate the content of a picture to blind and partially sighted readers by extracting the salient information from complex or fussy images. The objective is to ensure the experience of reading is about clarity of communication and for the reader to be able to make pictures in their mind's eye as they explore the tactile pictures. Of primary importance is the ability to create a comfortable and inspiring experience for the reader, as well as building confidence and increasing knowledge.

The content is designed to address different educational stages and, depending on the age, different subjects are covered. For example, for teenagers subjects include fashion, football and performing arts. Other subjects range from current events such as the Olympics, educational books on dinosaurs or images of cultural objects from the British Museum collection *A History of the World in 100 Objects* (MacGregor, 2012).

Each picture block is hand carved. Rather than depersonalising the experience, the designers carve the block by hand. They also understand the physical parameters of the block, for example the height and depth of the relief for fingertips and that the distance between one object or another can be no more than the distance of a hand-span; therefore the scale of the picture needs to be kept within the proportions of the human body. The pages are vacuum formed using a thermoplastic sheet over the carved block. The sheets are ring bound together with other pages to form a flip-book. This cheap method means that books can be reproduced, are widely circulated and are fairly robust.

Another aspect of Living Paintings is the large archive of tactile paintings and cultural objects. As part of the loan system, a box is sent to the reader through the post. Inside each box is a range of items including a book, audio narrative on USB or CD, with large print and textured illustrations. As exampled in Figure 5.24, the idea is to explore the contents with a sighted person, who can help as a guide in highlighting aspects of paintings or read the story, which the blind person can also take part by following in Braille or by feeling the raised pages. Figure 5.24 shows how the book has been disassembled and

Figure 5.24 An example of a braille page and illustrated page, overlayed with tactile images of works of art by Claude Monet. From the Touch to See book, *The Magical Garden of Claude Monet* by Laurence Anholt, published by Living Paintings (www.livingpaintings.org) (copyright Living Paintings).

then rebound including brightly painted tactile pictures and Braille sheets on transparent plastic, which overlay the printed text but which the sighted reader can also read.

Other ingenious ways of engaging on a multiplatform level is the project called Talking Tactile, which combines phone apps as the didactic and audio provider, a 2.5D printed tile with embedded sensors and a wearable finger device that picks up the sensors. Examples include a side elevation of a historic building or a map of a church. Each tile contains embedded sensors that, as the user touches and explores the panel, the sensors are triggered to describe an interesting fact (http://tooteko.com/) (Figure 5.25).

Methodologies for the translation of paintings into a bas-relief or interactive artworks are now being widely studied. Tactile representations of Old Master and contemporary paintings can provide an enhanced understanding for the user. Appropriate methods for image segmentation is necessary to distinguish different parts of a picture. Studies have been conducted whereby single point perspective paintings have been transcribed into a bas-relief (Furferi *et al.*, 2014a, 2014b). Using computer modelling, the scenes are sorted according to their relative positions in the picture; for example figures and buildings at the front plane represent the highest point and the background is placed at the lowest point (Figure 5.26).

Figure 5.25 Exploring a printed panel, Tooteko Talking Tactile project (courtesy: Tooteko srls).

Figure 5.26 Images are relative to the same case study *The Healing of the Cripple and the Raising of Tabitha*, Three-dimensional bas-relief model (meshed surface) obtained by a Delaunay triangulation. The final rapid prototyped model (UV cured resin) was obtained with PolyJet technology (size 240 mm × 460 mm) (Furferi *et al.*, 2014b) (courtesy: Rocco Furferi *et al.*).

Figures are inflated or curved and then using height mapping details are added, such as the drape of clothing or faces and hands. Other spatial arrangements such as oblique planes to suggest a three-dimensional building may include stepping to indicate depth or perspective of a building.

Other organisations working on bridging the gap between cultural heritage and tactile pictures is the project *Didú*, developed by Estudios Durero based in Bilbao in Spain (Figure 5.27). The studios in collaboration with the Prado Madrid have studied ways to make Old Master paintings accessible to partially sighted and blind people through the *Didú* relief printing method. Their designers worked alongside art specialists and with blind people to understand how to either exaggerate textures already present in a painting or to create new ones that would be visually relevant but needed to be exaggerated to be understood by blind participants.

In an exhibition entitled *Touching the Prado* (20 January 2015–18 October 2015), a series of relief printed Old Master paintings were on display. The museum selected works they considered could be translated into a surface texture. However as Fernando Pérez Suescun, curator of *Touching the Prado* explained, 'It is hard for a blind person to build a mental image of what these works are like so we looked for paintings that provided information but was clear.' The digitization of the relief surface of a work of art is a relatively new field of research: it can be used to both study and monitor the surface of an object or to rematerialise it in diverse forms, ranging from projections to the construction of exact facsimiles. The validity of this type of recording is dependent on the quality of the data gathered and its

Figure 5.27 Estudios Durero, example of the *Didú* relief painting technique that adds volume and texture to allow the blind visitor to create a mental image of a painting by feeling it (courtesy: Estudios Durero).

correspondence to the original surface (http://www.3ders.org/articles/20150311-3d-printing-allows-visually-impaired-to-see-priceless-paintings.html; https://www.museodelprado.es/en/whats-on/exhibition/touching-the-prado/29c8c453-ac66-4102-88bd-e6e1d5036ffa).

This case study has demonstrated the importance of well-designed tactile images that can assist in active participation. We would argue that an exact replica is not wanted for most of the applications and that carefully selected objects, figures or features are more useful to communicate a story than a busy textured replica. For example, confusion may occur where pictures, objects or concepts are just too big to touch across a hand-span. Many of pictorial illusions that artists use, such as perspective or vanishing points, and placement of objects to suggest depth (large objects in the foreground and small objects on the same plane), are pictorial concepts that have been learnt and understood. Alternative methods, whereby a story or a scene is played-out along one plane, such as the carvings on Assyrian temples, may present a much more useful pictorial narrative.

Case Study 4: Coins and Medals

The most enduring and portable examples of relief are coins and medallions. Most people have a jumble of coins in their pocket, and can quickly and easily differentiate their value by edge, size, weight and shape. The coins may include two different types of coloured metal as exampled in the one euro coin. They can be considered as both commonplace as well as highly valued and collected. The reason for choosing coins is the high degree of intricacy and precise repeatability.

To make a coin is to *mint* a coin, and in time-honoured tradition, from earliest Roman examples to twenty-first century computer-aided methods, the process of hammering a relief surface into a metal blank on to a two-sided die has changed very little. To produce a coin a variety of processes are required: making the blanks, making the dies and striking the coins. Two dies are cut in hard metal such as steel. Making dies for the production of ancient coins included iron or steel tools, burins, punches and small chisels to engrave into the surface of the metal die. The metallic content of the coin has also changed over time, from silver in the twelfth century to an alloy of metals that includes a copper-plated steel for a one-pence coin or nickel and brass for a one-pound coin. Over the centuries, intricacy and complexity has continued to be added to coins to compete with counterfeiters. Samuel Peyps provides a detailed description of the minting process and counterfeiting methods employed in the Royal Mint in the seventeenth century (http://www2.lawrence.edu/dept/art/BUERGER/ESSAYS/PRODUCTION6.HTML; http://www.pepysdiary.com/diary/1663/05/19/).

Making Coins

Similar to Josiah Wedgewood's methods for making cameo reliefs (Chapter 2, Figure 2.9), the traditional approach to making a coin involved the transcription of shaded tonal drawings made by the designer on to a low-relief plaster roundel, approximately 1000% larger than the size of the final coin (see Figure 5.28). Working on this scale enabled the sculptor to engrave the image at a higher level of detail and accuracy. A detailed account is given by D. Wayne Johnson on coin minting (https://medalblog.wordpress.com). A mould was taken from the plaster model and an electrotype (electroplated with nickel and copper) reproduction in metal was created. The surface of the large electrotype coin was traced using a reducing machine. Acting like a three-dimensional pantograph, the reducing machine, as it suggests, traced the outlines and surface relief of the large electrotype and the movements of the tracer were communicated to a rotating cutter, which engraved the lines at coin size into a steel block. Reducing machines continued to be used by coin makers until the early years of this century (Figure 5.29).

Over the last decade national Mints, such as the Royal Mint in the United Kingdom, have revised traditional technologies and replaced the reducing machine with computer-controlled scanning technologies. The surface of the plaster is scanned using a coordinate measuring machine (CMM) or touch trigger

Figure 5.28 One penny plaster carving on the left and version engraved into metal on the right, compared to the actual size of the die below (courtesy: The Royal Mint Experience) (http://www.royalmint.com/en/the-royal-mint-experience).

Figure 5.29 Royal Mint's reducing machines (http://www.royalmintmuseum.org.uk/collection/collection-highlights/minting-equipment/reducing-machine/index.html) (courtesy: Royal Mint Museum).

probes to precisely record the *X*, *Y* and *Z* coordinates of the object to achieve a greater level of accuracy. The original touch-trigger probe is based on a spring-loaded kinematic arrangement of rods and balls. These provide six points of contact to ensure precision movement, measuring and repeatability. The data generated can then be imported to 3D software. Although CAD methods are now used as part of the workflow, traditional methods are still employed that include creating a plaster or wax low relief.

One can also make comparisons to profilometers, which, as the term suggests, are used to measure a surface's profile. These range from contact to non-contact (optical) devices as used in areal photography to measure the topography of a landscape (Stout and Blunt, 2000) or road pavement profilometers that were used for measuring the roughness of roads (Sayers and Karamihas, 1998). On a micro level, profilometers are used for measuring the roughness of paper to study the properties of paper-coating formulations applications and their relationship to print attributes (Wu, 2008).

Although plaster disks are still in use by the Mint, an all-digital approach is now being included in the workflow. This is exampled in the work by designer Matthew Dent and Edwina Ellis who have worked with the Royal Mint on a number of major projects. Two commissions were the design of the *Shield of the Royal Arms* representing the United Kingdom for the one-pound, one-penny, two-pence, five-pence, ten-pence, twenty-pence and fifty-pence coins, which forms part of the Royal Arms (2008–2014), and the design of the two-pound coin to mark the 200th anniversary of the birth of Charles Dickens (2012); although both comprised digital components they were approached differently by Dent (Figure 5.30).

For the *Shield of the Royal Arms* commission, the sculptor John Bergdahl transcribed the seven coins into low relief designs on a plaster disk. At a larger scale, Bergdahl was able to add the minute details

Figure 5.30 Matt Dent, New Reverses (2005–2008) and Charles Dickens £2 (2010–2012) (courtesy: Crown Copyright and Matt Dent).

such as the shape of a ribcage on the dragon, the directional grain of wood on the harp or the curl of hair on the lion, which would probably be overlooked on a coin but contributed to an overall convincing look and feel of the composition. After the models were scanned, Dent worked with the engraving department using CAD software to add a texture to define the background of the shield. The textural background details served to increase the contrast between the individual relief elements of the coins and the background and to provide a harmonious balance across the six coins. The digital design process also enabled minute modifications to the curved path of lettering in order to work in the round. For the *Charles Dickens* coin reverse, Dent also designed the artwork on computer, but as the design was more typographical, he was able to supply the electronic file directly to the Royal Mint's engraving department, and using the CAD data the output could be scaled down and the CNC engraving machine was able to cut the design into the steel at the same size as the coin.

Another series of coins that was designed by printmaker Edwina Ellis demonstrated other methods of composition and design. For the *Bridges of Britain* composition (representing different bridges from England, Ireland, Scotland and Wales, 2004–2007) the designs were drawn and then traditionally hand-cut as a relief in lino by Ellis. This enabled her to consider the relationship of the bridge to the border design, alongside the contrast of tone and level of detail such as bricks. Ellis' experience of working with the sculptor who translated her prints described assumptions in assigning relief height according

Figure 5.31 Edwina Ellis, *Mind the Gap*, 2013 (courtesy: Crown Copyright and Edwina Ellis).

to light and dark areas to a coin was not as straightforward as she had anticipated. Ellis described what she thought was obvious in the relief heights (big white lines high, black areas low) but that was apparently not at all the case. Likewise, for the very geometric design for the 150th anniversary for the London Underground, the two-pound coin designed by Edwina Ellis shows the London Underground logo, which appeared on underground stations in 1908. The sculptor was able to achieve the illusion of shading and high contrast through very little relief (Figure 5.31).

The Royal Mint has also introduced relief colour printing on to collectors and commemorative pieces. For example, series of commemorative coins to celebrate the 150th anniversary of illustrator Beatrix Potter includes engravings by Royal Mint designer Emma Noble and four of the most well-known Beatrix Potter characters, which shows the animals in full colour (https://www.royalmint.com). Likewise, a commemorative coin produced by Royal Mint coin designer Thomas Docherty for the 2016 Remembrance Day was a wreath of red poppies with dark green leaves printed on to the relief surface. The printing process is a pad printing method that is used to print on to objects that are already fabricated. The soft silicone pad is able to conform to the surfaces and print detailed information. The shape of the pad can be customised to fit around different objects. Pad or Tampo printing is a three-stage printing process and, similar to offset printing, transfers a layer of ink from a flat plate to a soft silicone pad and then on to the object. In more detail, a gravure or engraved plate is inked and wiped clear of residue, the soft silicone pad is pushed into the plate and ink adheres to the surface of the silicone pad. The pad is positioned directly above the substrate, is pressed down and wraps around the object while stretching very little but with enough pressure to transfer the ink from pad to substrate.

Medals of Dishonour Exhibition British Museum

Although medals are usually associated with celebration or acknowledgement of merit, bravery or achievement, the British Museum host a large collection relating to so-called *Medals of Dishonour* (Attwood and Powell, 2009). These medals, some dating back to the sixteenth century and the most recent in 2009, communicate messages that are neither celebratory nor complimentary, but demonstrate social, political, warfare, financial, personal aggrievement and satire. The artists invited to produce medals for the recent collection (2009) and as part of their own practice have often demonstrated commentary on current social and political issues were asked to consider the medal's history of 'opposition, dissent, subversion, outrage and crude humour'. Similar to the themes of the previous 400 years, the artists in this collection also chose to use allegory, satire and horror through minute representation and observation. It is interesting to note that as the majority of the artists had not worked with metal before, they chose to use an alternative media to resolve their compositions. For example, artist Helen Gallagher layered and carved into thick watercolour paper. Casts were taken directly from the paper and therefore highlighted the carving techniques. Likewise, Grayson Perry, who is well known as a potter, worked in clay to develop his roundels. Perry worked at a larger scale of 205 mm diameter, as he wanted to obtain as much detail as possible. The roundels were then mechanically reduced. Situated in Birmingham's Jewellery Quarter, Toye Kenning & Spencer Ltd worked with Perry to strike the medals. Toye Kenning & Spencer, which is over 320 years old, produces a wide range of woven ribbon, gold cord, embroidery, insignia, badges, buttons and medals and is holder of a Royal Warrant to Her Majesty Queen Elizabeth II (https://www.telegraph.co.uk/finance/newsbysector/retailandconsumer/9192967/A-trip-back-in-time-to-our-industrial-heritage-at-Toye-Kenning-and-Spencer.html).

English Pop artist Richard Hamilton (1922–2011) was asked to make a medal. His medal, entitled *The Hutton Award* (2009), used portraits of Tony Blair and Alasdair Campbell (Figure 5.32). These

Figure 5.32 Richard Hamilton's Hutton Award (2008), CNC milled development model in polyurethane resin board, 72 mm diameter. *Medals of Dishonour*, the British Museum, 25 June–27 September 2009 (courtesy: CFPR, UWE, Bristol).

images were generated from greyscale 2D image files and digitally converted into a relief surface. Hamilton started by digitally manipulating the two sides, firstly converting the colour images into black and white, then adjusting the greyscale tones appropriate for a relief surface. The greyscale images were then assigned different tones to a relative height, and by this method, the lightest tones of the greyscale image produce the areas of highest relief and the darkest tones recede as the background. The Centre for Fine Print Research (CFPR) at the University of the West of England was invited by Richard Hamilton to contribute to the realisation of his medal *The Hutton Award.* CFPRs input included 3D modelling from Hamilton's original image files of the artwork, production of a series of development models by 3D printing, using 3D Systems Thermojet, Envisiontec Digital Light Processing-based rapid prototyping systems, and CNC milling into a dense polyurethane resin board using a Roland Modela MDX 540 3-axis milling machine, with a tapered cutter with a 0.25 mm tip. Working from the master patterns, Irene Gunston (Royal College of Art, London) produced wax master copies for casting into silver and bronze by BAC Castings Ltd (London), and these were hand-finished at the RCA Foundry. In this instance, for 2.5D printing the most relevant approach is through the second projection and a semi-relief of the first projection. For these approaches the background surface is primarily the support for the modelling (http://www.britishmuseum.org/about_us/news_and_press/press_releases/2009/medals_of_dishonour.aspx).

Case Study 5: Capturing Texture of Paintings for Museum and Heritage

Since the 1970s, museums have undergone a radical change, incorporating stunning new spaces to traditional buildings or the creation of brand new buildings. In this climate of 'new museology' and through increasing social change and education, their function is no longer a place for the elite, but about openness and social inclusion that appeals across disciplines. Although museums are in charge of caring and preserving a nation's cultural heritage, this new openness has enabled the curator to organise and present their cultural artefacts in new ways, by 'breaking down of established boundaries … to some transgressing of the boundaries of specific subject disciplines' (Ross, 2004).

Technology can provide us with new insights about how a painting or an artefact has been created. This can be of benefit for curators in bringing to life museum artefacts and engaging with visitors in new exhibitions. Technology can also be of assistance for conservators in choosing the most appropriate methods for restoration. Today, in the cultural heritage field, there are many available methods and technologies that are capable of capturing three-dimensional data from objects. Of all these techniques, there are only a few that meet the necessary requirements to satisfy the accuracy needed for the capture and reproduction of painted surfaces. Within a wide range of sensors only non-contact optical sensors can be considered due to the fragile nature of paintings. These are classified into passive and active methods. In active methods, a light beam or pattern is projected over the surface and the reflection is captured by an adequate sensor, which measures the location of the image in space at the illuminated point. Amongst these systems, structured light and laser triangulation are the most important. In passive methods, the light reflecting from the object and the illumination of the scene are computed to derive the shape information from two-dimensional photographs. Although not as accurate as the active systems, passive methods such as photogrammetry have shown potential within the cultural heritage field.

Visualising Surface Texture Through the Combination of 2D and 3D Data

A special thanks to Xavier Aure for providing detail on this section. Xavier is undertaking a collaborative PhD research project at CFPR and The National Gallery, London, funded by The Arts and Humanities Research Council.

Studying the texture of paintings and the surface qualities of many other cultural heritage objects is crucial for interpreting how an artist or a studio operated, and may also provide important clues to art historians and conservators about their condition. However, the display and presentation of three-dimensional information through two-dimensional reproductions or photographs, either printed or on computer screens, is a difficult task. In the past, the only approaches available to capture surface detail were either observing the object and drawing the details by hand or as contact methods such as moulds and imprints, but for obvious reasons of possible damage to the paint layer they cannot be used. The only existing non-contact method are photographs taken in raking light to emphasize the surface topography, although the two-dimensional aspect of the photographs cannot truly represent the texture.

With the recent developments in digital imaging presented above it is now possible to capture contact-free three-dimensional details from the surface of paintings. Advances in 2.5D and 3D printing also allow the production of accurate three-dimensional physical models. Both techniques present great potential within the heritage sector, and especially in the reproduction of the surface texture of paintings.

Using a Lucida laser scanner designed by Manuel Franquelo and custom built by Factum Arte, The National Gallery in London has recently been testing the potential of 3D scanning technology to record and measure the surface of paintings. The Lucida scanner outputs 3D information as 2D greyscale depth-map files, which can then be used to generate both 2D shaded renders of the textured surface in addition to the traditional 3D triangular meshes.

There are, however, issues with these two outputs when it comes to visualisation. On the one hand, greyscale rendered images can only display surface texture illuminated from a range of defined positions, they are individual static images that cannot convey texture information as a full interactive 3D model. On the other hand, generated triangular 3D meshes of detailed surface texture require a vast amount of computing power to visualise and interact with the final processed 3D model.

A new workflow is being developed that combines 2D images and 3D data captured at The National Gallery with RTI images to facilitate the efficient visualisation of the surface texture of paintings (Figure 5.33).

Workflow: 3D Scanning and Processing

The 3D scanner projects a beam of red light on to the surface. The deformation of the beam is recorded by two cameras as it moves across the surface at a resolution of 100 μm. All files are saved as raw black and white video data and processed as a tone depth-map. The scanner records individual tiles (480 mm × 468 mm), which are then edited, stitched and merged to create a whole record of the surface. After scanning and processing the 3D data, a final 32 bit greyscale depth-map file is obtained. The different greyscale tones in the file define the depth of each pixel recorded of the surface of the painting (Figure 5.34).

Generation of Colour RTI Images

The Lucida software uses the depth-map file to generate a controlled number of shaded rendered images lit from different directions and a light position file that indicates the normalised position in space of each light source used in each image. The images and the LP file can then be processed with the RTI builder or the PTM software to generate a final RTI interactive image.

Because the shaded renders are only in greyscale, the RTIs can be further explored by registering existing colour, infrared reflectography, elemental XRF maps and X-ray images of the painting to the shaded renders to visually compare the relationship between the textured data and the other image-based examination techniques.

Registration is a two-step process: first, a perspective match transformation in order to correct size, translation, rotation and perspective distortion differences between the two images; second, converting all images to Lab colour space and substituting the 'ab' colour channels into the greyscale shaded renders. This results in the same series of images with surface data information, but with the added benefit

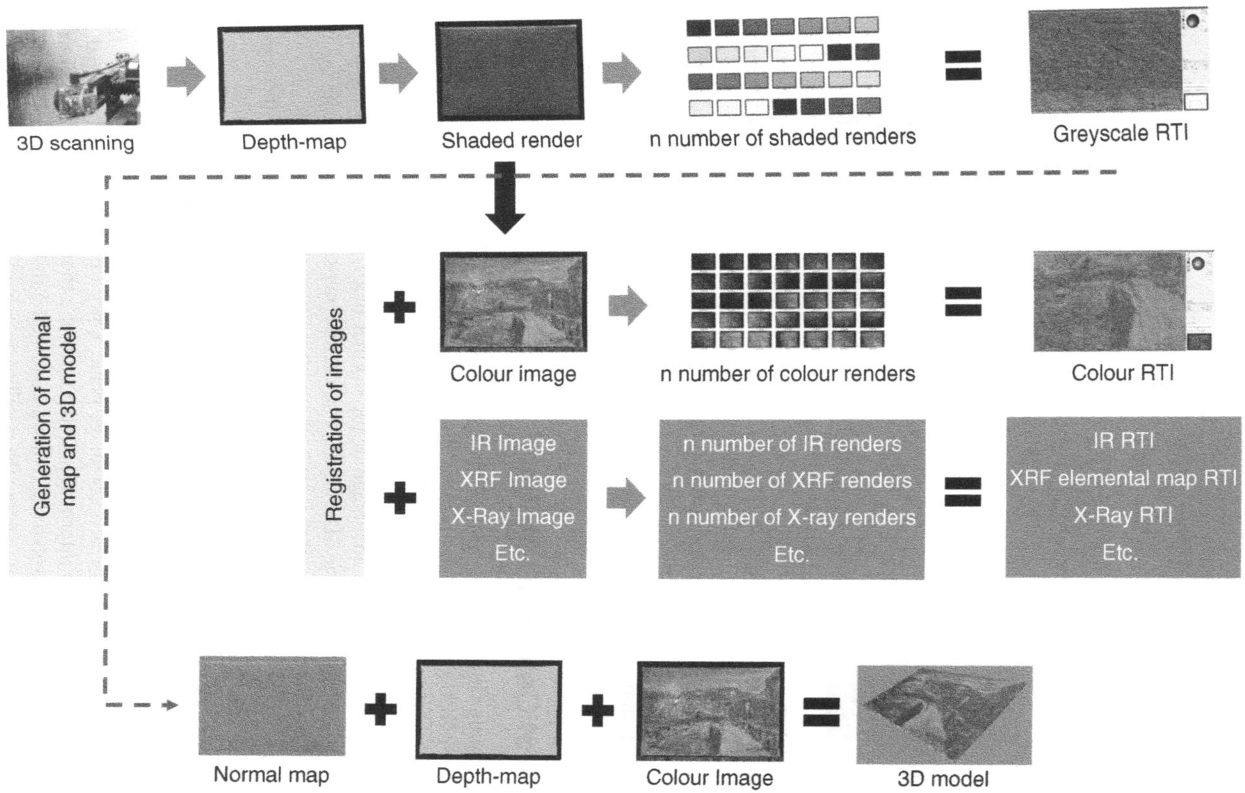

Figure 5.33 Workflow of the scanning methods to generate a normal map and 3D model (courtesy: Xavier Aure, CFPR).

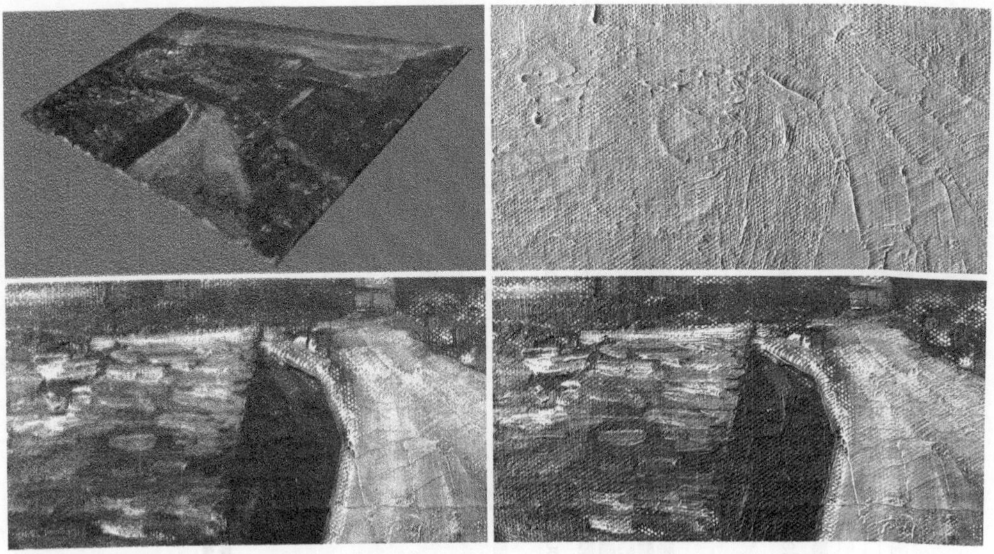

Figure 5.34 Showing a complete render of the 3D model of the painting (top left). A detail of the painting showing a greyscale render of the 3D model with the normal map (top right), a render of the 3D model without the normal map but with the colour image (bottom left) and a render of the 3D model with the normal map and the colour image (bottom right) (courtesy: Xavier Aure, CFPR).

of having colour, infrared or X-ray information. These operations are efficiently done with Nip2 open source software.

Generation of Colour 3D Models

Visualisation of highly detailed meshes of painted surfaces requires the use of professional workstations with powerful GPU cards that can process the model in real time. This ideal scenario is highly unlikely even within important heritage institutions. In order to overcome this issue the proposed workflow borrows some of the techniques used in the gaming industry to produce highly detailed 3D models of the surface of paintings. We take advantage of the normal maps generated with the RTI or PTM software and incorporate them into the 3D models. The normal maps allow us to create the illusion of a highly detailed surface texture without the strain of a multimillion polygonal mesh.

Normal maps are 2D images used to change the angle at which light reflects off the surface at per pixel level. Each pixel in the image stores the *X*, *Y*, *Z* coordinates of that angle in relation to the point of origin at the surface of that pixel. The coordinates are represented by RGB values. With this information, the 3D software can create the illusion of height and shape when the image is illuminated with much more precision than those normally extracted from polygons.

To generate the 3D model the same greyscale depth-map used to produce the shaded renders is imported into the 3D software blender as a 2D image plane. Once imported, a number 'modifiers' are applied to the plane to extract the 3D information in the form of a triangular mesh. In order to keep the model manageable the plane is only subdivided into around half a million polygons. This provides the model with an overall shape sufficient to show the possible deformations of the canvas or wooden support and enough detail to display the thickest impasto on the surface of the painting as the finer surface details will be later provided by the normal map.

Once the modifiers are applied, the normal map and the colour image are mapped to the mesh. As the colour image has been previously registered to generate the coloured RTI and the normal map has been extracted from that same RTI, there is no need for further adjustments when mapping the images. The final stage involves adding the correct type of lights to illuminate the model and modifying the diffuse and specular properties of the various maps to achieve the appropriate appearance of the surface of the painting. When the model is ready it can be both interacted with instantly within the 3D software or it can be rendered using current physically based rendering standards to produce 2D images or videos of the painting. The models can also be exported for online visualisation.

An RTI dome is now being built to allow for the generation of high-resolution colour images and normal maps. This will improve the quality of the current normal maps that have been so far extracted from the 3D data at maximum resolution of 100 μm (254 dpi). The resulting images will be stitched and blended together so they can be mapped to the 3D meshes. By incorporating the improved normal maps and the colour images without shadows and specular highlights it will be possible to examine the surface texture at greater accuracy.

The generation of 3D information as 2D greyscale depth-map files facilitates the production of shaded renders that can be used to create interactive RTI images. By registering these with existing 2D images to generate further RTI images provides a way of visualising possible relations across paint layers and whether hidden layers actually show on the surface of a particular painting. By combining existing 3D data, colour images and normal maps to generate realistic 3D models of the surface of paintings, we overcome the problem of relying on extremely powerful computers to interact and visualise the models.

Case Study 6: Textiles

Textiles are composed of fibres, which are single filaments that, when spun or plied, create longer strands or yarns that can be woven, knitted or meshed together (Cook, 1984a, 1984b). One can also consider yarns as the molecular units of a fabric. Fibres can be natural (silk, wool, bamboo, cotton) mineral (fibreglass, asbestos) and synthetic (Nylon, Kevlar, Polyester). Each has special or inherent properties, such as absorbency, fire-resistance, warmth, softness, drape, stretch, pliability strength and durability. Fibres may also be combined with metallic alloys to create functional or smart materials that respond to human touch or the environment.

An important aesthetic consideration of a garment is a material's tactile and visual texture such as colour (natural), ease and unease of dyeing, lustre and texture; the pattern, weave, fibre content, decoration and how a material looks and feels, protects and drapes around the body. Decoration may be obtained by a relief pattern, for example as part of weaving or knitting, or during the finishing processes, for example through embellishments, embossing, embroidery or printing. Digital textile printing has been slow to develop and still represents a tiny market share. According to reports in 2015, digital textile printing accounted only for a 2–3% of the textile printing market. Nevertheless, there have been ground-breaking improvements in digital textile printing during the last few years that will open the market to new digital alternatives. According to Smithers Pira, a total of around 650 million square metres of fabric were digitally printed in the year to mid-2014. It is estimated that the total digital textile printing will rise to 1.6 billion square metres annually by 2019. This includes displays, soft signage and direct-to-garment printing (DTG) (http://www.smitherspira.com/news/2014/october/digital-textile-printing-set-to-grow-by-20-1-from).

There are many traditional methods that are integral to the manufacturing process or at the finishing stage that can radically change the appearance of the material. Some processes may change the gloss (optical effect) of fibres and fabrics, such as mercerising or glazing to create a lustrous or smooth finish. Other processes are used to produce raised textures, for example, *devoré* is a chemical process where an acid paste is printed as a pattern on to velvet, thus dissolving fibres to create a flock appearance. Other methods include embossing, pleating or crimping the surface with heat and pressure to create permanent relief or by using abrasive tools to create a suede-like or flannel-like surface, such as emerising using rollers covered with emery powder or napping using wired brushes.

Some interesting textile products have evolved using 3D printing or additive manufacturing. In the 1990s, the company Freedom of Creation, based in the Netherlands, was an early user of additive manufacturing to make 3D printed objects including lampshades. Designer Jiri Evenhuis, in collaboration with Janne Kyttanen of Freedom of Creation, was the first to consider the idea of using 3D printers to create textiles. As explained by Evenhuis, 'Instead of producing textiles by the meter, then cutting and sewing them into final products, this concept has the ability to make needle and thread obsolete' (http://www.ecouterre.com/are-3d-printed-fabrics-the-future-of-sustainable-textiles/3d-printer-fabric-6/).

Other more recent examples include Danit Peleg, who designed the dress for Amy Purdy for the Paralympics Opening Ceremony in Rio 2016 (http://danitpeleg.com/paralympics-dress-amy-purdy/#sthash.JKdOBjHd.dpuf). Peleg is a fashion designer based in Tel Aviv, Israel. In collaboration with Gerber Technology (http://www.gerbertechnology.com) and using Accumark and Blender and printed on Witbox2, Peleg used a flexible filament called Filaflex, which is produced by Spanish company Recreus (https://recreus.com/en/) (Figure 5.35).

Other examples include Shape Shifter, which are modular garments made by Angelene Fenuta (https://www.angelene.net/concepts-01/), who was a prototype winner of the Lexus Design awards 2016 (http://www.lexus-int.com/lexus-design/lexus-design-award). Dutch designers Anouk Wipprecht, with body tech designs (https://www.anoukwipprecht.nl), and Iris van Herpen, with organic-inspired clothes and shoes (http://www.irisvanherpen.com), both worked with high tech companies as well as the 3D printing innovator Materialise. The limitations of these so-called textiles is the appearance, tactility of the material and the chain mail construction to obtain a material that is flexible enough to drape and conform to the shape of the body. The geometric complexity of these materials lies in the need to produce and support thousands of tiny elements.

3D processes are already in place in other areas of the fabric industry, such as weaving and knitting. Knitted or woven textiles in 3D are used mostly for industrial purposes and less for decoration, where the structural properties prevail over aesthetic aspects, at least for now. These include high tech applications including fire-safety (http://www.dupont.co.uk/products-and-services/personal-protective-equipment/nomex.html), radiation resistance and aerospace; for example firefighters, military personnel, police officers and auto racing teams are protected by clothing made with Nomex®, which is resistant to extremes of heat and fire.

There are a range of emerging trends in textile fabrication, as well as in 3D printing, to develop adaptive processes for a wider set of materials and technologies that will lead us to combine 3D weaving, knitting, embossing, printing and more into aesthetic and functional objects. Examples include Open Knit (Figure 5.36), Knitic (Figure 5.37) and Disney Research Pittsburgh (Figure 5.38). Computerised knit technologies are also being explored and developed for 3D printing for sports shoes, for example Primeknit by Adidas (http://www.knittingindustry.com/flat-knitting/adizero-primeknit-how-its-made) and Klyknit by Nike, that digitally knits the upper part of the trainer in just one piece. These highly

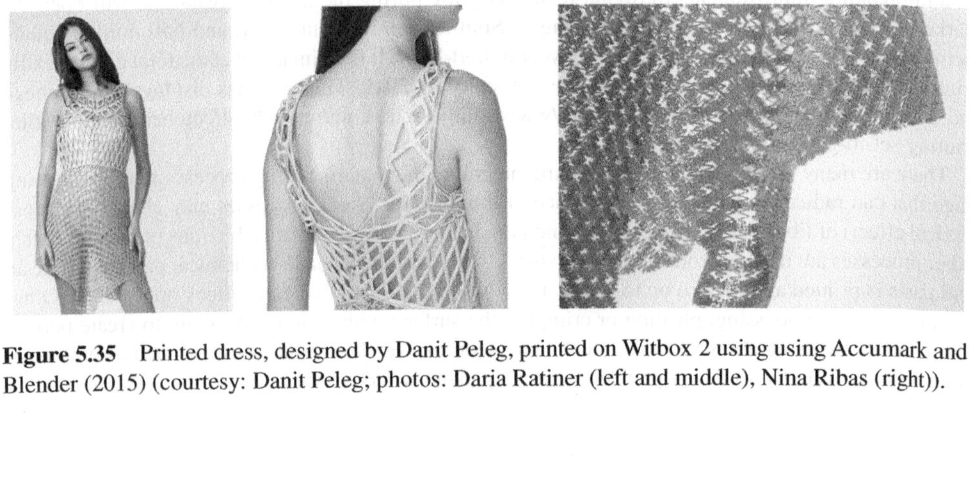

Figure 5.35 Printed dress, designed by Danit Peleg, printed on Witbox 2 using using Accumark and Blender (2015) (courtesy: Danit Peleg; photos: Daria Ratiner (left and middle), Nina Ribas (right)).

Figure 5.36 Open Knit (courtesy: Open Knit) (www.openknit.org).

Figure 5.37 Knitic knitting machine (courtesy: Knitic) (http://blogs.elpais.com/arte-en-la-edad-silicio/2013/04/trabajos-de-punto-y-pixel.html https://www.varvarag.info/circular-knitic/).

Figure 5.38 Disney 3D printer for teddy bears (courtesy: Disney Research) (https://www.disneyresearch.com/publication/printing-teddy-bears-a-technique-for-3d-printing-of-soft-interactive-objects/).

automated methods present a challenge to traditional manufacturing, but also offer opportunities for increasing the variety of materials that can be used (Economist, 2017).

Modern looms, weaving and knitting machines are a mechanical version of traditional implements such as needles and crochet hooks. Until the industrial revolution manual labour in textiles was very common and still is, but it has switched domains. Manual labour nowadays is more common in assembling and finishing of garments and in operating textile machines than in the actual making of the yarns and fabrics. A significant impact on the textile industry was the Jacquard machine (Figures 5.39 and 5.40), created in the beginning of the nineteenth century by Joseph Marie Jacquard. It simplified the creation of complex patterns such as damask and brocade by using a series of punched cards that controlled a sequence of operations for a given design (http://www.nms.ac.uk/jacquardloom). Each row of punched holes in the card corresponded to the row being woven. The Jacquard loom is an important precursor of computing programming and, in a more indirect way, digital weaving. Traditional weaving is still in high demand, for example Humphries Weaving (http://www.humphriesweaving.co.uk), established in 1972, undertake a range of conservation restoration and replication of fragments of materials, as well as contemporary production methods that incorporate new materials and weaving construction. Other working heritage mills include Sir Richard Arkwright's Masson Mills in Derwent Valley, Derbyshire.

Quite a different company that works in the field of fine art is Magnolia Editions, which was founded by Donald Farnsworth in 1981. He works with Flanders Tapestries in Belgium using an adapted 2.1 m wide double-headed electronic Jacquard loom that interfaces with a digital weave file to create high-resolution tapestries (Jacquard Weaving and the Magnolia Tapestry Project: https://www.magnoliaeditions.com/exhibitions/weaving-in-belgium/) (see Figure 5.41). The double head meant that

Figure 5.39 Jacquard loom, showing punched hole cards, National Museum of Scotland, Edinburgh (via: WikiCommons/Creative Commons 3; credit: Ad Meskens) (https://commons.wikimedia.org/wiki/File:NMS_Jacquard_loom_2.JPG).

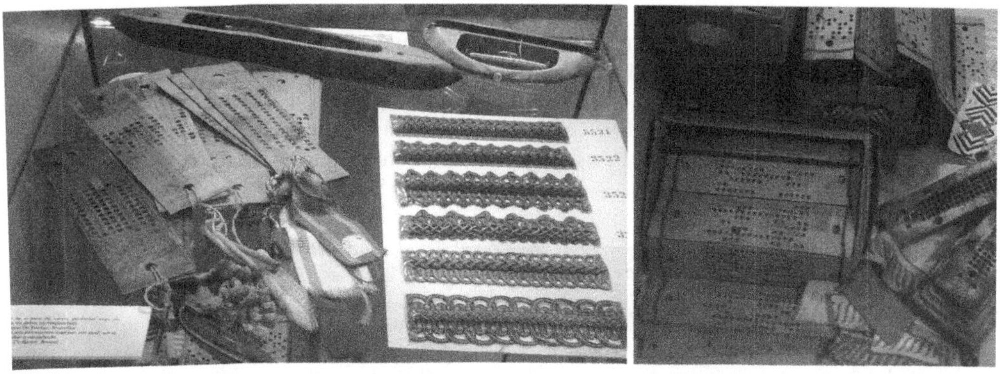

Figure 5.40 Samples produced on a Jacquard machine at La Fonderie, Brussels Museum of Industry and Labour.

he could create images using twice the number of threads, thus creating a highly dense weave. Farnsworth worked with artist Chuck Close to create high-quality woven wideformat colour images woven from cottons and viscose (see Figure 5.41). Adam Lowe, director of Factum Arte, has also worked with a number of artists, including, for example, Grayson Perry and Marc Quinn, to convert their work into digital weave files (Sheets, 2012).

As presented at the Ars Electronica Festival in 2016, *Project Jacquard* (https://atap.google.com/jacquard) introduces the idea of fabric as interactive surfaces. Led by Ivan Poupyrev, the objective of *Project Jacquard*, in collaboration with Levis and Google ATAP, is to add function to clothes by introducing interaction through touch and gesture. These ideas have evolved by studying the relationship of user gesture and electrical energy generated through touch, such as tapping, rubbing and sliding (Seung-Chan, Israr and Poupyrev, 2013; Karagozler *et al.*, 2013). Working with Google's Advanced Technology and Projects incubator programme, the group are investigating interactive materials that can be made using standard looms by weaving and combining thin metallic alloys with natural and synthetic yarns such as cotton, polyester and silk.

Analogous to the patterns created on a Jacquard loom is a printed example developed at the CFPR using a Rostock 3 axis printing approach to material and pattern fabrication that creates gauze-like structures (O'Dowd *et al.*, 2015). Using a Deposition Modelling 3D Printer, a filament is drawn out of the nozzle faster than usual and extruded along a spiral cylindrical path. As the filament reaches just

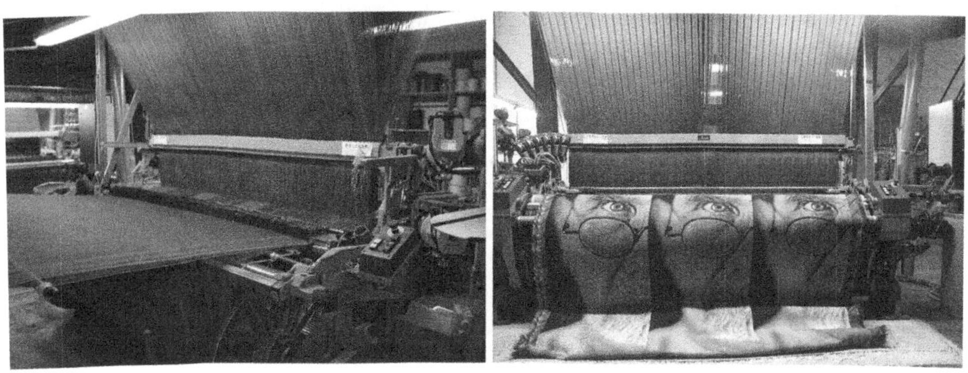

Figure 5.41 (left) 17 000 warp threads and lifters (right) testing different greyscales of a woven portrait of Chuck Close at the mill in Belgium (courtesy: Magnolia Editions, Oakland, CA).

melting point, tails of molten filament are stretched and angled in the direction of travel, bonding to the next available point to form a pattern. The pattern is obtained by mapping a greyscale bitmap image to the surface of a cylinder. The image is printed as a cylinder and then cut vertically to open up as a flat panel (Figures 5.42 and 5.43). *Project Silkworm* is an open project group exploring methods similar to weaving (http://projectsilkworm.com). Silkworm is a plugin that enables users to translate their 3D files into G-Code, which can be used to drive different printing platforms to extrude pastes and filaments.

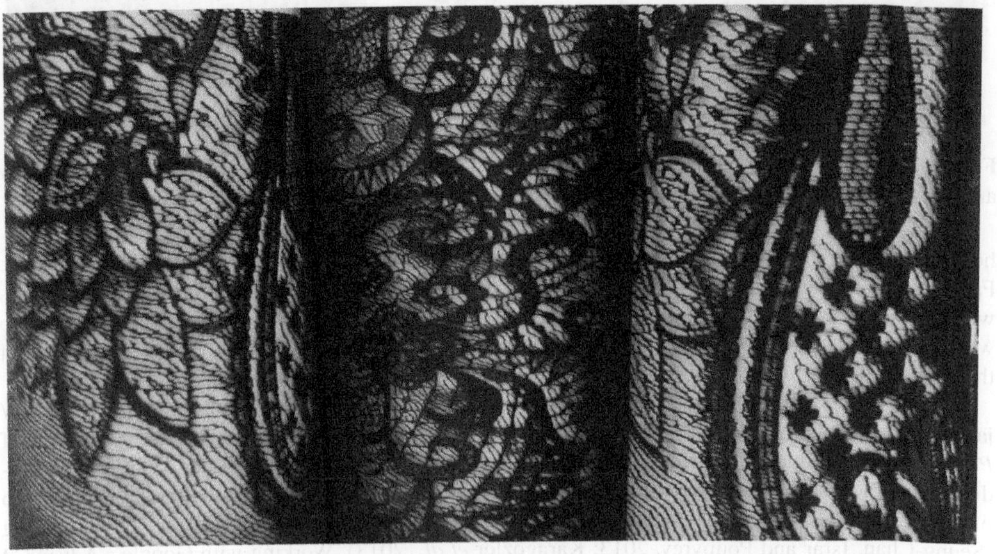

Figure 5.42 Paul O'Dowd, 3D printed lace using a Rostock 3 axis extruder (2016), Centre for Fine Print Research, University of the West of England, Bristol (courtesy: Paul O'Dowd).

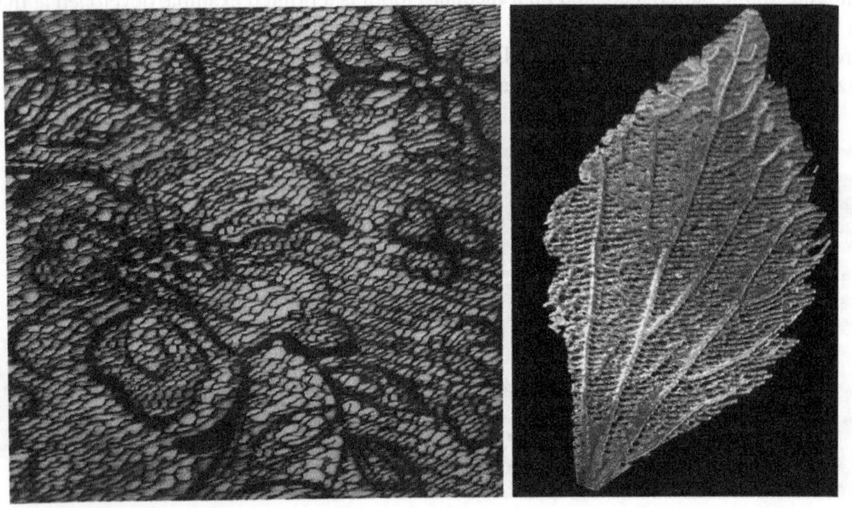

Figure 5.43 Paul O'Dowd, detail of 3D printed lace and leaf Centre for Fine Print Research, University of the West of England, Bristol (courtesy: Paul O'Dowd).

Whether relief or 3D printing is ready to be used to create flexible and wearable fabrics is still debatable, but as new materials are beginning to emerge, 2.5D printing for textiles presents many new opportunities. We can be sure that when digital printing increases its pace in the fabric industry there will be a need for richer textures and forms that 2+ printing technologies may be able to address. Therefore studying the 'analogue' techniques (embroidery, embossing, mercerising) used today to create textures may prepare the future for digital textile printing with new dimensions.

Case Study 7: *Trompe l'Oeil*

The art of mimesis, from the Greek word imitate, could be considered as the means to represent nature rather than directly copy it. In his introduction, *Lives of Painters*, Giorgio Vasari (1511–1574) describes how painters' perception of the world could achieve a spatial deception (*piacevolissimo inganno*) in their paintings through relief (*relieve*) and depth (*fondo*). On a material level, artists working in the context of still-life painting, as demonstrated by Hans Holbein, Charles Bell and Robin Eley, can engage the viewer to an intimate display of the minute characteristics of many hundreds of different surface textures (see Chapter 2).

On a larger scale, by changing the cues that we normally use to perceive depth, leads us to misinterpret size, distance, dimensionality or colours of objects within a scene. Not only have artists consciously included these effects in their creations but also architects and designers have taken advantage of our *deficiency* to detect them. It is the ability of artists to fool the viewer into thinking that a casually placed object on a shelf or leaning against a wall is a real object, or that a small room has achieved a further dimension such as a window or a hole in the ground. There are so many examples that it is difficult to choose, so the following examples are based on *trompe l'oeil* on a *grand scale*, and sometimes is dependent on where we know to look (for example see Figure 5.44).

A theatrical example is the *Teatro Olimpico* (1580–1585) in Vicenza (Figure 5.45), designed by the highly influential architect Andrea Palladio (1508–1580). Although not completed in his lifetime, architect and writer Vincenzo Scamozzi (1548–1616) oversaw the completion of the project. Working from Palladio's designs, Scamozzi remodelled the staging, the ornate entrance gates through which the actors

Figure 5.44 Church of St Anastasia, Verona. The artist has combined sculptural relief and *trompe l'oeil* grissaille paintings.

Figure 5.45 Inside the *Teatro Olimpico*, Vicenza, Italy.

entered, and behind the gates he created the illusion of the long vista of radiating streets flanked by buildings, even though the stage was but a few metres deep. The illusion of distance was enhanced by the illumination of the buildings, the exaggerated upward slope of the street and decreasing scale of the buildings. The marble and sky are painted effects that mirror the plaster and wooden façade of the entrance gates and the surrounding of the theatre.

A similar theatrical effect is explored in the life-scale frescos on canvas paintings entitled *Devotees* in Gallerie dell'Accademia in Venice by Giovanni Battista Tiepolo (1743–1745). In these theatrically illuminated scenes, the transfixed onlookers lean out over the sides of the balcony to obtain a better view (http://www.gallerieaccademia.it/sale/sala-xi). Tiepolo has illuminated the stonework of the four corner balconies, where some of the faces of the devotees as they lean forwards are enhanced by the light. The upper carvings of the balcony are subjected to a dazzling light whereas the background and some sleeping figures remain in shadow. The effect and size of the work, which is 4 metres in height and hangs high in the Gallerie forces the viewer to look up, thus increasing the visual impact of the painting and aura of the scene.

Such *trompe l'oeil* – a trick of the eye or the illusion of reality – has been used to great effect when the objects and scenes are in proportion to their surroundings. The most successful or convincing works are those painted on a grand scale, such as the stunning dome in the Jesuit Church in Vienna. The fresco is painted on a flat ceiling, but when the viewer looks directly above, they perceive a concave-shaped dome (Figure 5.46).

Likewise, the painting *A View Through a House* (1662) is a large still life illusionary work by the Dutch painter Samuel van Hoogstraten (1627–1678) (Figure 5.47). Placed at the end of a corridor in Dyrham Park near Bath, the painting transforms the Baroque seventeenth century English country house into a seventeenth century Dutch interior. Van Hoogstraten cleverly created a scene that suggests the viewer has arrived at a crucial point of the drama. This particular placement of the painting may raise confusion for the viewer as to which is the real interior and which is the painted. For example, the bird-cage appears to be hanging in the hallway in front of the painting, and the illumination from the windows in the corridor in the house is similar to the light in the painting. The corridor in the painting is lit from both sides, the viewer can perceive the objects and social interactions and there is a glimpse of activity at the far end of the hall and rooms beyond. Van Hoogstraten is possibly best known for his illusionistic works of art, which required a balance of lighting, mimesis of materials, colour and

Figure 5.46 Fresco with *trompe l'œil* dome painted on low vaulting, Jesuit Church, Vienna, Austria.

perspective to achieve the overall effect. As exampled by *A Peepshow with Views of the Interior of a Dutch House* (c.1655–1660) in The National Gallery, London, Hoogstraten paints objects, including a dog, chairs and a broom, that are placed within a series of rooms, on to the inside of the box. Through the clever use of false perspective Hoogstraten has transformed the interior of a three-sided box into a three-dimensional multiroomed domestic interior. All corners of the rooms are perfectly lit and balanced so as to be able to perceive all the architectural details and objects. As described by Brusati, 'The device graphically demonstrates Van Hoogstraten's understanding of how the eye is deceived both by the painter's art and by the act of seeing itself. As such it offers an invaluable presentation of the assumptions about the nature of vision and picture making that inform Van Hoogstraten's art and writing' (Brusati, 1995).

More contemporary examples include the work by artist Kurt Wenner, who has incorporated the pavement into his canvas (Figure 5.48), He uses one-point perspective or anamorphic projection to create his 3D illusional artworks. Like Hoogstraten, Wenner uses depth cues, oblique perspective, subtle changes in illumination, and mimesis of materials and surfaces to reinforce the illusion.

The addition of fake houses may be used to bring uniformity to a street. For instance, streets in Paris incorporate façades to hide unsightly services and utilities (Figure 5.49). Building 145, Rue la Fayette is a constructed façade to hide a giant ventilation shaft belonging to the metro. Likewise the end-of-row house on 29, Rue Quincampoix also hides a ventilation shaft, but the exterior walls are painted to resemble the other windows in the street. It even includes a man practising his violin and a woman standing in profile in the window (http://parisisinvisible.blogspot.co.uk/2013/10/three-imaginary-homes.html).

Figure 5.47 Samuel van Hoogstraten (1627–1678), *A View Through a House* (1662), oil on canvas, 2642 mm × 1365 mm (reproduced with permission: National Trust Images); (right) photograph taken during a visit to the house during its restoration.

Superwideformat digital printing is being used to great effect to cover and wrap buildings, for example during building and restoration (http://www.signindustry.com/banners/articles/2003-09-10-Heritage-LB.php3#story), for marketing and advertising or to promote sporting events, as art projects (https://weburbanist.com/2015/04/12/wrap-it-up-10-creative-building-scrims-scaffolds) and *trompe l'oeil* (https://www.thinktank.org.uk/blog/1694-iconic-building-gains-trompe-l-oeil-wrap.php). These decals can be used to blend with the existing building to create surprise effects and maximise impact, for example an image of a gigantic football that is bursting out of the windows of an apartment block (https://i.pinimg.com/originals/bd/5b/02/bd5b02454540a58e85179a1cbef97143.jpg), or, similar to the Paris facades of Figure 5.49, 39 George V, also in Paris, which is wrapped in a giant photographic image that is based on the original building underneath but has been warped to create a wobbly appearance (https://www.deceptology.com/2010/05/creepage-distances-multiply-spectacular.html. Wideformat printed decals can be applied directly to walls and windows, which can be used as decoration, to fill in gaps and losses and to adhere to brickwork, stone or glass. Aside from advertising, these can work well as temporary creative site-specific interventions that can inject visual delight and surprise.

Figure 5.48 Kurt Wenner, Eurostar Publicity Event in Brussels, Belgium (2007) chalk on pavement (courtesy: Kurt Wenner) (http://kurtwenner.com).

Figure 5.49 145, Rue La Fayette and 29, Rue Quincampoix each conceal ventilation shafts, WikiCommons/Creative Commons 3: (left) credit: Geralix (https://fr.wikipedia.org/wiki/145,_rue_La_Fayette#/media/File:Immeuble_145,_rue_La_Fayette-Paris.jpg), (right) credit: Tangopaso (https://commons.wikimedia.org/wiki/File:Rue_Quincampoix,_29.jpg).

Case Study 8: Marble

Marble is an ornamental stone often found in churches, wealthy homes and palaces, and because of its aesthetic and enduring qualities has long been associated with luxury and permanence. Since ancient history, attempts to mimic the qualities of marble can still be seen, as demonstrated in the frescoes of faux yellow marble with red veining at Pompeii. Importantly in this case study, these qualities are still desired and copied in the twenty-first century. Whilst the desired appearance of marble is as a flat, smooth and polished surface, it is the micro subsurface such as grain shape, grain size, grain boundary and crystalline qualities that define its 2.5D qualities. This case study undertakes to demonstrate the different methods that have been used to capture the appearance of marble.

Marble is probably the most consistently used sculptural material across the centuries. It is more porous than other stone and soft enough to work with chisels and hand tools, and can also be polished to a high gloss (Figure 5.50). Marble also demonstrates many translucent, polychromic properties and patterns that were used to decorative effect, for example for Roman statuary to create contrast between the rich coloured robes, to mimic a pattern of the fabric by using a grain of the marble, or to incorporate different coloured marble to differentiate other fabrics, pleating, breast plates and medals.

Marble is a metamorphic rock (when limestone is subjected to heat and pressure) of dolomite and calcite crystals, which contains other minerals such as quartz, mica, iron oxide and pyrite to create hundreds of different combinations of patterns and colours. The display case (Figure 5.51) represents a selection of a much larger collection of seven hundred marble tablets in the *Kings Gallery* in The British Museum, London. Each of the drawers contains samples of different patterned and coloured stone marble, each measuring 9 cm × 6 cm by 1.6 cm. This collection, also known as *The Tolley Marbles*, was gathered by an English collector Henry Tolley. It is a database of geological samples that were collected from quarries in Italy and used as reference material by Tolley for the study of Greek and Roman sculpture and buildings. On each tile there is a label that describes the name of the rock, details the place of collection, date and time. Many different marble patterns and colours have been widely studied and, as

Figure 5.50 Example of different marbles, Palazzo Nuovo, Musei Capitolini, Rome.

Figure 5.51 *The Tolley Marbles*, The Kings Gallery, British Museum in London.

a result, have evolved methods to mimic and reproduce the appearance of marble (see Figures 5.52, 5.53 and 5.54).

Solid slabs of marble were traditionally imported from Italy and used extensively in large English country houses in the eighteenth century. Although it is a robust hardwearing material, it is also highly expensive and heavy to manoeuver. Therefore, due to the cost of importing, the possibility of delay and damage during transit, alternative and cheaper methods were sought. Unless considered fashionable, it was unlikely the house owner would have drawn attention to using a *faux* marble, and so great efforts were made to mimic their qualities. *Scagliola*, a method of imitating marble, became a popular method

Figure 5.52 Contemporary coloured hardwearing resin-based samples that can be machined to exact specification for worktops and sink-tops in kitchens and bathrooms.

Figure 5.53 These columns are around the first floor staircase in Berrington Hall, Herefordshire. The plaster colour was mixed to resemble marble from Sienna. (Detail) flat panel of *Scagliola* showing the small marble chips and the base colour.

in the latter half of the eighteenth century. The name derives from the Italian word *scaglia*, meaning pieces or chips. These small pieces are added to pigmented plaster. The colour of the plaster can then be coordinated with existing marble, or to mimic a marble from a particular region. A rough base plaster is applied to a wooden or steel base, and then layers of smoother plaster were added, with the final *Scagliola* as the top-coat. Using woollen cloths, the surface is polished and finished with boiled linseed oil. The *Scagliola* method is still used today. It is also employed on an industrial level where chips of marble are mixed with concrete and then highly polished.

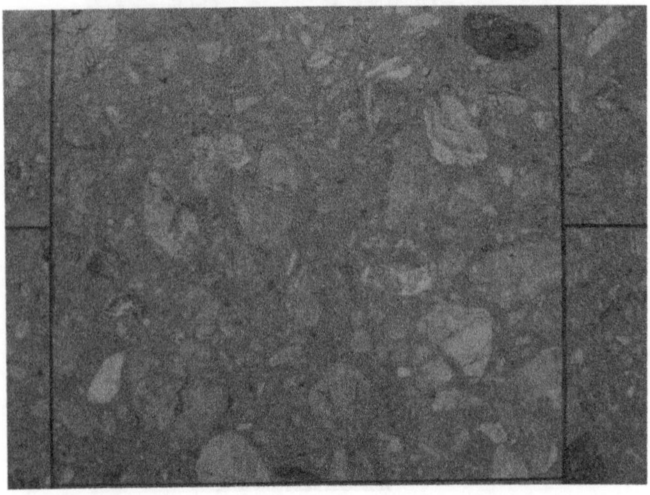

Figure 5.54 An example of a composite industrial marble.

The second approach is to create the appearance of marble using paint. Particular examples are painted columns and panelling in old houses, and scenery for theatres. A non-drying varnish is applied to a white ground, which maintains the fluidity of the subsequent layers of paint and to prevent them drying out. Then several layers of different colours are applied using sponges, brushes and feathers. The choice as to range of colours will depend on the desired effect: similar tones and colours will achieve a subtle rich texture whereas contrasting colours will result in a vibrancy of colour. To obtain the delicate veining of the marble, a final colour is applied using a fine paint brush and softened using a feather.

Depending on the viewing distance, different contrast effects may be employed, for example for theatre scenery, designers may use less detail but incorporate stronger highlights and shading to create a bolder three-dimensional surface that is visible to the audience. However, for a domestic interior where the elements are viewed at close proximity, a painter may use subtle colours and softer edges to achieve a more detailed appearance. These painted marble panels, columns and fireplaces were painted to resemble a generic marble or to blend with existing marble. Examples of this include mock-marble columns in seventeenth century Dyrham Park (Gloucestershire) (Figure 5.55, left) and eighteenth century Castle Coole (County Fermanagh, Belfast).

On a more contemporary level, in collaboration with the BBC, historians and specialist painters worked with The National Trust and Avebury Manor (Wiltshire) (Figure 5.55, right) to reconstruct nine rooms, each reflecting a different period of time originating from the time when the house was first built in Tudor times to include Georgian, Queen Anne, Edwardian, and early twentieth century. The specialists and restorers explored the history of the 450 year old house and its inhabitants, and explored ways the public might imagine how the families lived and the impact of the different periods and changing way of life (Evans, 2011). A room that was decorated as an early eignteenth century styled room included striking orange painted marble panels with red borders.

Early publications for artists provided practical advice on pigments, brushes and painting techniques. For example, Il Libro dell'Arte (*The Craftsman's Handbook*), written around 1390 by Cennino Cennini, offers advice on how to make paints and brushes and how to apply varnishes. These skills

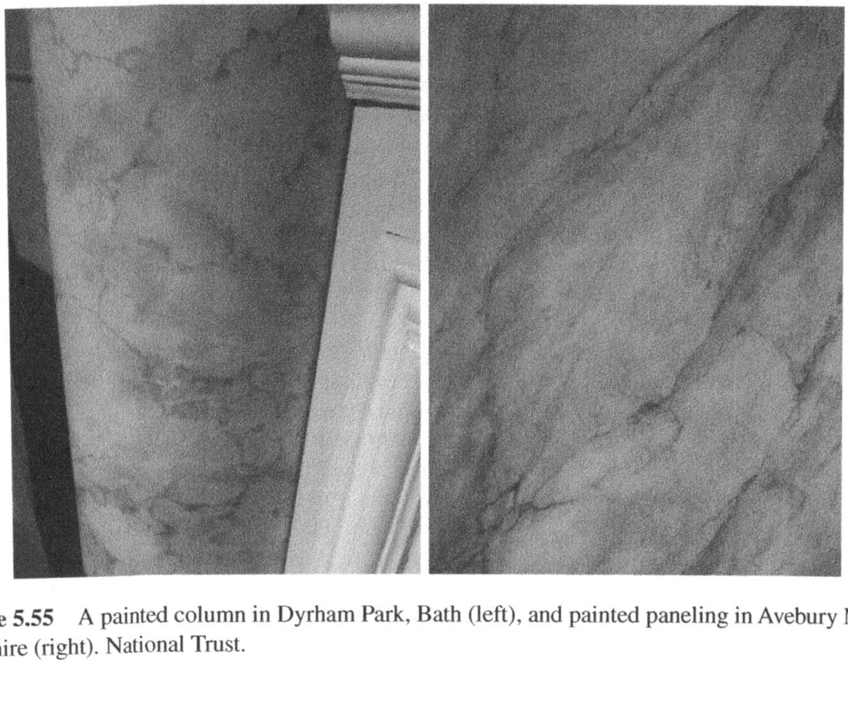

Figure 5.55 A painted column in Dyrham Park, Bath (left), and painted paneling in Avebury Manor, Wiltshire (right). National Trust.

Figure 5.56 A combination of original delft blue tiles and marble columns with painted sections to recreate a fireplace. The pediment and the flattened column against the wall have been painted to look like the marble column. Rijksmuseum, Amsterdam.

would traditionally have been taught by a master craftsman, but in the nineteenth and twentieth centuries these methods were replaced by *how to* books or at art schools. It is increasingly popular to learn about the craft and techniques on the Internet, such as through YouTube tutorials or Instructables. These films are often made by amateurs and enthusiasts (http://www.instructables.com/id/Faux-marble-concrete-vanity-top/step6/Special-effects). Michael O'Regan has studied and recorded artisan methods of decorative finishing techniques by interviewing and filming artists at work (https://asogenius.wordpress.com/about/; https://asogenius.wordpress.com/2015/02/28/vert-de-mer-faux-marble-a-tutorial-by-laurent-hissier). The National Theatre scenery department released a series of short films demonstrating how backdrops and decorative techniques such as marbling and woodgraining are created (National Theatre: How to create a *trompe L'oeil* painting, used for stone mouldings, https://www.youtube.com/watch?v=c_H7T-bLKEk&list=PL39C7B0CCE67102E3&index=2).

Likewise, a site by Hip Hut Deco provides more domestic examples and design inspirations (http://hiphutdeco.com/blog).

Some successful examples of the subtle appearance of marble, that can blend with all the other decorative elements in an interior, are shown in Figure 5.55. Another example is the fire surround in The Rijksmuseum in Amsterdam (Figure 5.56), which originally would have been complete in a domestic house. Aside from the delicate blue and white Delft tiles that would have surrounded the fire, the only real components of the display are the marble columns at the front of the fireplace. The pediment and the columns against the wall are painted to blend with the marble columns and are probably a reconstruction by the museum. Once one begins to notice, then of course these neither look nor feel like marble.

The third category comprises artworks made by artists to capture the appearance of objects in marble using paint. Figure 5.57 is one of a pair of paintings by Gerard de Lairesse and hangs in the Rijksmuseum,

Figure 5.57 Gerard de Lairesse (1641–1711), *The Allegory of the Sciences* (c. 1675–1683), oil on canvas, 289 cm × 161 cm (courtesy: Rijksmuseum, Amsterdam) (https://www.rijksmuseum.nl/en/collection/SK-A-4177); (right) detail.

Amsterdam. *Allegory of the Sciences* is a *trompe l'oeil* painting of marble that depicts a haut sculptural relief carving. These *grissailles* – pictures painted in shades of grey – show veins of marble running diagonally across the figures as if they have been cut from a solid block of marble. Like the domestic interior decorator, these paintings are remarkable for their subtlety in tonal range and *sfumato* – softness of edges. In both paintings, Lairesse has positioned the figures in a shallow alcove. He was very interested in pictorial illusion, and where and how paintings should be hung. One can imagine these positioned on either side of the room in a high-ceilinged Amsterdam canal house.

The fifteenth century master painter of material appearances, Jan van Eyck, used rich glazes of pigment and flicks of white highlights to increase the dynamic range and texture. The *grisaille* painting, *The Annunciation Diptych* (https://www.museothyssen.org/en/collection/artists/eyck-jan-van/annunciation-diptych), in the Thyssen-Bornemisza National Museum in Madrid, Jan van Eyck recreates the appearance of different types of materials, such as stone, marble and wood. These artworks are rendered and modelled using a subtle range of light and shade that convey a highly convincing three-dimensional effect and realism. The statues appear to be standing in a highly polished black marble niche, and to the right the reflections of the statues appear in the black marble.

A fourth category is the reproduction of marble digital through print. There are many examples of paper decorative laminates using offset, screenprint, and inkjet that provide a cheaper alternative to traditional marble. Classic materials such as Formica laminates were patented in 1913 as a brighter, cheaper and highly durable solution and was used extensively in diners, cafes and nightclubs. It became widely archetypical as the material of the mid-century café culture and *modern* domestic interior (https://www.formica.com/en/ca/about/our-history. As well as tabletops and work surfaces, contemporary laminates are used for a range of furniture, especially flat-pack furniture such as wardrobes and shelving units.

Figure 5.58 demonstrates a facsimile reproduction of flooring that is placed over a delicate mosaic floor to prevent further erosion or wear and tear. Instead of covering the floor with a different floorcovering and thus potentially changing the atmosphere of the place, these mats are colour-matched facsimiles of the existing floor that blends with the environment (http://www.eyemats.co.uk/eyemats%20Conservation%20Flooring%20Brochure.pdf). The process of capturing the entire floor involves using a lighting rig set at a fixed height and illumination; a series of images are then captured from a fixed height and stitched together. The image is printed on to a soft composite material of a knitted polyester surface bonded to a nitrile rubber backing, which protects the ground (mosaic, marble, wood, ceramic or encaustic tiles) from the impact of the many visitors' feet.

Significant developments in the ceramic printing industry have meant that inkjet printed images can be created on a grand scale using flatbed, superwideformat inkjet printing to produce hardwearing ceramic tiles and surfaces, which are used, for example, in hotels and reception areas to create the appearance of high-quality highly polished marble floors and walls. Inkjet ceramic decoration printers use piezoelectric inkjet printheads and ceramic powders. Because the wideformat inkjet hardware is able to print on such a large scale and at high resolution, there is no discernable repeat pattern. The Italian ceramic company Domus can print and produce tiles that range from 80 cm × 80 cm to 160 cm × 320 cm and a thickness of 6 mm, which can be cut to any size and shape and can be used on both walls and floors (http://www.domustiles.co.uk/whats-new/2016/02/domus-magnum-collection/;

Figure 5.58 Eyemat covering a mosaic floor at Tyntesfield Church. The actual mosaic floor can be seen in the top right-hand corner.

http://www.xaar.com/en/MediaDocuments/Xaar-Ceramic-Guide.pdf). The convincingness of these huge panels is demonstrated in the ability to accurately mimic the appearance and colour of many different colours and patterns of marble and surface finish to a uniform high gloss. These wideformat printing methods also minimises the waste involved in extraction of stone marble from the ground.

The reproduction of marble has evolved over many centuries. As the need for accurate reproduction has continued, likewise the enduring obsession for maintaining an aura of luxury and opulence has resulted in the development of many different methods of construction and materials to suit demands.

Case Study 9: Gold

Gold has been a constant material in the embellishment of illuminated manuscripts, paintings, relief surfaces and creating functional and decorative objects. Gold was an important component of sacred texts (http://www.bl.uk/onlinegallery/sacredtexts/golden.html; http://www.khalilicollections.org/all-collections/islamic-arts), and was considered to be a bridge between humans and the spiritual world (Behr, 2010); therefore books, words, texts and illuminated manuscripts provided that corporeal link. Craftspeople embellished paintings by adding gold to panels and relief surfaces, as exampled by relief sculpting in *pastiglia*. Artists have also captured the appearance of intricacies of gold on canvas and panel-work and have explored the dynamic range of lustre, shine, reflection of metalwork or glints of thread-work through pigment. Gold has been used extensively as decoration on Japanese and Chinese panels (Figure 5.59). The following images demonstrate different gilding effects by either the application of a thin gold leaf, as gold paint on to carved relief, as an outline into which colours are flooded, or engraving into panels of gold.

Cennino d'Andrea Cennini (1370–1427) wrote extensively on methods for gilding onto panel, stone, wool and velvet. He describes the methods for creating egg tempura and the level of humidity

Figure 5.59 Karamon (Chinese style gate) (1651) Toshogu Shrine Tokyo, Japan (top-left), Toshogu Shrine in Nikko (top-centre and bottom-centre, bottom-left), and Nijō Castle, Japan (right: top and bottom).

necessary for laying down gold leaf. He describes how to create decorative elements that will add surface effects to the *ancona* (panels or pieces of an altarpiece).

> 'When you have burnish and finished your ancona, you must start by taking the compasses; swinging your crowns or diadems; engraving them; tapping in a few ornaments; stamping them with tiny punches, so that they sparkle like millet grains; embellish with other pinches; and do stamping if there are any foliage ornaments … This stamping which I am telling you about is one of our most delightful branches. And you may do all-over stamping, as I have described; and you may model in stamping, so that, with imaginative feeling and a delicate touch, you may work out foliage ornaments, on a gold ground, and make little angels and other figures so thay [sic] they show up in the gold' (Cennini, 1954, pp. 85–86).

Possibly our most evocative encounter of gold and polychromic paintings is in a darkened room that is illuminated by a flickering candlelight. The craftsperson knew well the psychological effect of combining the yellow light from tallow candlelight, an oil lamp or stained glass upon gold and polychromic panel paintings. Under museum lighting, artworks are illuminated to provide a uniform experience of the painted panel and a gold surround. However, in original places panels were illuminated by lamplight. Artists added tiny radiating lines and complex patterns in the gold, so that, when illuminated, the patterns, lines and pin-pricks became dazzling, resulting in a contrast between the halo and the painted figures. One can imagine the psychological effect and impact on a devotee when confronted by the gilded artworks in a darkened church. Their perception of the relationship between the surrounding of gold and the painted figures, as the gold halos or frame shone brightly, would have brought the faces and subjects into sharper focus and the experience would have been far more evocative than today.

The demonstration surface relief and physicality in a painting was sublimely exampled in the work by Carlo Crivelli (1435–1495) in the fifteenth century. He was well known for illusionistic painting of materials and objects, and, for example, of *cartelino*, or small pieces of paper that were painted as if attached to a wall. Crivelli's paintings were also highly decorated with gold and in many cases heavily embossed with ornament. *Pastiglia* is a bas-relief that is made from a paste of gesso (chalk, white pigment and animal glue as the binder), which was either used as a ground for gilding on paintings or to make moulded picture frames. The *pastiglia* was applied to rigid surfaces and the relief, in combination with gold leaf, was used to give a surface effect, imitate precious metalwork and rich velvet brocades. As the gilded *pastiglia* caught the light, it enhanced the illusion of three-dimensional forms. For example, the decorated beads in relief around the neck of Saint Stephen (1476) in The National Gallery, London (https://www.nationalgallery.org.uk/paintings/carlo-crivelli-saint-stephen) appear smooth and glassy, whereas the gold background is stamped (as similarly described by Ceninni) and appears to be textured and woven. The cloak around St Peter (1476) (https://www.nationalgallery.org.uk/paintings/carlo-crivelli-saint-peter) is edged with gilded rope. In the *Triptych of the Virgin and Child with Saints Peter and Paul*, Crivelli has incorporated many different relief elements. The example in Figure 5.60, and the image of the artwork taken at an angle, shows the height of the relief of the rope-work, the clasp around St Peter's cloak, and the keys hanging from his gloved hand. The heavy looking keys are in the highest relief and appear as if they could swing free from the picture. St Peter appears to be physically weighed down by these gilded objects.

One can now begin to imagine the mix of pigments and maybe materials that would be required to create such an effect in a reproduction, assuming the necessary workflow and the gamut of the printing device can handle the original effects of the painting. Gilding on its own is a delicate process, and to record and mimic the details achieved by the incisions and punches on the surface of the gold leaf, which seem to 'play' with the light as it strikes them, would certainly be a complex task. Moreover, the technique of water gilding goes even further by creating gold-red reflections due to the transparency of the gold leaf and the redness of the clay used as the base. This technique can also be seen in Fungai's work in the yellow inner dress of the Virgin. Again Cennini describes the process, 'if you want to make

Figure 5.60 (left) Carlo Crivelli (1435-1495), *Triptych of the Virgin and Child with Saints Peter and Paul* (1482) (courtesy: Pinacoteca di Brera). (right) The surface relief is captured at an angle, showing the detailed pastiglia elements of Saint Peter.

a cloth of gold, lay the gold with bole, and scratch in the folds of the drapery … then if you want to make the cloth red, lay in vermilion over this burnished gold' (Cennini, 1954, p. 86).

Possibly, where we see the most compelling examples of the illusion of gold in paintings is when artists have used pigments to create a subtle yet highly effective appearance of this precious metal. There are different reasons for mastering certain effects on a flat media, which could either give the illusion of depth, perspective and volume, replace the cost of gold with inexpensive pigments, or to mimic the appearance of materials and match the illumination of the painting in a site-specific environment. In the work by Bernardino Fungai (1460–1516), *Virgin and Child with Cherubin*, in The National Gallery, London (Figure 5.61), there are sections where the artist has combined gold leaf and pigment to create a gold dynamic range, firstly by imitating gold in the shadow regions using pigments, which also serves to enhance the brighter gilded sections. This can be seen in the white drape of the Virgin, such as the brighter areas across the knees and the sleeve on the left, where the embroidery is gilded, whereas in the shadow sections, the embroidery is painted in a brown-orange colour to simulate a darker gold.

Fungai modified the classic *Sgraffito* gilding technique to achieve a better rendering of shadows that gives fluidity and a three-dimensional appearance to the drape. Before gilding, he separated the drapery into highlights and shadow regions. The gold leaf was applied only to the highlights and a brown-orange pigment on top of a greyish background in the shadow areas. When the painting is observed from different angles one can see the transitions between the metallic reflections of gold and the dull reflections in the shadows, which give movement to the folds in the fabric. Under candlelight, the effect would be more startling: the glints of the gold would have shimmered, thus creating the appearance of chiarscuro and three-dimensional modelling.

Other examples include *The Annunciation* (1434–1436) by Jan van Eyck (Chapter 2, Figure 2.23), who used pigment to simulate the threads of gold on the cloak of the Virgin, and the painting by Willem Claeszoon Heda (1594–1680), who explored reflection and surface decoration of gold and other metals. Heda has accurately captured the ornate engraving into metal, and the gleam and lustre of different objects (Figure 5.62).

Figure 5.61 Bernardino Fungai (1460–1516), *The Virgin and Child with Cherubim* (circa 1495–1510), Tempera and oil on wood (https://www.nationalgallery.org.uk/paintings/bernardino-fungai-the-virgin-and-child-with-cherubim) (reproduced with permission: National Gallery, London).

Figure 5.62 Willem Claeszoon Heda (1594–1680), *Still Life with a Gilt Cup* (1635), oil on panel (88 cm × 113 cm) (courtesy: Rijksmuseum, Amsterdam).

The last example is the work by Cosimo Tura (c.1431–1495) we see similar effects applied to textured fabric. The impression of velvet and gold in A Muse – Calliope (1455–1460) in The National Gallery, London (https://www.nationalgallery.org.uk/paintings/cosimo-tura-a-muse-calliope), demonstrates even without gilding, the illusion of materials and complex reflective patterns. The velvet parts of the sumptuous brocade of the sleeve are created by a raised pattern with shades of red lake and the highlight of the golden areas is achieved with a layer of pure lead-tin yellow. We can compare this polychromic example with colour management workflow methods for rendering devices, where we attempt to approach a given colour by proposing a colorimetric or a perceptual solution that is made up of a combination of existing colours (pigments, filters) in the system. Let us remember that colour in rendering devices is usually defined for a single viewing geometry and does not take into account other reflection functions such as gloss. Therefore, even if we can achieve colour accuracy in some cases (usually under a 45/0 geometry), the reflections of different viewing conditions would be hardly guaranteed by this workflow, especially highly interreflective and complex surfaces such as gold, silk and velvet.

Printing Gold in the Digital Market Place

Whilst e-readers are enjoying widespread popularity and have addressed the problem of a luggage full of paperback books, there is a renewed interest in collecting ornately illustrated hardback books. Publishers have dispensed with the book jacket to display cloth bound books with gold-embossed designs. Likewise, in a highly competitive market, the addition of gold, embossing and decorative features to packaging are being used to enhance a luxury product.

The majority of gold effects are created using blocking, hot and cold foils and lamination processes (http://www.kurz.de) (Figure 5.63), which are incorporated as part of the finishing of prints. A gold ink layer can also be added as a spot colour, and printed as an extra plate to the CMYK process.

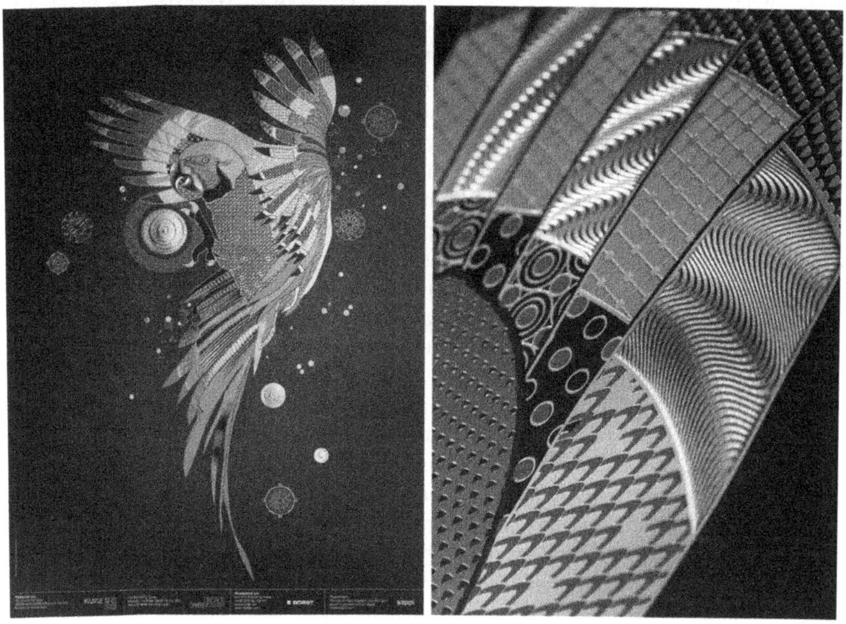

Figure 5.63 Different gold effects using blocking, hot and cold foils and lamination (courtesy: Kurz).

However, for short run and digital printing, foiling still remains expensive, due to the production costs of blocks and dyes. The current speed and quality of inkjet printing gold onto artworks is slow and less convincing, due to the need to build a thicker deposit and pigment size. Digital inkjet solutions include: printing a silver layer and applying a CMYK colour on top to create different colour lustre effects (Roland DG), a gold *dry* ink is the last layer (Kodak NexPress) to be added, to create different gold effects; printing on to a silver or pearl card or foil (HP Indigo ElectroInk, Epson); or UV cure printing on to metal surfaces (Mimaki, Roland).

An alternative printing approach, by combining gold pigments and lustres, can be used to create different surface effects. For example, Merck (http://www.emd-performance-materials.com/pm.country.us/en/images/W578281_EMD_MERCK_EFFEKTE_15_02_tcm2080_146166.pdf?Version=) have introduced VE3D, a virtual 3D embossing process that uses flexographic printing to create the illusion of an embossed surface. The effect is quite realistic and tricks the eyes into believing the surface is embossed, but to the touch the surface is smooth. The process employs four steps: first, an image or background colour is printed; the second step involves the application of a layer of Merck Iriodin Silver or Gold varnish; the third step uses a patterned polymer plate, which is applied to the still-wet and uncured pigment varnish; fourth, the plate is lightly (kiss) printed, which despite the lack of pressure, creates the impression of a raised surface. Once the surface is exposed to UV light, the illusion of embossing or surface relief is apparent and both positive and negative embossing can be combined in one image (https://www.merckgroup.com/en/news/schmid-rhyner-ag-cooperation-3d-printing-02-05-2017.html).

The Stafford Hoard and Gold Printing at The Jewellery Industry Innovation Centre, Birmingham Jewellery Quarter

3D scanning and printing has presented new opportunities for the recreation of facsimiles and replicas, and to gain new insights as to how artefacts were made, and their complete appearance before they were damaged. One example is the recently unearthed Staffordshire Hoard (http://www.staffordshirehoard.org.uk/; http://www.tctmagazine.com/tct-events/formnext-powered-by-tct/recreating-the-staffordshire-hoard-with-3d-technologies), found by a treasure seeker and his metal detector whilst walking across farmland. Of the 1662 objects that were eventually recovered, most are gold, some are silver and many decorated with precious stones. The seventh century Anglo-Saxon metalworkers' knowledge of such intricate soldering and wirework tells us a lot about their skill, but according to Frank Cooper at The Jewellery Industry Innovation Centre (http://www.bcu.ac.uk/research/-centres-of-excellence/centre-for-design-and-the-creative-industries/jiic) in the Birmingham Jewellery Quarter (BJQ), they can only guess as to how these craftsmen achieved such high-quality metal work. This hoard demonstrates how Anglo-Saxon people were making the finest examples, and which would be difficult to surpass. 'The quality of the craftsmanship displayed on many items is exquisitely delicate given their generally warlike nature' (http://www.bcu.ac.uk/cmsproxyimage?path=/_media/docs/frank-cooper-report.pdf). Of course, many of these artefacts are damaged, broken, or parts are missing. When asked to reconstruct a gold cross, different CAD programs to undertake 3D scanning, reconstruction and prototype printing, alongside consultation with contemporary jewellers and goldsmiths and historians was necessary to gain insights as to how the cross was originally made. As a religious object or an object associated with battle, the question for Cooper and his team was could they adequately recreate the aura of this historically significant artefact? *The making of the replica crosses needed to give a visualization to the modern, contemporary observer so that peoples from two entirely different ages could share the same vision. To achieve this we needed to make the historical connections between the seventh century Anglo Saxons, the nineteenth century Birmingham Jewellery Quarter and the twenty-first century BJQ, and link them together with craft skills that have spanned these ages and the most up-to-date technologies available to us. By using traditional observation and analysis methodologies and coupling them to the available CAD and prototyping technologies, we attempted to build a bridge that spanned this vast gulf of modern time, linking their 'then' to our 'now'* (http://www.tctmagazine.com/tct-events/formnext-powered-by-tct/recreating-the-staffordshire-hoard-with-3d-technologies).

Case Study 10: Exterior Decoration Tiles and Ceramics

Relief has been an important element across architectural styles, interior and exterior decoration. We can see many examples, influences, styles and genres across Europe. The following examples include the cities Sevilla, Barcelona, Lisboa, Porto, Milan and Vienna, which demonstrate Moorish architecture of the first centuries, and to Art Nouveau at the end of nineteenth century, as well as the incorporation of relief elements as decoration in both interiors and exteriors. Made in relief with plaster, clay or wood, Moorish architecture is rich in geometrical forms. Between the eighth and fourteenth centuries during Islamic rule of the Iberian Peninsula in Spain, we can see many examples. During this period many Alcazares (Moorish castle or palace) were built in the peninsula, especially in the region of Andalucía, where we find the most notable examples.

Not only relief and carved shapes were prominent in Moorish art, but also rich gilded designs are found inside mosques and palaces. In the cathedral of Cordoba, a former mosque in the Moorish period, there is an exquisitely decorated horseshoe arch that serves as the entrance of the *mihrab* or prayer's niche (Figure 5.64). The arch is decorated with gold, glass and colour tesserae, containing dazzling combinations of black, blues, browns, reds, yellows and gold, depicting intricate calligraphic bands and plant motifs. Above the *mihrab* there is a richly decorated ribbed vault enclosing a patterned small dome, also adorned by golden tesserae. Similar artistic expressions are found in the *Alcazar* in Seville,

Figure 5.64 (left) Mihrab, in the Mezquita-Catedral, the Great Mosque of Cordoba. (right) Moorish architecture in the Alhambra (Granada).

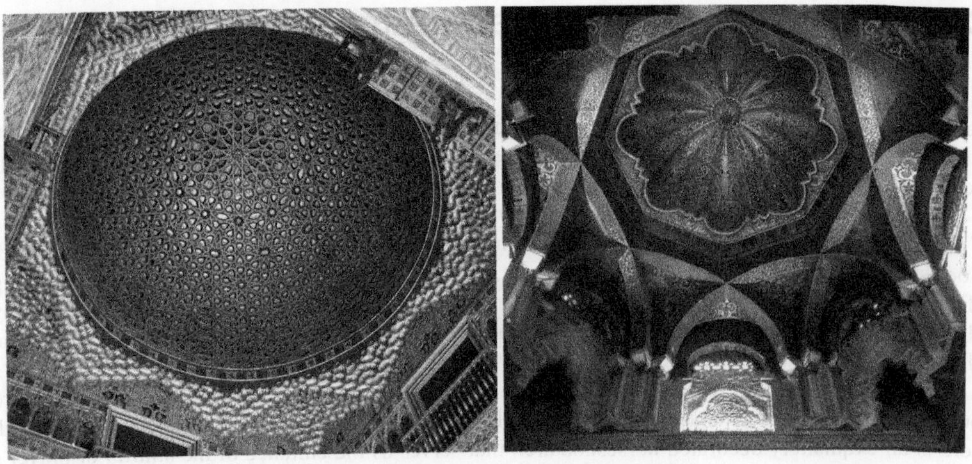

Figure 5.65 (left) The stunning gold ceiling of Alcazar in Seville. (right) Dome of Mezquita-Catedral, the Great Mosque of Cordoba (gold ceiling of Alcazar) (credit: James Gordon (CC BY 2.0; http://creativecommons.org/licenses/by/2.0), via Wikimedia Common).

a former royal palace, where we can see a wooden dome richly ornamented. The dome is so bright it gives the impression of being back-lit (Figure 5.65).

Towards the last centuries of the Moorish era in the Peninsula, the use of tiles gained in popularity (see Figures 5.66 and 5.67). Different styles of tiles or *Azulejos* are still seen in the facades of downtown Porto, exhibiting distinct methods of production (both artisanal and industrial) (Figure 5.68). *Azulejos* demonstrate the balance between the looked-for duality of ornamental and functional objects. Since the thirteenth century, tiles in Portugal and Spain, as well colonies in Central and South America, were used to decorate facades and to control the temperature of houses. Due to a large demand from

Figure 5.66 Casa dos Bicos (House of Spikes), Jose Saramago Foundation Lisbon. Built in the early sixteenth century with a Renaissance and Manueline (Portuguese late Gothic) influence.

Figure 5.67 Arista tiles (pre-moulded tiles, made by pressing clay against carved wooden moulds) in the Alcazar in Seville, sixteenth century, influenced by Islamic art and the Renaissance period.

the Americas, the production of tiles continued through the seventeenth and eighteenth centuries. Polychrome, blue-and-white tiles and a transfer-print method became fashionable in Europe during the nineteenth century, where industrial methods produced simple, stylised designs. The National Museum of Tiles in Lisboa provides visual delight as well as a fascinating historical background

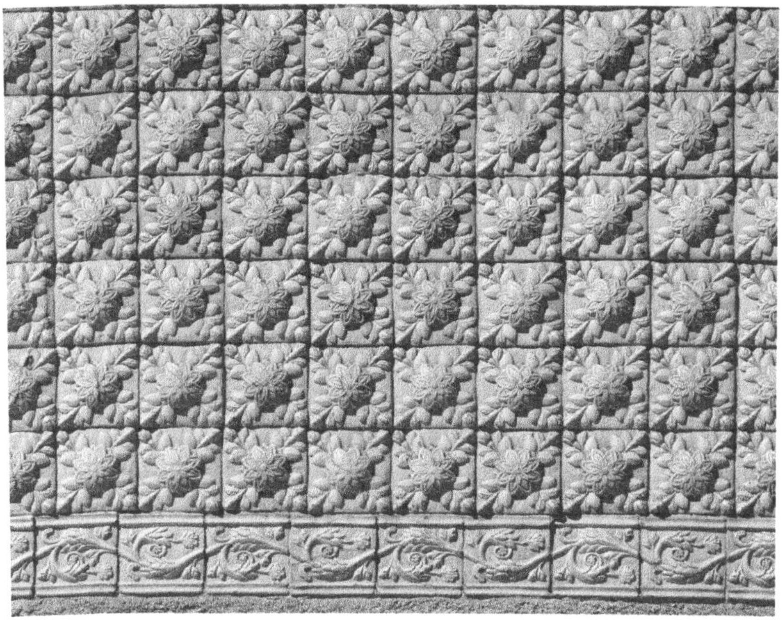

Figure 5.68 Modern Azulejos in relief from the last decades of the nineteenth century (Manufactura de Massarelos, Porto) (licence: Adobe ST © Karol Kozłowski).

(http://www.museudoazulejo.pt/en-GB/default.aspx). Interesting examples of Portuguese contemporary relief glazed tiles can be found at Mambo Unlimited (http://www.mambounlimitedideas.com/?id=60).

The Art Nouveau style towards end of the nineteenth century extended the stylistic design and use of tiles on building facades. Well-preserved examples of Art Nouveau and Art Deco and their derivations (Secession, Liberty Style) of the end of the nineteenth and beginning of the twentieth centuries can be found across North America and Europe (see Figures 5.69 and 5.70). Art Nouveau in particular was an important demonstration of rich decoration in relief that was afterwards ‘flattened’ during the period of Art Deco as a way of producing cheaper and streamlined products that were more suited to the more austere wartime decades. Art Nouveau artists adopted new materials and machined surfaces to achieve dynamic, undulating and flowing lines in a response to the maturity of the industrial revolution.

Sgraffito is a decorative technique for creating wallpaper-like facades on exterior walls that has been in use since the fourth century BC (Figures 5.71 and 5.72). It was prominent during the Renaissance in Italy and Art Nouveau architects revised the process in the twentieth century in Catalonia and Vienna. The process involves two (or several) layers of paint or plaster of differing colours. A negative repeat

Figure 5.69 Secession Building, Secessionsgebaude (1897), Vienna, Austria. Specifically ‘secession style’, a branch of Art Nouveau. (right, licence: Adobe ST ©dbrnjhrj).

Figure 5.70 Majolikahaus, one of the finest examples of Art Nouveau in Vienna, designed by Otto Wagner in 1898. The facade of the house is created with colourful ceramic tiles decorated in floral motifs. (WikiCommons/Creative Commons 3; photo credit: by Thomas Ledl) (https://commons.wikimedia.org/wiki/File:Majolikahaus_Detail_2.jpg).

Figure 5.71 Examples of exterior wallpaper in Segovia.

Figure 5.72 Examples of exterior wallpaper in Barcelona. Casa Amatller (left) and Casa Battló of Gaudí (middle) on the Passeig de Gràcia, and Plaça De Les Olles (right).

pattern is etched into the external layer (similar to the appearance of Devoré in textiles), revealing the coloured layer below.

Ceramic tiles in Spain are also combined with relief to suggest ornamental frames, patterned tiles, and hand-painted advertising. The following example is a pharmacy on a street corner in Madrid; the exterior is extensively clad with decorative tiles and hand-painted ceramic signs advertising different remedies that are available in the shop (Figure 5.73).

Twenty-first Century Ceramic Tile Inkjet Printing

Digital inkjet printing is beginning to significantly impact the ceramic tile industry. Traditional industrial methods of pattern printing on to tiles have involved screenprinting colour separations on the surface of tiles. The major disadvantage of screenprinting is consistency of mixing and matching colour between batches and print runs, and therefore the overall printed and fired colour may appear uneven. Other factors include the overall appearance of the tiles. Colour separation from an original design may result in desaturated or flat images, and due to the constraints of pattern size and variation of patterns may result in an overly repetitive appearance. Lastly, during the screenprinting process, as the squeegee and screen come into contact with the tile, any uneven pressure or uneven tiles may result in breakages.

Inspired by their Spanish tile heritage, Valencia-based Entic Designs have reinvestigated the nineteenth century manufacturing processes and are incorporating traditional patterns and contemporary designs to create a range of coloured cement tiles, or encaustic tiles Figure 5.74. Their contemporary tile collection includes the *Keidos* range, which is a hexagonal cement tile, and *Drops*, which is inspired by Japanese Koi Carp and uses a concentric circle motif to suggest ripples on water. Entic Designs also offer an online bespoke pattern builder, and the user can create a pattern, choose colour. The software can also simulate the overall appearance of the tile pattern as a complete wall (http://enticdesigns.com/?lang=en).

The term encaustic means *to burn in*, which describes the method of inlaying coloured clays into the surface of tiles. The tiles are built upside down, the colours that are laid during the first stage then appear on the top. The pattern grid is similar to a metal cookie-cutter that is divided into different sections, into which the different colours are poured or pressed. A layer of cement is then applied on top of the mould, smoothed and subjected to hydraulic pressure. This forms the base of the tile. The frame is

Figure 5.73 Painted ceramic tiles in Madrid, showing a combination of lettering, relief and advertising.

removed to reveal a smooth multicoloured tile (https://www.youtube.com/watch?v=1RsIEGKNNmU). There are many artisan and heritage companies involved in this labour-intensive process, both for the production of new tiles, as well as the restoration of churches and nineteenth century buildings. The colour, density of the pigments and richness of the glaze cannot be surpassed. The company Craven Dunnill Jackfield has been producing ceramic wall and floor tiles for more than 140 years, and at the heart of the Iron Bridge Gorge in Shropshire and part of a World Heritage Site is the oldest surviving purpose-built tile factory in the world (http://www.cravendunnill-jackfield.co.uk).

As opportunities for digital methods for the printing of ceramic pigments (see Case Study 8: Marble) is beginning to increase, contemporary tile production has brought about new and exciting opportunities for the designer. Digital printing has also increased the variety of patterns, so that any number of different patterns can be created and printed, therefore reducing the obvious and ubiquitous repeat pattern. The wood effect unglazed ceramic tiles, produced by Johnson Tiles, are particularly effective in both creating the appearance and texture of aged and distressed wood panels (http://www.johnson-tiles.com/catalog/products/loft/beige-mix/natural/tile). Italian company Mutina have also showcased a range of textured and bas-relief tiles (http://www.mutina.it.) Resembling the encaustic tiles, designer Patricia Urquiola has designed a contemporary Azulej pattern. These tiles are produced from a combination of hydraulic cement, porcelain stoneware and digital printing technologies. The patterns range from the pixelated, geometric and organic floral to lace compositions (Figure 5.75).

The creation of three-dimensional or photo-relief pictures has its roots in nineteenth century photomechanical print processes, in particular the Woodburytype process, which used gelatine to create a continuous tone surface topology (McCallion, 2017). Lithophanes, Woodburytypes and relief tiles rely

Figure 5.74 Hydraulic tiles or encaustic cement tiles (courtesy: Entic Designs).

on the varying thickness of the material in relief to obtain a greyscale picture as light is transmitted. Inspired by the glazed portrait ceramic tiles by Stoke on Trent painter and designer George Cartlidge (1868–1961), the team at the Centre for Fine Print Research (Stephen Hoskins, David Huson and Paul Thirkell) investigated the process of relief decorative ceramic tiles using digital technologies (Huson, Hoskins and Thirkell, 2005). The fine examples of Cartlidge's work can be seen in the Potteries Museum and Art Gallery, Stoke on Trent. Similar to the coin and medal designers, Cartlidge translated images and portraits into a physical relief by hand-sculpting the layers. However, unlike the coins and medals, which employs a range of bas and etch relief to obtain low relief, Cartlidge's method was different. The characteristic relief of Cartlidge's portrait tiles are more akin to lithophanes (https://www.lithophanemuseum.org), where the different depths of the ceramic correspond to different shades of grey. The lithophane image was originally carved in wax, from which a plaster mould was obtained. A slip porcelain was poured into the mould, dried, removed from the mould and kiln-fired. When the relief panel is held up to the light, a composite picture generated from the varying thicknesses of the ceramic is revealed. Cartlidge adapted this method by flooding his tiles with a translucent coloured glaze that, when filled with coloured glaze, the deepest areas of the relief correspond to the darkest parts of the image and the highest

Figure 5.75 Examples of contemporary encaustic tiles, designed by Patricia Urquiola, inkjet printed on to porcelain stoneware.

parts of the relief correspond to the lightest parts of the image. This effect is similar to receding steps in coloured water, as observed and described by Johann Wolfgang von Goethe (1749–1832), which he called 'intensification' (Goethe, 1970).

Working from Cartlidge's hand sculpting of relief, David Huson and colleagues at the CFPR explored the use of digital technology for the creation and conversion of images into relief surfaces. The digital method translated the different tones of a greyscale image into 256 different heights. The 2.5D relief image is CNC milled using G-Code that plots a tool path for the cutting tool, which removes material from modelling board to create a mould. Following the mould-making process, a ceramic slip is poured into the mould and allowed to dry. Once dry, the ceramic is biscuit fired (first stage). The low relief tile is filled with a lightly coloured glaze and fired again. The result is a tile where the continuous tone image and glaze are integral to the tile. The image in Figure 5.76 is an example from an exhibition entitled *Altered Images* (2004), which shows tiles by contemporary artists alongside the works by George Cartlidge (Thirkell, 2004). However, the process that we developed went through three different stages: first we machined model board, took a silicone rubber impression and made a plaster mould

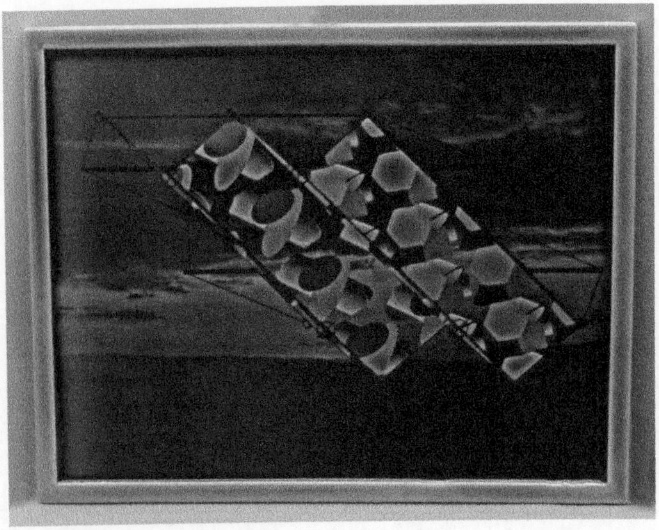

Figure 5.76 Stephen Hoskins, glaze relief image tile kite design (2004) glazed ceramic (244 mm × 19.8 mm) (courtesy: Stephen Hoskins, CFPR).

from this for slip casting. We moved on from this and cast a plaster block, which was machined directly on the CNC. Finally, at the end of the project we moved on to machining directly into dust-pressed ceramic tiles.

Creating multicoloured images using these processes has remained problematic due to the accurate registration of multiple layers of colour. Researcher Peter McCallion at the CFPR, has explored different nineteenth century plastographic processes and in particular the continuous tone qualities of the Woodburytype. The traditional Woodburytype used lead as a matrix from which a gelatine image was cast. McCallion has undertaken a reappraisal and has used a combination of CNC milling to create the master mould and silicone from which the gelatine is cast. In order to create a full colour image, he needed to make three silicone moulds for the magenta, cyan and yellow channels, where each gelatine layer is coloured, cast and then applied wet-on-wet using a highly accurate pin-registration method (Figure 5.77). When the three layers are first cast, a significant relief can be observed but, once dry, these layers flatten to create a full colour image.

Figure 5.78 shows a detail of the print created by combining the three silicone plates (Figure 5.77). The detail on the left shows how the colours are more vibrant when wet, and the gloss and relief is visible. As the print dries the relief flattens, the image sharpens but the colours are less saturated. Although the Woodburytype is not yet a suitable medium for commercial use, the significance here is that by translating photographic tones into a physical height presents alternative ways for translating images

Figure 5.77 Peter McCallion's silicone printing plates in preparation for wet-on-wet gelatine three-colour printing. Left to right: cyan, magenta, yellow (2016) (courtesy: Peter McCallion, CFPR).

Figure 5.78 Detail (left) of a wet-on-dry layer print showing the last layer in relief and, once dry, the final image (right) (courtesy: Peter McCallion, CFPR).

that move away from halftoning and stochastic screening towards creating permanent images with materials, such as glass and other translucent materials.

Case Study 11: Microstructural Texture

A viewing environment is a way to describe the conditions under which any object is observed. The environment comprises the illumination, the observer and the position of the object in relation to the direction of light and the observer. The distance between the observer and the object determines how well details and textures are perceived (macro- and microstructural textures). As suggested, the macrostructures provide the larger, visible components, whereas the microstructural textures are composed of tiny details. There are established standards as to how macro- and micro-structures are understood in relation to the viewing distance under a microscope. For example, according to metallurgy, a structure is considered to be micro only if it is visible by magnifications of 25× or greater, and macro if it is revealed by magnifications of 10× or less, but there is not a clear definition between the two magnifications.

In some cases the unit of a microstructure may be the same as the minimal unit of the surface (a pixel, texton, atom) but in other cases the microstructure unit may be composed of several minimal units (a sort of molecule). When considering what constitutes a relief, it is important to differentiate between near-planar surfaces and printed, machined or extruded layers. We could suggest these parameters are a function of viewing distance, application and intent. If we think about reproducing microtextures and macrotextures, we need to consider the relationship between these structures and the physical capabilities of the resolution of the device. For example, based on the tolerances of a machine tool or a printed layer, there may be some microstructures that cannot be reproduced due to the specifications of a particular machine.

At the Rolex Learning Center, at the École Polytechnique Fédérale de Lausanne (EPFL) (known also as the Swiss Cheese Building), there is an interesting example of near planar surfaces in metal that contain a relief (Figure 5.79). The installation is composed of four tilted silver square plates, that at first glance appear to be a smooth surface. Under closer inspection we notice some subtle aberrations on the surface, and with the addition of sunlight on the plates, reveal four faces that are reflected upwards on to the underside of the roof of the building. The discrete variations on the surface make use of the 'etched' effect seen in water, glass and metal reflections. The reflected faces contain enough detail to actually recognise the features. In optics, a *caustic* or a caustic network is a light phenomenon where

Figure 5.79 Solid installation (EPFL). Reflections of Marie Curie and Alan Turing can be seen on the underside of the roof of the building. Science shapes light to create a work of art, Sandy Evangelista (http://m.epfl.ch/news/25838/).

light patterns are created by liquids, glass and metals by refraction or reflection of light, usually considered as a random phenomenon.

The Computer Graphics and Geometry Laboratory (http://lgg.epfl.ch) in EPFL has developed a method to control such etched light patterns to form a given shape, by optimising the geometry of reflective or refractive surfaces. They have developed an algorithm that computes the shape of surface, so that light reflected and refracted off the surface results in a desired light pattern. The algorithm calculates the bumps and dents needed in a surface to produce an image as a function of parameters such as the direction of the light source, the focal distance and some material properties. Basically, it concentrates light in bright regions of the image and diverts the light in the darker regions. Once the surface has been computed using the algorithm the plates are fabricated. Following the instructions driven by the algorithm, a high precision CNC milling machine removes extra material from a metallic block. To obtain the final etched image, the plate is polished to remove any rough parts (http://solid.rayform.ch/).

Comparisons can also be made with the intriguing so-called *Magic Mirrors* or diaphanous mirrors, which were made in China and Japan, and used in Shinto shrines and Buddhist temples. Bronze mirrors were produced in large quantities during the Han Dynasty (206 BCE–24 AD) in China. In Japan, bronze mirrors are known as Magic Mirrors, or *Makkyo* (魔鏡) or Ghost Cave (Figure 5.80). The mirror is

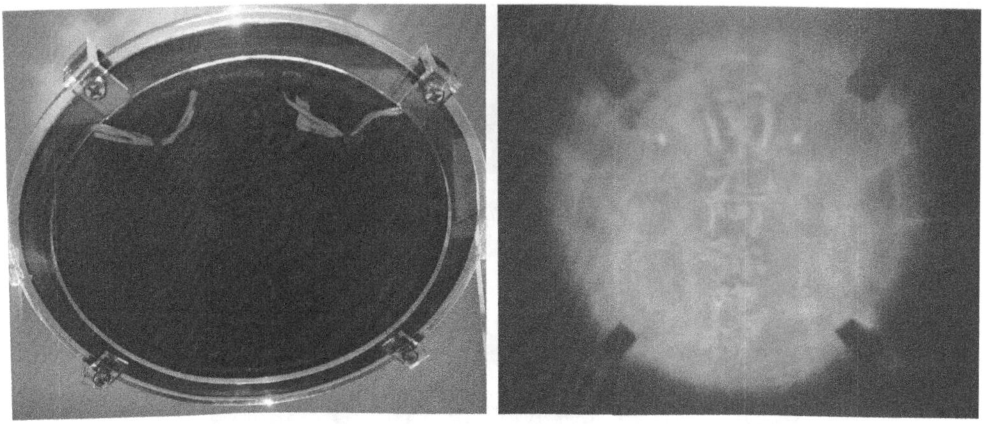

Figure 5.80 Makkyo or Magic Mirror, Tokyo National Museum, Japan. A bright light is shone on to the smooth side of the disk, but the image that appears on the facing wall is of the Japanese characters that are etched into the reverse.

made out of thin polished bronze of 5 mm with a thick outer rim. One side is smooth and highly polished, while an embossed design is cast into the reverse side. When light is directed on to the polished face of the mirror, the image from the reverse is magically reflected on to a wall. Although the metal is solid, the reflected image creates the impression of translucency. The secret is in the thick Kanji characters that are cast on to the reverse. When the smooth side is polished, this causes the mirror to become very slightly deformed to create minute convex variations that correspond to the outline of decoration on the reverse. As there is no focal point, the reflected image becomes larger as the distance from the mirror screen increases.

Another reflection-based example of subtle surface changes is found in stained and cathedral glass applications. Glass is one of the world's oldest materials. The use of coloured glass dates back to the Romans and Egyptians, who developed methods to apply glaze to objects. In churches and monasteries, stained glass windows have been present since the fourth and fifth centuries, gaining popularity during the medieval period in the form of narrative windows with panels to teach religious stories and decorate churches.

Glass can be coloured by adding metallic oxides and salts to molten silica sand. The mix is melted at high temperatures (1700 °C) and, as the molten silica cools, it is *floated* to obtain flat sheets of glass. The coloured sheets of glass are cut into small pieces, arranged into patterns or pictures and then held together by lead strips or *cames* and are soldered at the joints. Another way of colouring glass is to paint on the glass using vitreous enamel, which is then fired at a high temperature in a kiln, thus increasing saturation, depth and dimensionality as light shines through.

Glass reflects, transmits and refracts light. A unique aspect of stained glass is that its appearance changes according to the weather, season and time of the day. There is also a pleasing contrast between the colour pieces and the dark lead cames that hold the pieces together as a mosaic arrangement. The cames cement the different shapes together, but they also work as a perceptual framing that creates a contrast to the colours. The stained glass designed by Antoni Gaudí (1852–1926) for the Sagrada Família are cut pieces, which, when composited, form an abstract pattern of colours and shapes (Figure 5.81).

We can also see a texture on the glass itself (known as cathedral or architectural glass, often incorrectly applied to windows in cathedrals). Cathedral glass is a sheet of uncoloured glass with a certain texture on the surface (wavy, bumpy, etc.) that is used for decoration and privacy (Figure 5.82). The texture can be obtained by using moulds or by mechanically manipulating the hot glass sheet while it is

Figure 5.81 Antoni Gaudí (1852–1926) Stained glass in The Sagrada Família, Barcelona.

Figure 5.82 Antoni Gaudí (1852–1926) Glass window at Park Güell, Barcelona.

still in a pliable state; for example, ripple textures are created by stretching the glass sheet before it cools.

The texture in cathedral glass can be understood as a set of tiny lenses that reflect light in different directions. Measuring such lenses and controlling their creation during fabrication would be another challenge for the relief and 3D printing workflow.

Few labs are working on fabricating or printing glass. Achieving the target degree of transparency with 3D printers requires changes to the chemistry of glass and a procedure to find the right recipe. The transparency of kiln glass depends on the size of the particles of glass powder or frit. Even though fine frit produces better details in the glass object, it also traps air bubbles in the object that scatter the transmitted light and turn the object opaque.

Working on a microscale, the Dutch company Luximprint adds a new dimension to the 3D printing landscape by introducing 'Optographics' – a unique combination of optical and full-color 3D printing (Figure 5.83). The company offers optics design and manufacturing services for fabrication of functional, inspirational or decorative technical plastics. Most of the printed parts are used for demonstration purposes, fabrication tool validation, functionality testing, or art and decoration applications. When combined with daylight or artificial light, the printed structures are transformed, and can be used as an engineering tool or artwork.

Scientist Susanne Klein, when working at HP Labs, was part of the project RAGNAROK (Research on Advancing Glass and Nonorganic Applications to Recreate Objects and Kinetics) and also worked on 3D printing of glass. Thanks to printing, the mould for firing objects in glass becomes obsolete but the resulting object does not show the same transparency as achieved with traditional methods due to issues during extrusion of the frit and its sizes (Figure 5.84). Not only transparency is affected by the frit size but also the colour may shift during heating. This shift is unpredictable, differs from batch to batch, and gives inconsistent results when measured in transmittance and reflectance. This area however, certainly presents a potential for further research. (Klein *et al.*, 2015).

Figure 5.83 3D printed functional and decorative optical plastics (courtesy: Luximprint) (www.luximprint.com).

Figure 5.84 The sample is made from particles with an average diameter of 700 nm. On the right, the average particle size was 50 μm (Klein *et al.*, 2012) (courtesy: Susanne Klein).

At a microstructural level, and even at nano level, there are many security applications that are possible thanks to new printing technologies. For instance, at the laboratory in South Dakota School of Mines and Technology (*Porous-Wall Hollow Glass Microspheres for Security Printing Applications*, Forest Thompson and Grant Crawford, South Dakota School of Mines and Technology, and George Wicks, Applied Research Center, USA), porous-wall hollow glass microspheres are used as storage vessels for functional materials for the development of security inks. The laboratory has demonstrated the feasibility of developing and delivering security inks by using glass microspheres that are loaded with functional materials. So far, they have been able to load the spheres with gold nanoparticles and print them through aerosol jet deposition systems. Such encapsulation methods demonstrate the potential for to a new generation of security features and applications. The same laboratory, in collaboration with the Research Institute of National Printing Bureau in Japan (*Structural Color for Security Printing: Patterned Robust Colloidal Crystal*, Takaharu Kobayashi, Research Institute of National Printing Bureau, Japan, and Mark Owens, William Cross, Jon Kellar, and Grant Crawford, South Dakota School of Mines and Technology, USA), have been studying the influence of microstructures in colour. Structural colour is the production of colour through the use of micro structures that can interfere with light. The particles self-assemble into colloidal crystals that are then suspended in printable inks. Given their unique colour and gloss effects, the printed colloidal crystals are difficult to reproduce by traditional printing methods. These specialty inks carrying capsules or microstructures, will not only enhance security features but will also bring appearance reproduction to the next level.

Microstructures are also important in medical fields such as biomembranes, dental restoration and external and internal prosthesis. Besides finding and certifying biocompatible printing materials, the challenges from the functional and aesthetics points of view are a perfect example of how the current 3D printing priorities and concepts will at some point merge with those of 2.5D printing. Even internal prosthesis need to be concerned about (micro) surface aspects more pertinent for the moment to relief printing, such as roughness and porosity.

Even though studies about the user experiences of the finished products are not yet clearly defined, there are certainly biocompatible materials available (https://www.theguardian.com/technology/2015/oct/31/3d-printing-plastic-replacement-teeth; http://www.stratasys.com/materials/polyjet/dental-material; http://formlabs.com/products/materials/dental-sg/) that allow 3D printed teeth to be developed as a medical product, and with the potential for a range of future applications. Recently, Stratasys have launched multi-material dental 3D printers (http://www.stratasys.com/en/Dental) that replace the typical plaster models and can print directly from digital scan data to create dental models and surgical guides for implants. Likewise, Renishaw are working on advanced rapid manufacturing methods for Direct Metal Laser Sintering (DMLS) of medical-grade cobalt chrome (CoCr). These types of osteophilic (bone-loving) materials can be directly used as dental frameworks or for craniofacial implants and orthopaedics (http://www.renishaw.com/en/metal-3d-printing-for-healthcare--24226). Other materials such as glass ceramics and zirconia are increasingly used for the dental industry. For example, Roland DG dental mills offer 4 to 5 axis milling machines specific for dental restorations (http://www.rolanddg.co.uk/products/dental).

Aesthetically speaking, micro- or mesostructure modifications of the print surface also have an impact on appearance attributes such as gloss and roughness. An important aspect of print quality is the perceived gloss level across the print. Gloss levels can be affected by the type of inks and the set of print parameters such as ink deposition time, colour separation (amount of each ink in a mix to produce a given colour), halftoning screening, print swath size, curing time and settings, among others. Variations and combinations of these parameters, which can be referred to as 'print modes', lead to a variety of surface appearances that are usually considered as a trade-off between print quality and print speed in traditional 2D printing. These parameters can affect the roughness of the print surface on a micro- or mesoscale level (i.e. changing the optical properties of the printout and therefore the perceived gloss) depending on the printing system. The different combinations of print parameters, along with the use of relief printing, can be characterised as a feature instead of being just a side effect. Figure 5.85 demonstrates the specular gloss levels as measured on uniform patches, with percentages of ink coverage ranging from 0 to 300%. The curve called 'standard mode' shows how the gloss level increases with the ink coverage in patches printed by a print mode provided by the print maker; something that is not actually intended to happen and is known as differential gloss. Other combinations of printing parameters can compensate for differential gloss by modifying ink is deposition, and depending on the coverage of

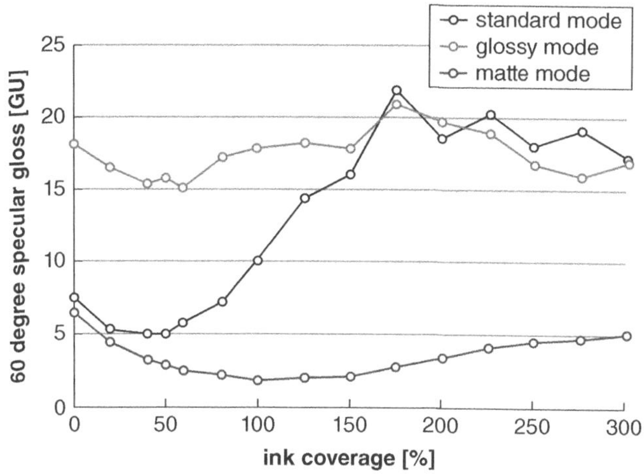

Figure 5.85 Relationship between ink coverage and measured gloss of a standard print mode, together with two print strategies obtaining either a full gloss or full matte appearance of the printout, independent of the local ink coverage. Gloss variation through multilayer printing.

ink that is required. For instance, in order to obtain a uniform gloss (or matte) appearance, independent of the ink amount, a new print mode can be created to achieve a nearly constant gloss (or matte) appearance across a print surface, as shown in the blue and red curves of Figure 5.85.

Besides the print parameters, we can also modify the topography of the print surface by changing the substrate–ink interaction, making use of the multilayer capabilities of a 2.5D printing system or by using pretextured media (Figure 5.86). Using a multilayer system, we can either print different layers of ink to reduce the substrate–ink interaction and achieve a flatter (and therefore glossier) surface, or we can print small structures on the substrate underneath the final image, to create a rough surface that provides a more matte appearance.

The print mode in Figure 5.86 (b) can also be used to generate different levels of gloss through the manipulation of the printing time (1 to 10 seconds) in between the two sub-layers of white ink. The glossiness of the print decreases as the ink drying time between deposition of the two white layers increases (Figure 5.87). This is due to requiring more time for the ink to dry, as well as allowing more time for droplets to spread to create smoother surfaces.

When printing with a UV-based inkjet printer, each layer of ink is polymerised by UV light. The intensity of UV light also influences the adherence of the ink to the media and the roughness of the ink layer. Ink that is not cured directly after deposition is cured by UV exposure during following passes. The extra time in between passes (as shown in Figure 5.86 (b)) and a lower UV light intensity will allow

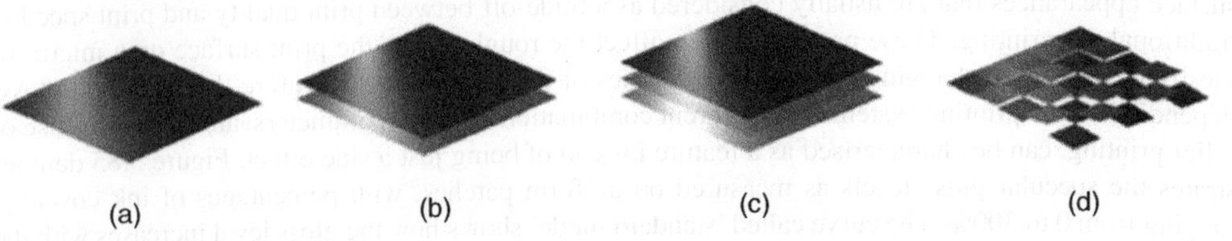

Figure 5.86 Schematic representation of multilayer strategies, where a desired coverage of ink is either printed in (a) one pass or on top of (b) one or (c) more layers of white ink to isolate the ink from the substrate and create a flatter and glossier appearance. A matte appearance can be obtained by creating a rougher surface either by printing (d) ink droplets on neighbouring locations in different passes or by printing the final image on top of small structures or pre-textured media.

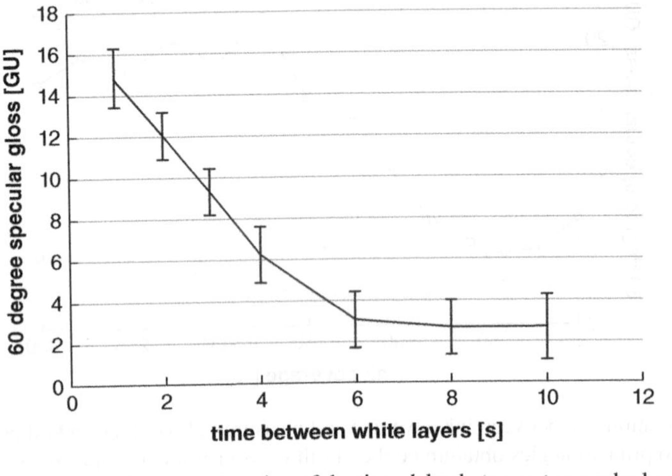

Figure 5.87 Measured gloss level as a function of the time delay between two under layers of white ink.

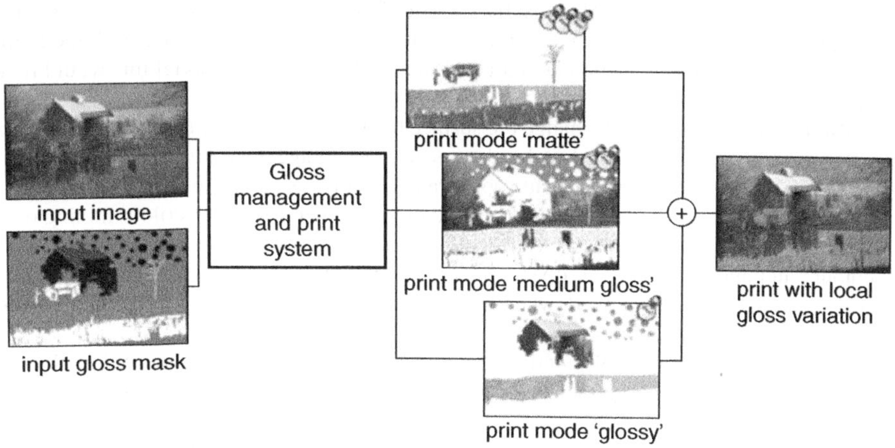

Figure 5.88 Gloss management workflow to obtain a print with local gloss variations. The input contains local colour and gloss information. The content of each intended gloss level is printed with a corresponding print mode determined by the gloss management workflow. The outcome is a print that presents different levels of gloss.

more time for the ink to spread on the print surface before being cured, resulting in a smoother and glossier surface.

By adding specialty inks such as varnish, the appearance of gloss can also be changed. Although the use of varnishes is not a straightforward process, given the different chemical and physical behaviour of varnish compared to standard inks, it is still possible to achieve a number of gloss levels by modifying the size of the varnish droplets in a system with print heads offering variable dot size technology.

Once the printer's capabilities are assessed we can think about generating a desired level of glossiness in different regions of the reproduction by using different print modes. Analogous to colour management for traditional printing applications, we may capitalise on gloss levels by creating specific combinations of gloss and print modes to create a 'gloss management' workflow. Figure 5.88 is an example of how this workflow could be achieved for an image by using three levels of gloss. A gloss mask would be used to codify the level of gloss assigned to each pixel or zone in the image. Each level is individually processed by incorporating the corresponding print parameters, resulting in an print with the desired levels of gloss.

The relationship between roughness and gloss has been widely studied in the last decades, but the dependence on colour and the impact of these three parameters in the context of complete appearance is yet unresolved. In the printing industry gloss meters are used to estimate the glossiness regardless of the fact that the mapping in between measured gloss and perceived gloss is not linear. For systems that can produce relief or 3D prints, a gloss meter falls short of explaining gloss as it neglects the influence of mesotexture and macrotexture and colour. All we have to do is watch an observer evaluating gloss to realise the drawbacks of these measuring devices. Observers move an object around when assessing the gloss, and by comparing the amount of reflection at different angles, they assess the local variation of the gloss. For highly glossy surfaces even binocular vision plays a role, where one eye perceives a different amount of surface reflection than the other.

Case Study 12: Painting Machines

Artists working with traditional drawing and painting media have an abundance of tools and materials, whereas the digital artist is restricted to a few. There are many digital painting applications that mimic the appearance of drawing and painting tools, but the correlation between on-screen representation and how they might be reproduced as a print still remains unsatisfying. This is because each medium used in painting and drawing conveys particular characteristics and to be able to model this would require a significant library of data.

Artists also tend to mix mediums, change the surface texture or add embellishments, by incorporating unusual materials such as glitter, glass beads or sand, or add agents to pigments to increase fluid dynamics, surface texture, glossiness or matte appearance. Additives can be used to modify the drying speed to change the behaviour as one colour is applied on to another. Likewise, colour mixing of paints and pigments may well affect the subsurface of colour mixtures, for example, whether they are mixed as solid colours and applied as a thick opaque layer, or applied as thin layers of glaze to build up a colour. Lastly, how materials are applied, whether by brush or sponge or palette knife, will also contribute to changes in appearance.

There are a significant number of painting programmes by which images can be digitally created using graphical interfaces, drawing tablets or directly working on screen. What is the relationship between a brush stroke that is made by an artist working on a screen to a mark that is made by a machine? Can machines make marks that look like brush stokes? Is it possible to create machines that mimic the behaviour of an arm that handles a brush or a drawing tool? How does this relate to the painter's art or one that the machine made? How might we bridge these gaps?

There are also highly mature and advanced commercial sectors of CGI that can simulate objects and animate in such a convincing way that it is difficult to tell apart the real from the digitally generated. It is now possible to bridge the virtual and physical, digital and analogue. This section will consider the relationship between images that are generated on screen, and the physical representations using painting and drawing materials on paper and canvas.

Computer Simulations for Representing Materials and Objects

Since the introduction of the computer interface, there has been a long history and interest in digital modelling of materials, rendering of the surface qualities of objects, and furthermore how animated figures and objects appear to move in a realistic way. Each point in 3D space is generated as a numerical model, and connected by lines, curves and triangles (Sayim and Cavanagh, 2011), which is stored in the computer as an objective description. Surface effects evolved from limited algorithms to describe reflective, smooth surfaces (Phong, 1975) in the 1970s, to contemporary approaches that can describe highly complex organic, asymmetric and animated materials. All these variables require an incredible amount of processing, and methods are being sought to reduce resolution through downsampling and based on the photographic methods of antialiasing (Fenlon, 2014). The earlier models sought ways to explore and test different modelling and simulations. *The Utah Teapot*, developed by Martin Newell in 1975, presents both a complex 3D object to model and as a test to render in different materials (ceramic, metal, glass), and was extensively used as a reference standard. Newell made a sketch of his own teapot on graph paper to create the original coordinates (http://www.computerhistory.org/revolution/computer-graphics-music-and-art/15/206/556; http://www.computerhistory.org/revolution/computer-graphics-music-and-art/15/206/554). Since then, modelling, simulation methods and reference standards have evolved to incorporate many different objects and textures. Huge efforts and advances have been made in the computer animated simulation of objects and modelling of material qualities, as demonstrated by Pixar (*Toy Story*, 1995; *Monsters, Inc.*, 2001) and Walt Disney Pictures (*Jungle Book*, 2016), whereby scenes and moving animals now appear super photo-real.

Of course, these images are created for viewing on screen-based media, but the use of point clouds and vector approaches used here are highly pertinent as transferable data for printing. The difference here is in the animator's goal to create objects that are photo-real, whereas the goal for the artist is to create images and surfaces through the strategic placement of brush strokes and, in computer terms, could be considered in terms as non-photographic rendering, thus reflecting on painterly methods as so well demonstrated by painters throughout the centuries. An interesting visual reference to the *Utah Teapot* can be found in

Pixar's RenderMan software tutorial. The tutorial has used this historical teapot as the object for different rendering and lighting effects (https://simplymaya.com/forum/showthread.php?t=39768 https://dbql2bd8kqpp.cloudfront.net/network6/storage/7/attachments/1909-lighting-basics-coverpage-01.jpg).

Direct Manipulation

With the introduction of the personal computer, artists have found the computer could become both the tool and the medium. In 1973 Dick Shoup at Xerox Parc was working to develop SuperPaint. He borrowed heavily from traditional painting, graphic arts and animation techniques to create a colour digital painting system. The software included a '*jaggie* removal' program (now called antialiasing) whereby smooth straight lines could be drawn at any angle, without the jagged stair steps always associated with raster computer displays at that time. In the 1980s, 'direct manipulation' (a term first used by Ben Shneiderman in the 1980s) was achieved through the revolutionary method of being able to point at the screen with a mouse and Mac Paint (developed by Bill Atkinson, which was revolutionary in being designed to interact with the user, interpret mouse and keyboard requests and decide what was to be drawn where) enabled artists to use a range of *tools* to paint, draw, spray and fill with black and white patterns. Andy Warhol used an Amiga 1000 system including a drawing pad, camera and drawing software such as the GraphiCraft painting system (http://www.computerhistory.org/atchm/lost-and-found-andy-warhols-amiga-artworks).

In the twenty-first century 'direct manipulation' is achievable using graphics tablets, touch screens, Wifi drawing pens; there are many physical tools by which artists can create images. Today, picture editing software such as Photoshop provides us with 'hundreds of tools' (Harrod, 2015, p. 226) and options, and using a range of digitally mediated methods has had a profound intervention in the way we think and work. Working on touch-screen devices or drawing tablets, by painting on a touch-screen the user can scroll their finger to select colours and brushes and manipulate paint. An interesting example, recently developed at Rochester Institute of Technology (RIT), is a touch-screen paint mixing application, Tangipaint, that creates the appearance of gloss and the texture of paint (Blatner, Ferwerda and Darling, 2011; Ferwerda, 2015; Flores, Ferwerda and Gu, 2012). A recent collaboration between Adobe and NVIDIA, called project Wetbrush, is a virtual oil paint that adds texture to digital painting (https://blogs.nvidia.com/blog/2016/07/26/adobe-wetbrush/). Further information on the evolution of electronic painting system programs over the last 50 years is described by Maureen Nappi (Nappi, 2013).

These examples highlight interesting opportunities for manipulating virtual paint, but do not indicate how images will appear when printed. The assumption is that images remain screen-based or are shared between mobile platforms. However, without touch or an engagement with materials, 'What are the exploratory and manipulative aspects of touch as opposed to passive sensations?' (McCullough, 2010, p. 313) As Margot Lovejoy describes how in the *Age of the Mechanical*, the 'hand of the artist' was considered as 'doubly precious' (Lovejoy, 2004, p. 73). McCullough goes on to explain that the 'practiced digital hand' humanises computer-aided design and in return digital technologies have provided a currency for the tacit knowledge of craft skills.

Non-photorealistic rendering (NPR) uses computational methods to interpret digital formats (captured visual scenes or original works) and renders them in an alternative stylistic expression (Kasao and Miyata, 2006). Typical source formats are pixel-based images, although NPR can also be derived from 3D models and determined algorithmically without a prior source. They can also be used for the stylisation of images, videos and 3D animations (Lu *et al.*, 2013). The utility in NPR can be functional, for instance 'unrealistic' shading schemes aid in the visualisation of complex parts drawn in computer-aided design (Gooch and Gooch, 1999), or methods of illuminating objects that are inspired by Old Master paintings (Zhang, Chan and Constable, 2013). The utility of NPR can also be used to provide a purely artistic interpretation from a source or to provide interactive software tools to aid in the creation of stylised digital art works. An interesting example that achieves a separation of image content from style is based on an artificial neural system to transform a photographic image using different stylistic representations of well-known artists such as Vincent van Gogh (1853–1890), Wassily Kandinsky

(1866–1944) and Edvard Munch (1863–1944) (Zhang and Yu, 2013; Gatys, Ecker and Bethge, 2015), thus allowing to recast the content of one image in the style of any other image. In most cases the majority of the rendering is based on an optimisation that creates an overall look of an image.

Stroke-based rendering methods (SBR) is a subset of non-photographic rendering processes that simulates the stroke or line of paint. Depending on the complexity of the stroke, the brush stroke may range from the simplest such as a stipple to a more complex stroke that emulates a painterly style. Studies have ranged from generic rendering effects and painterly strokes (Hartsmann, 1998, 2002, 2003; Hartsmann, Interrante and Lum, 2002; Hedge, Gatzidis and Tian, 2012) to specific rendering styles, such as Impressionism (Kasao and Miyata, 2006), Pointillism (Yang and and Yang, 2008), Cubism, oriental ink painting (Cheoh, Lim and Tan, 2007; Lee, 2001; Ning *et al.*, 2011) or pen-and-ink illustration (Winkenbach and Salesin, 1994), wax crayons (Rudolf, Mould and Neufold, 2003) and pencils (Yamamoto, Mao and Imamiya, 2004). The study on Pointillism is of interest here as it highlights the link between the colour studies on complementary colours and the spatial arrangement of colour by Michel Eugène Chevreul (1786–1889) and Ogden Nicholas Rood (1831–1902), who was of influence to Seurat's Impressionist colour palette.

Other studies have modelled the characteristics of tools such as the physical qualities of the shape or bristles of a brush (Xu *et al.*, 2004), the viscosity and fluid dynamics of different mediums (Curtis *et al.*, 1997) and the interaction of paints on different substrates. Other studies into the physical simulation of media have included oil (Baxter, Wendt and Lin, 2004), water-colours (Chu and Tai, 2004) and approaches to blending colours such as Microsoft Fresh Paint (https://www.microsoft.com/en-gb/store/p/fresh-paint/9wzdncrfjb13). A further subset of NPR is a data-driven approach, an example-based approach (Lu *et al.*, 2013) and the ability to synthesise painterly virtual strokes on different substrates or an interactive approach (Kaplan, Gooch and Cohen, 2000), where a range of artistic effects are due to the fact that almost any type of mark made on paper or canvas can be imitated. According to Lu *et al.* (2013), Real Brush is a modifiable painting system where the user can create their own brush strokes and reproduce different paint effects. This study includes the behaviour and appearance of paint as other layers are applied or if it is smudged or smeared. Smudging is particularly useful for the application of paints, charcoal and graphite, which can be, manipulated and blended on the paper over existing applications to achieve a softer appearance. Another simulation painting system is Wet Brush (Chen *et al.*, 2015), which models the interactions of the brush and the fluid qualities of paint on to the canvas. It also demonstrates fine surface details and interaction of paints to suggest a three-dimensional quality. These examples highlight significant opportunities for interactive and real-time computer-based painting systems, but these tools and methods have yet to be used as exportable data that can be used for physical surface deposition or 2.5D printing.

Translating Brush Strokes into Painting Machines

Returning to artists' approaches to drawing and painting: How does an artist hold a brush or a pen? How can we record the movement of the wrist and arm to make a mark? Likewise, characterise different processes and materials as resistance of the tool on paper or canvas, the fluid dynamics of the material, the relationship of medium on substrate, how the medium dries or how colours or marks are overlaid – all present highly unique tactile experiences resulting in a myriad of complex elements and appearances. How does this translate from a drawing tablet or, if mediated by software, to paper and canvas? Does current hardware and software genuinely capture the experience or appearance?

There is a significant body of research dedicated to computer-controlled machines that are able to transcribe artists' methods of drawing and painting to produce artworks. Some of these machines appeal to our creative consciousness through the ability to mimic an expressive abstraction of lines, shapes, colours and texture; they also convey human characteristics. In some cases these machines are able to replicate gestural marks in a highly spontaneous and lifelike way. The most interesting are those that explore notions of authorship, subjectivity and objectivity. The debate between whether these are effectors of creative art or are dumb transmitters for making marks – points, lines and planes, which amount to mere daubs without the numerical coding of the artist – is an important point and is explored in detail by Nick Lambert (Lambert, 2003). Can computers assist creativity? Furthermore, do we as humans, as suggested by Sennett, learn about what it is to be human through the machine interaction?

Most famously, the artist Harold Cohen (1928–2016) was regarded as a lifelong practitioner in artificial intelligence and the arts. Since 1973, when he started working on an art-creating program called *AARON*, Cohen collaborated with his painting machine *AARON*, teaching it to perform increasingly complex tasks – Cohen wrote programmes, the computer made the paintings. Cohen describes his project:

> 'AARON is an autonomous intelligent entity: not very autonomous, or very intelligent, or very knowledgeable, but very different, fundamentally different, from programs designed to be "just" tools. Electronic paint boxes, for example. And its use is equally different from the way computer artists use electronic paint boxes: I don't work with the program, that is to say, I work on the program. The goal is the program's autonomy, not the making of a better – orthodox – tool' (Cohen, 1986).

In the 1987 interview between Harold Cohen and Ray Kurzweil, whilst discussing Cohen's *AARON* project – a programmable painting machine – Kurzweil asked Cohen whether the computer was being creative. Cohen replied 'yes', but to a relative degree. He explained that although the images were different each time they were drawn, the computer was still performing functions according to a set of algorithms, which were written by Cohen. As Cohen points out, 'for a programme to be fully creative in a more complete sense creative it has to be able to modify its own performance and that is a very difficult problem'. (The Computer Museum in Mountain View, California, USA, provides a rich resource on the evolution of the computer. See also the interview between Cohen and Kurzweil, http://www.computerhistory.org/revolution/computer-graphics-music-and-art/15/231/2306).

More contemporary examples such as *Paul* is a robotic hand–eye system that is dedicated to the activity of analysing, interpreting and drawing the key features of human faces. It begins by identifying the face and then draws the salient features and incorporates a shading process to add tonality. The drawing still appears to be machine made but conveys a sensibility for the line. Tressett explains that although machine-made, the materials he uses are important and that an artwork is an object that has to be perceived as rare, collectible and of high quality. He also highlights the fact that he has years of experience as an artist and is insistent about details such as materials and the weight and surface finish of the paper. Tresset believes that, like Cohen, the creation is in the process or as a system that generates drawings (Tresset and Fol Leymarie, 2013, p. 358).

A similar project, *eDavid* (Deusssen *et al.*, 2012; Lindemeier, Pirk and Deussen, 2013), is an industrial robotic arm programmed to simulate a human painting process, incorporating a visual feedback system to iteratively analyse, generate and place brush stroke marks to approximate a source image. The machine watches itself whilst painting and then decides where to add new strokes where they are needed. The subject matter is also important and provides semantic information as to how it will be painted: '… a tree will be painted in a different way than a face even if the colours are similar. Such semantic information can be obtained from semantic image analysis of the given target image' (https://www.youtube.com/watch?v=NvifGpmACJE; http://graphics.uni-konstanz.de/eDavid/).

The *Painting Fool* is a computer algorithm and 'aspiring painter' and on the website (http://www.thepaintingfool.com/galleries/you_cant_know_my_mind/index.html) is interestingly written in the first person, suggesting that maybe it (the computer) will be taken for a *serious* painter one day. The computer programme paints portraits and in the project 'You can't know my mind' interacts with the sitter, asking them to pose, change their expression or tilt their head in a certain way. The computer will indicate what sort of mood it is in and then paint a portrait, and, when complete, explains its reason for painting and whether it was 'pleased' or 'annoyed' by the result (Colton, 2010; Colton, Valstar and Pantic, 2008). The above examples are evaluated as a whole by how well they appear painterly or stylised in their output. These examples also demonstrate an interest in representation. Other forms of generative art that explore the relationship between CNC, the dynamics and interaction of tools and materials presents an alternative understanding of machine interaction to create more abstract painting.

The *Interactive Robotic Painting Machine* by Benjamin Grosser demonstrates painting and colour mixing as performance (http://bengrosser.com). The machine comprises an x and y axis flat bed and a z axis to raise and lower the brush and uses a four-paint colour system. The brush is attached to a circular rotating mechanism for dipping and swirling the brush in the paint pot. It uses a genetic algorithm (GA) to drive the system that applies lines, dots and curves according to any stimulus it gains from the ambient noises around it, as well as the noises it makes. Dragan Ilic's new work *DI-2K4* is a collaboration with a large multiaxis industrial robot (2016) and could be considered as more performance *with* the machine (than collaboration). On a simple level, the robot draws with a large array of pens, creating circular images according to programmed commands. On a more perilous level, Ilic is strapped to the robotic arm and as the robotic arm moves across the plane, he applies paint to the surface (https://www.youtube.com/watch?v=The6OE5IOpA). Kai Franz' *Dual-Axis Precision Deposition System*, or *Plopper*, is a vector-based drawing programme, which interprets the digital drawings and converts them into a machine CNC code to drive the Plopper. The printhead deposits various materials including sand, resin and liquid paint. The sand is mainly used to create a texture, but also binds the resin and paint. The machine operates as a mediator between the precision lines of the CAD drawing and the more unpredictable and fluid behaviour of the sand and liquid. Both Ilic, Franz and Grosser's machines demonstrate the relationship between the input as (fixed) code and the output, whereby stimuli and materials are dynamic and unpredictable. As demonstrated so well by Franz, the layering and textural qualities are crucial to the finished work (http://www.widewalls.ch/kai-franz-exhibition-rockelmann/).

On a commercial level, there are some fun examples of painting machines, such as the Kickstarter funded project for Sylvia Todd (http://sylviashow.com/tags/watercolorbot) in collaboration with Mad Evil Scientist to make a WaterColorBot (Figures 5.89 and 5.90), which is a specialised desktop pen plotter that uses a set of watercolours. Other inventions by the Mad Evil scientist group include the AxiDraw (http://shop.evilmadscientist.com/productsmenu/846), which is designed to precision draw on (almost) any flat surface with a range of drawing tools that can extend beyond the parameters of its own frame. Another product is EggBot, which, as well as eggs, can draw and engrave on to other spherical objects, including ping-pong balls and light bulbs. These types of hardware and projects, by working with open source communities, provide vital contributions to makerspaces, hackerspaces and online communities around the world. Whilst neither give texture dimensionality, these products demonstrate both the open opportunities for development of ideas and methods for moving from two to three dimensions.

Methods of Converting Images into Brushstrokes

It is generally acknowledged that although there has been remarkable progress in digital imaging software, it is still difficult to mimic the appearance of marks made through real painting and drawing. Returning to the questions at the beginning of this study on how to convert a brush line on-screen to a brush mark on canvas, there are broadly two ways to consider how an image might be converted: firstly, using automatic methods whereby presets and filters are used to convert an existing image using filters to create painterly effects and, secondly, using an autographic approach by which an image is generated on-screen using drawing tools. These approaches present subtly different methods of constructing and understanding how an image is created. Of course, there are many overlaps and file formats that can assist in generating and converting images. Images captured using cameras or phones are represented and stored as pixels. A pixel is one point, dot or element that represents the smallest unit in a picture or on-screen. Other variations are the texel, which is used to describe a texture element in computer graphics, or a voxel, which is used to describe a volumetric point as a coordinate in 3D space. As an alternative to using pixels, vectors may be a more appropriate approach to image construction.

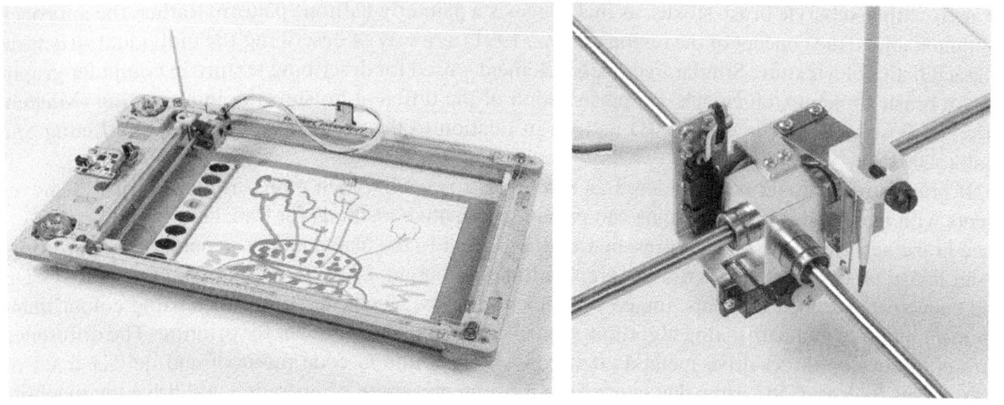

Figure 5.89 Sylvia's WaterColorBot and detail of the brush holder at http://www.evilmadscientist.com (courtesy: Windell H. Oskay).

Why vector? In drawing software programmes, vectors describe lines or curves that are connected by points: vectors, splines or Bezier curves, that mathematically expresses points in space and time. These lines can be straight or curved and can be orientated in two dimensions or three dimensions. A path can be stroked to obtain different thicknesses or colours, or used as input for other elements such as pressure, gradients in height and blends to include opacity and translucency. A vector-based drawing philosophy can be considered as significant to gaining an understanding of alternative approaches, which break away from pixels (digital screen resolution) and halftones (analogue print resolution). Furthermore, one can consider brush strokes as analogous to vector lines, which are represented by a series of movements with many different magnitudes and directions. By observing the brush strokes of painters, images are generated through the use of lines, subtle modulation of similar strokes and a repetitive overlayering of paint. Artists such as van Gogh and Seurat apply paint in a methodical way, yet each brush stroke is different. The objective is not to replicate painterly styles and populate images

Figure 5.90 Deluxe EggBot, drawings on eggs and baubles at http://www.evilmadscientist.com (courtesy: Windell H. Oskay).

or areas with pixel-style brush stokes, as this suggests a painterly halftone pattern. Rather, the approach is comparable to the concept of the texton (Julesz, 1991) as a way of describing the individual structural characteristics of a texture. Similar to the texel element – used for describing texture in computer graphics – a brush stroke texton creates a representation of the different brushworks in a painting (Maaten, Eric and Postma, 2010) as well as 3D textons in relation to the appearance of materials (Leung and Malik, 2001).

If brush strokes could be considered as analogous to vectors, then representing a series of movements with magnitudes and directions can provide alternative methods for translation. The key difference in the approach is the data representation that is sent to the machine. For general digital printing, raster image processing (RIP) is used in preparation print output, takes file formats such as EPS or PDF and converts the embedded fonts, images, graphics, colours, blends and greyscales using colour management modules and half-toning algorithms into a continuous-tone bitmap for printing. The difference here is to develop direct drive methods that convert SVG into G-code protocol and deliver a set of instructions to drive CNC paths that control the movements, speed of operation and drive the machine and tools in a direct way, whereby the vector path on screen becomes the tool path on paper.

It is interesting to note that the earliest computer graphics, as developed by Ivan Sutherland in the 1960s for his program Sketchpad (1963), also used vector graphics. It has only been a result of the use of photographic images that dots, pixels and bitmaps became the more expedient method to define an image. However, by building on traditional computer-aided design (CAD) practices, CNC milling and machining, there are increasingly more opportunities for artists to work in vector formats. It is important to recognise at this point that whilst a vector approach (digital format) is highly machine repeatable, the resulting individual painted brush strokes (analogue) are not.

Brush Strokes and Painting Machines

The research undertaken at the Centre for Fine Print Research (CFPR) (the research was funded by the Arts and Humanities Research Council, 'The development of novel inkjet inks', May 2014–April 2015, Grant: AH/L015277/1) makes a distinction in the specific emphasis of approach: to develop a machine capable of producing texture in the added dimension in the form of relief on the 2D surface (2.5D). The objective is not to produce a painting machine that emulates human expression, but instead to produce a digital machine capable of consistently producing textured prints. Therefore the research also looks to develop accompanying software and data representation to produce a cohesive workflow toward future 2.5D printing technology and its creative applications. The study has included the creation of non-uniform but harmonious painting effects across paper substrates through the actuation and placement of the brush and the flow of paint. The research at the CFPR, as the name suggests, is predicated in the fabrication of printed artefacts and the relationship between on-screen images, development of colour separation algorithms and computer-controlled methods for applying paint to paper. The objective in the context of 2.5D printing is based in a highly pragmatic field in developing technologies and methods that can assist the user to apply inks, paints and viscous materials to create low relief or 2.5D printed surfaces. The research has so far considered alternative approaches to image making and printing that move from the on-screen representation of images and painting applications to the physical generation and methods for surface deposition or 2.5D printing.

The focus here is within a digital context and the opportunities to render colour and texture information as a combined printing method to create a complex pictorial composition. The term 'combined' is emphasised here to describe how colour and texture are applied at the same time, and not a printed texture upon which a halftone over-coat is applied. The research has investigated the ability to physically print combined colour and texture on to a 2D substrate. The motivation is to explore human-analogous gestures for deposition of paints, inks and fluid materials toward a digital machine capable of physically reproducing painterly styles and to produce visual effects that exploit the character of the materials used. The inspiration is the work of Old Masters of painting who were able to describe complex visual details with relatively few strokes. The approach to image segmentation methods of coloured images into stroke-based rendering has considered two different image-making routes (see Figures 5.91 and 5.92).

The autosegmentation approach converts pixels in a picture/photograph (bit-for-bit, raster graphics, pixels, points, image, jpeg) such as photos, screengrabs ans iPad art into vector paths that can be used to drive a paintbrush or pen on the printing machine. The algorithm automatically separates the image into coloured layers. The layers are determined by edge strength, texture direction, *x*/*y* pixel adjacency and hue/saturation/brightness. This approach is inspired by the non-photorealistic rendering algorithms for texture directionality (Kasao and Miyata, 2006; Kasao and Nakajima, 1998).

The autographic process provides the user with a higher degree of freedom. The artist can use drawing tablets and pens, and working on drawing software such as Adobe Illustrator or Inkscape, where vector-based graphics (paths, points, handles, anchors, Bézier curves and smooth lines), such as AutoCad Illustrator, Freehand and Inkscape, can be exported as vector-based images (line work, *x*-*y*-*z*, postscript, svg, animation). Layers, brush dimensions and lengths, characterisation of brush strokes can

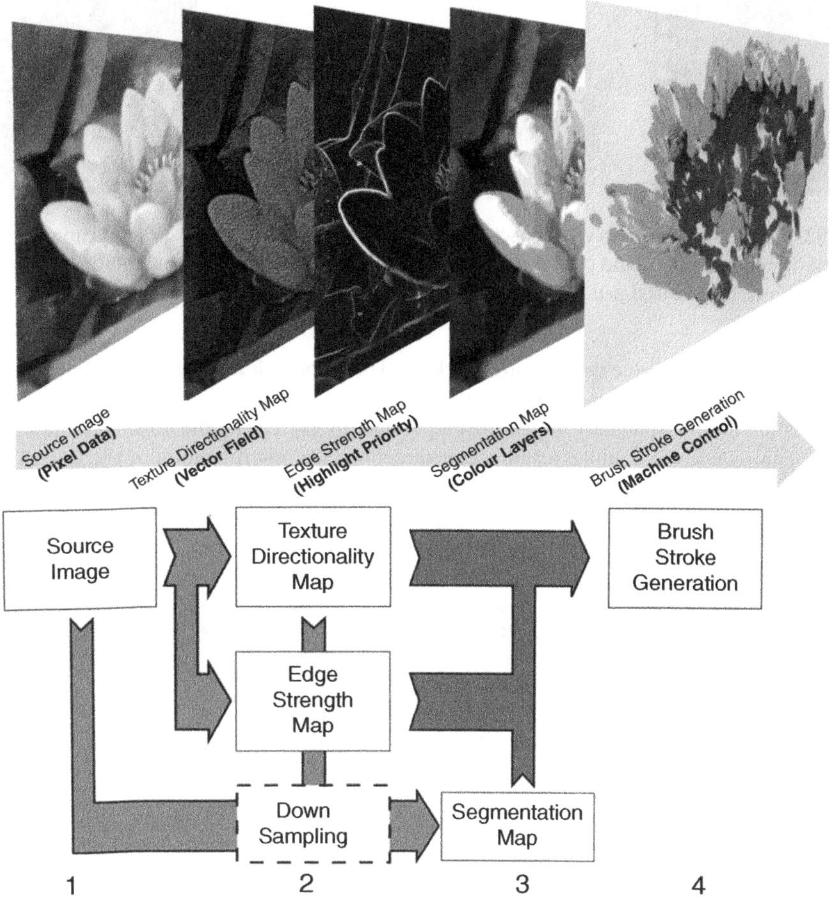

Figure 5.91 From left to right: (1) source image is used (JPEG, PNG) as a colour reference and pixel position; (2) texture directionality creates a vector field to guide brushstrokes in 2D and an edge strength map creates a priority of brush strokes in time; (3) segmentation map creates colour channels; or layers; (4) steps 2 and 3 are combined to generate brush strokes per layer and are then sent to the machine (courtesy: Paul O'Dowd, CFPR).

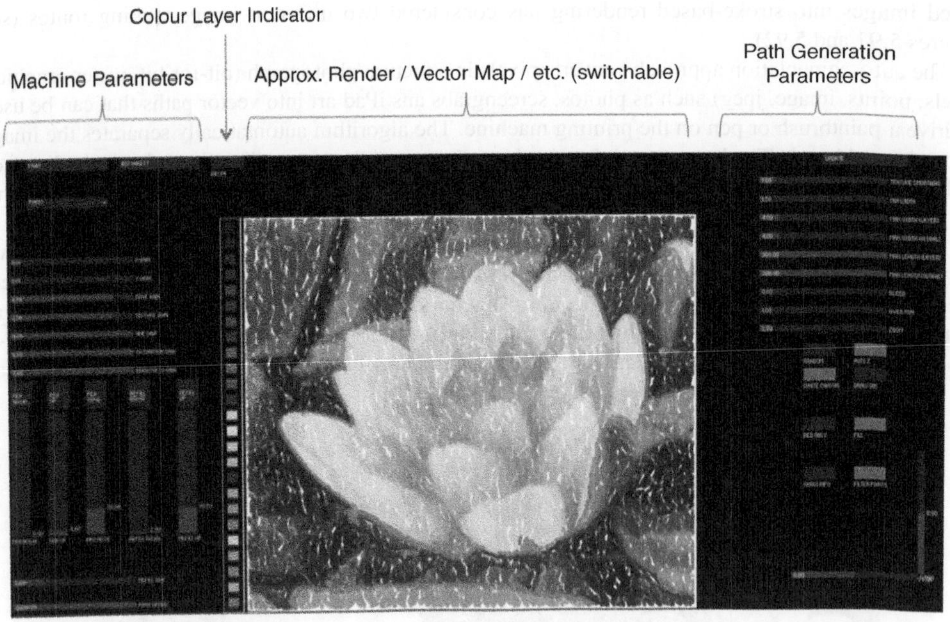

Figure 5.92 The interface enables the user to modify a picture that has either been captured as a photograph or generated by hand on a drawing tablet or screen and manually adjust many different settings to achieve the desired printed output (courtesy: Paul O'Dowd, CFPR).

be maintained and used to drive the painting machine. This approach is inspired by Direct Write printing processes (Lewis, 2006; Clem and Bell, 2013).

Slider bars are used to manually adjust the output and, by increasing the texture smoothing setting from 0 to 4, can reduce the complexity of vector direction. The interpretation of the waterlily photograph into a vector-based image has the following example characteristics. For a print of 310 mm × 260 mm, a minimum stroke width of 1 mm and maximum width of 20 mm, the image has a total of 4826 brush stroke marks; the colour distribution is shown in Figure 5.93.

Figure 5.93 (left) Segmentation of the digital image, (right) showing 25 layers and colour swatches (courtesy: Paul O'Dowd, CFPR).

To obtain the 2.5D movement of the brush stroke, the width of a point on the stroke is used to determine the *Z* component. The *Z* height is constrained to the lowest point at which the point of the brush adequately meets the paper to form a stroke (Figure 5.94). This is not yet characterised as the user must predict the range of width and height. Figure 5.95 shows examples of paintings produced using the 2.5D software and printing machine.

This project is beginning to explore 2.5D printing methods with the aim to gain deeper insights into the appearance of texture in both real-world and computer-based image processing, by documenting the relationship of textured appearance of objects as seen in the real world and how these are interpreted by artists and rendered as paintings and drawings. The end result is not to replace artists by painting machines but to explore a more meaningful interaction between marks made on a tablet or screen and

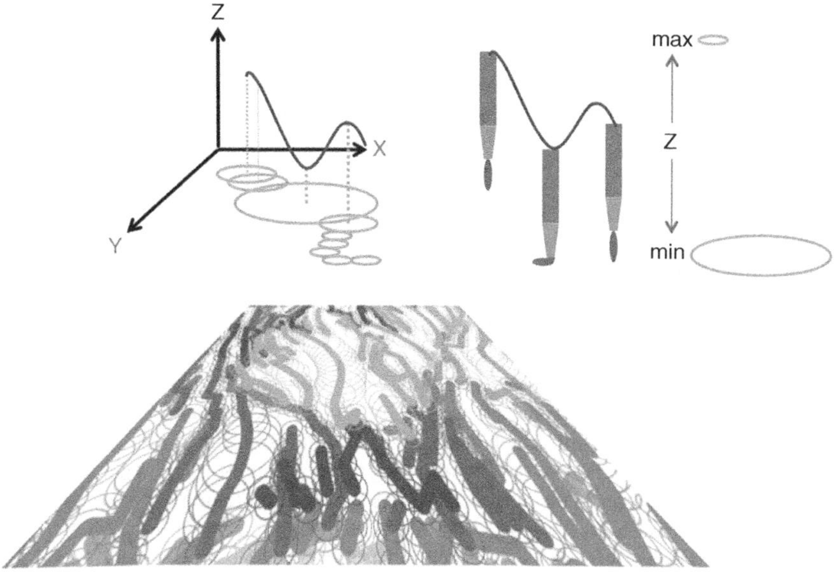

Figure 5.94 Demonstrating the machine and path parameters and the *Z* height of the brush (courtesy: Paul O'Dowd, CFPR).

Figure 5.95 Watercolour and acrylic paintings made by Paul O'Dowd at the Centre for Fine Print Research, Bristol (courtesy: Paul O'Dowd, CFPR).

how gesture as defined by the movement, angle and pressure of a hand and body can be transcribed using real materials and tools.

Case Study 13: Analogue Printing Methods

Many contemporary artists combine digital, photomechanical and analogue methods to make prints. The photomechanical process, as developed by Fox Talbot, employs the use of light and a photo negative or positive to produce an image. The plate is printed on a press using inks and paper. The following descriptions are a glossary of common analogue printing processes (also see Case Study 2: Wallpaper Design for other printing methods). There are three primary printing processes – intaglio, relief and planographic – from which have evolved a range of other processes, including screenprinting, rotogravure and flexography. The following brief description of the primary processes are illustrated by mini-prints from the CFPR archive (http://www.uwe.ac.uk/sca/research/cfpr/MINI%20PRINT) alongside microphotographs showing the microtexture of the print. The microimages are useful in demonstrating the richness and variety of the print processes that, albeit on a micro scale, indicate the fundamental relationship of ink and paper and the physical 2.5D qualities and their different textures.

Intaglio

Intaglio is a printing process that describes a range of cutting or etching techniques by incising a plate with a sharp instrument or acid to create an image. Early methods similar to wood engraving involve very sharp tools that cut away the metal to obtain a very fine line (see also the engraving by Hendrick Goltzius in Chapter 2, Figure 2.31). The process of etching employs a sharp needle on to a copper or zinc plate that has been coated with an acid-resistant wax ground. The needle is used to draw lines through the ground, to obtain outlines, detail or cross-hatching to create tones. The plate is dipped into acid and where the surface has been scratched by the needle the acid 'bites' into these exposed areas. The wax ground is removed. The ink is pushed into the groves and cells of the paper and surplus ink is wiped away. Paper is pressed on to the surface and runs through an etching press under pressure. The lines of the ink from the plate are transferred on to the paper (see Figure 5.96).

An **aquatint** method may also be used, which is similar to adding tones to an image. It involves covering the surface of the copper plate with a fine resin powder. The plate is heated so the powder melts and adheres to the surface to create a ground. Analogous to making a watercolour where individual layers of thin paint are applied to increase the tonal range, in order to obtain a series of tones the

Figure 5.96 Louise King, *Owains Raven* (2005), etching (courtesy: Louise King).

plate is dipped into acid incrementally. In order to protect the lighter areas of the image, a resist paint is applied to the surface, called *stopping out*, and then returned to the acid until a full range of tones are achieved. Figure 5.97 demonstrates a combination of drawn lines into the ground and fine aquatint layers. The background, eye sockets and nose cavity show a deeply etched aquatint surface.

Photo etching differs from photogravure and is similar to commercial gravure printing in that all the tones of the image are etched to the same depth (Figure 5.98). Commercial gravure printing is used for

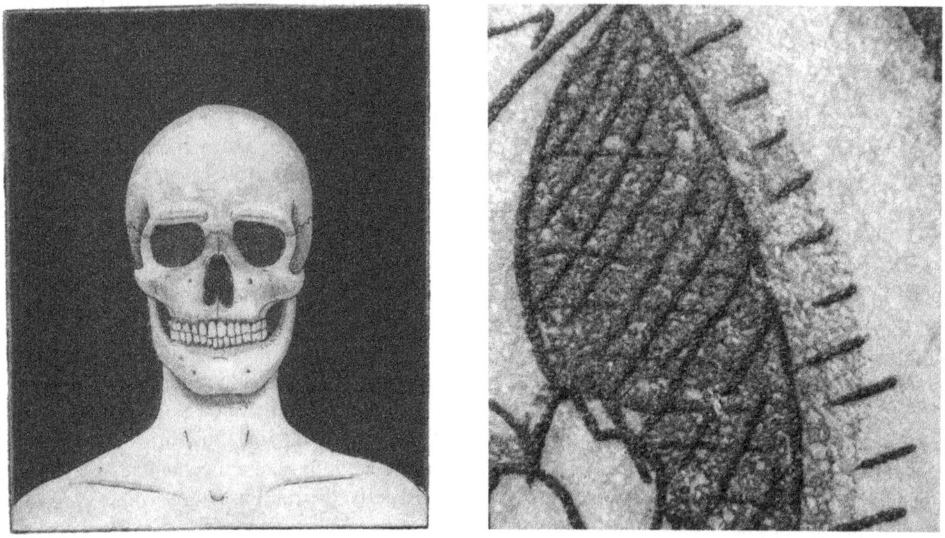

Figure 5.97 Roxanne Goffin, *In My Head* (2015), etching and aquatint (courtesy: Roxanne Goffin).

Figure 5.98 Akiko Takizawa, *Fresh Noodles* (1998), photo etching (CFPR Editions).

high-quality print runs, including postage stamps, magazines, fabric and wallpaper. The image is colour separated and broken into halftone printing elements or cells are engraved on a cylinder, where each cell holds the ink – the larger the cell the larger the dot. Before printing the cylinder is coated with ink and then the surface is wiped clean so that only the engraved cells contain the ink. Each cylinder needs to be in registration with the next cylinder. As the roller rotates and comes into contact with the ink, excess ink is then wiped off with a doctor blade, leaving the ink only in the cells. As the paper comes into contact with the roller, pressure is applied from the roller and the cells of ink are transferred to the paper, the paper then moves on to the next colour and so on. Gravure produces high-quality images for a range of applications including wallpaper and stamps.

Photogravure is a photomechanical intaglio process whereby a thin paper is pre-coated with gelatine and red pigment (used to be carbon black) and sensitised with potassium dichromate. After the tissue is applied to a copper plate, it is exposed to a photo-positive image and is then washed in warm water to remove the more soluble gelatine and to develop the image. Once dry, the plate is aquatinted to create a ground, which is then etched using varying strengths of ferric chloride to obtain a tonal range. The plate is inked and printed according to the intaglio method. A comparison between Figures 5.98, 5.99 and 5.100 shows differences between the halftone. In Figure 5.98 the photo etching demonstrates a typical halftone pattern, the flexo print (Figure 5.100) illustrates a very subtle halftone, whereas the photogravure shows none (Figure 5.99). The flexo print and the photogravure also show thick deposits of ink in the dark areas.

Flexographic plates can also be used for achieving a high-quality photo intaglio print. The plate is processed and inked using a similar method to etching or gravure. A significant aspect of flexo printing is the ability to use stochastic halftoning as a method of improving the visual qualities of a photomechanical print. Stochastic technology can be considered as the beginning of a mechanical simulation of nineteenth century continuous tone. The term stochastic is derived from the Greek word *Stokastikos*, meaning to predict or guess, and is used in mathematics to analyse and predict the movement of particles through liquid. In printing, it is a method that applies the placement of a more random dot FM (frequency modulated) screening. Unlike the traditional halftone system, the process employs no fixed grid or screen angles and the size of the dots and spacing between the dots is variable. Therefore more

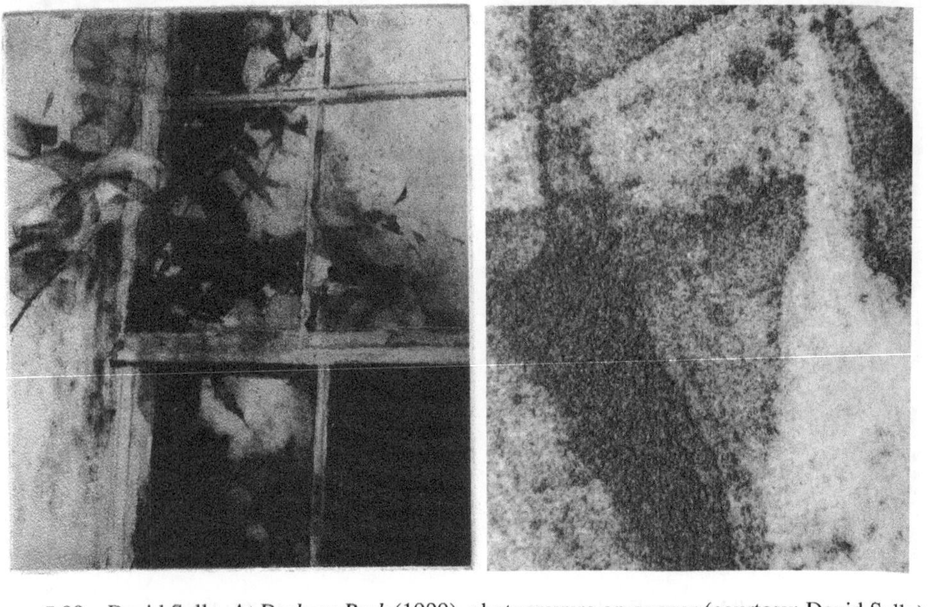

Figure 5.99 David Sully, *At Dyrham Park* (1999), photogravure on copper (courtesy: David Sully).

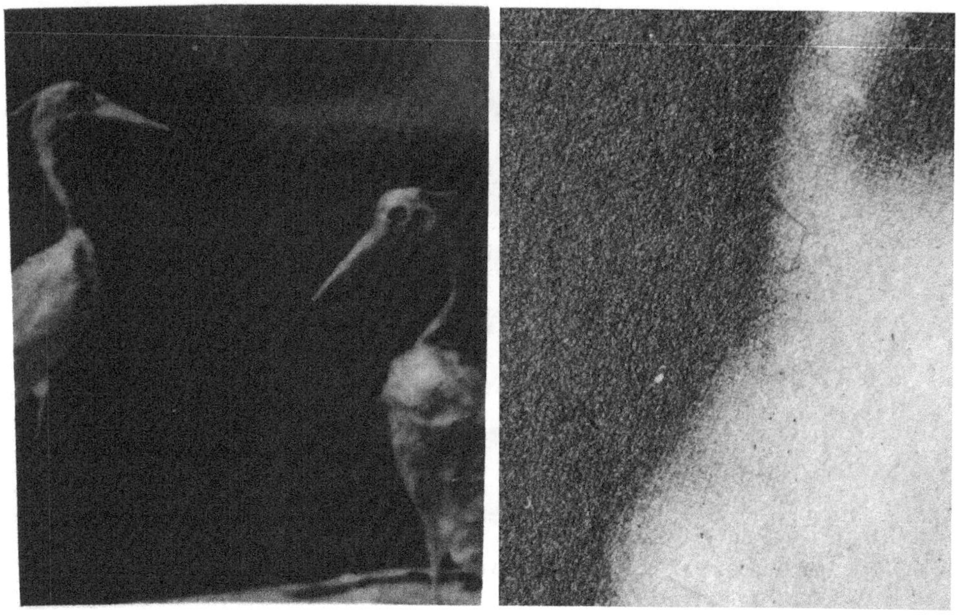

Figure 5.100 John Brennan, *Lear* (2001), intaglio flexograph (courtesy: John Brennan).

dots can be placed in a specific area in order to achieve a greater concentration of colour. Also, due to the removal of screen angles formerly preventing moiré effects, the stochastic system enables more than four colours to be printed. Extra colours have been added to the process colour range, which include process orange and green. This has created a greater depth of colour, range of tone and improved contrast.

Printmakers reproducing paintings used mezzotint because of its ability to create chiaroscuro images, smooth gradations in tone and rich, deep velvet blacks. The mezzotint process is possibly considered the subtlest of printing methods. The mezzotint is a process by which the engraver roughens the entire surface of a copper plate with a curved ended very sharp cutter with a serrated edge ('rocker'). The lines are incised into the surface at many angles, which is uniformly roughened so that the surface 'ground' has tiny prickly burrs. If the plate is inked at this stage the print taken from the plate will appear completely black. The engraver then works back into the plate to remove portions of the burr or burnish areas to hold less ink, to obtain a range of dark areas, mid-grey areas and light areas. During the inking process, the ink is wiped into the plate and is caught by the roughened surface. Paper is pressed on to the surface and run through an etching press under pressure.

Relief Printing

Linocuts, woodcuts, letterpress and ink stamps are common relief methods (Figures 5.101 to 5.104). The surface is uniformly raised from the plate and the ink is applied to the raised areas. Sharp cutting tools and different profiled chisels are used to incise and remove unwanted areas from the block. These areas will remain white when printed. When printing using a letterpress, there is often a discernable embossing in the paper.

Flexo relief is a contemporary version of surface relief or letterpress where the printing elements are in relief at a constant height (Figure 5.105). A light-sensitive photopolymer plate is exposed under a halftone image on transparent film. The unexposed/unhardened sections of the plate are washed, leaving the exposed/hardened sections and a relief image. The rubber or plastic print plate is flexible and is

Figure 5.101 Ben Goodman (2015), wood engraving (courtesy: Ben Goodman).

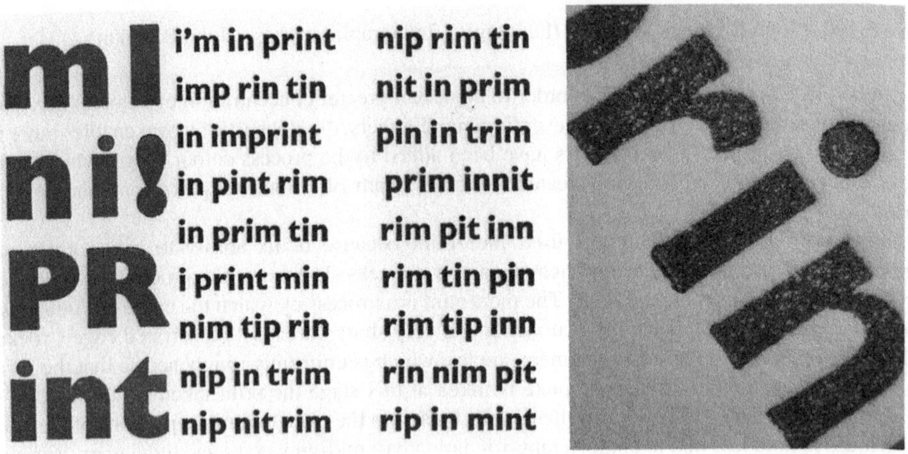

Figure 5.102 Lucy Guenot, *Mini Print* (2014), letterpress.

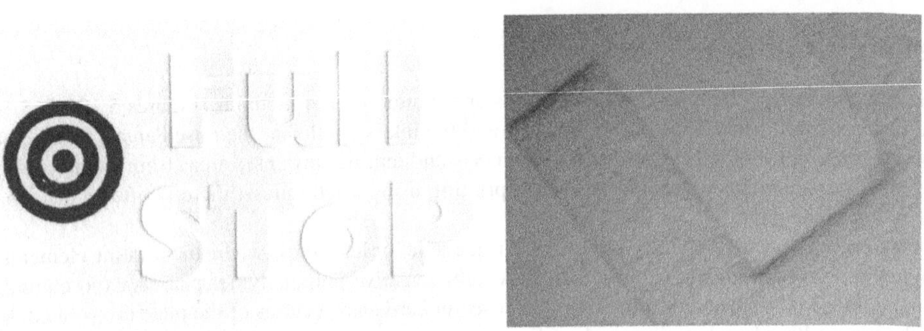

Figure 5.103 Cath Ingram, *Full Stop* (2014), blind emboss and relief print (courtesy: Cath Ingram).

Figure 5.104 Carinna Parraman, *Along the Lines* (2013), hand-cut linocut.

Figure 5.105 Carinna Parraman, *Holiday Memories* (2001), flexograph relief.

wrapped around a printing cylinder. The ink is then transferred on to the substrate through pressure from the cylinder. It is a very versatile industrial printing technology that can print at great speed on continuous rolls of paper or other substrates such as cardboard, paper and thin wrappers.

Helio relief is a method that employs a thin photopolymer film stencil used for sandblasting images into hard surfaces such as wood, glass and metal (Figure 5.106). The photopolymer light-sensitive film, which has a clear plastic support backing, is exposed to a photo opaque image. The unexposed areas are washed away leaving the exposed and hardened stencil layer that is supported by the clear backing film. The block of wood is coated with a layer of glue and the stencil is applied face down on to the block, pressure is applied and then the backing film is removed. The unexposed areas (that were washed away) reveal the bare wood, and the hardened areas serve as a barrier to protect the wood. The exposed areas of wood are then sandblasted. Once sandblasted to the required depth, the protective emulsion is removed to reveal the wood grain. The surface is then inked and printed.

Lithography

Lithography is a planographic method of printing based on the antipathy of oil and water. A positive image is applied using a greasy ink called *tusche* (Figure 5.107). Other drawing materials include a wax crayon or a pencil. The print matrix is either a grained flexible aluminium plate or porous limestone block. The surface is then *etched* using a solution of weak nitric acid HNO_3 and Gum Arabic. As the surface of the printing plate is rolled with ink, the greasy marks (the positive image) on the plate attracts the ink and repels water; the blank etched areas (negative areas) are hydrophilic, which attracts water but repels ink. When printing, the stone or plate is kept damp with water to maintain these conditions. Paper is placed on to the surface and the stone or plate is run through a press under pressure. The resulting image is a mirror image of the image on the stone or plate. Offset lithography employs a second step where the ink from the plate is transferred to an intermediary roller or blanket and is then deposited on to the paper facing the same way as the original.

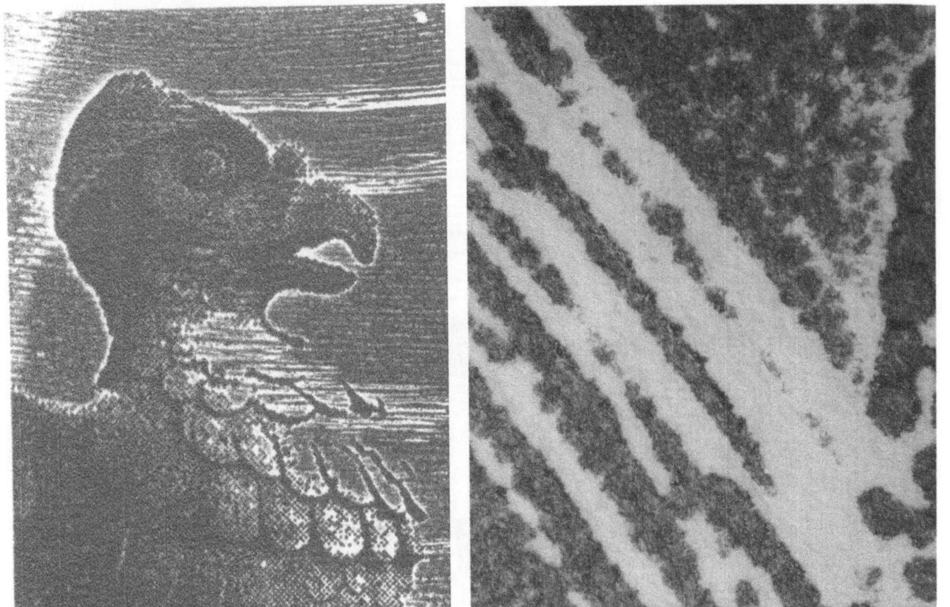

Figure 5.106 Richard Anderton, *Eagle* (2003), helio relief (courtesy: Richard Anderton).

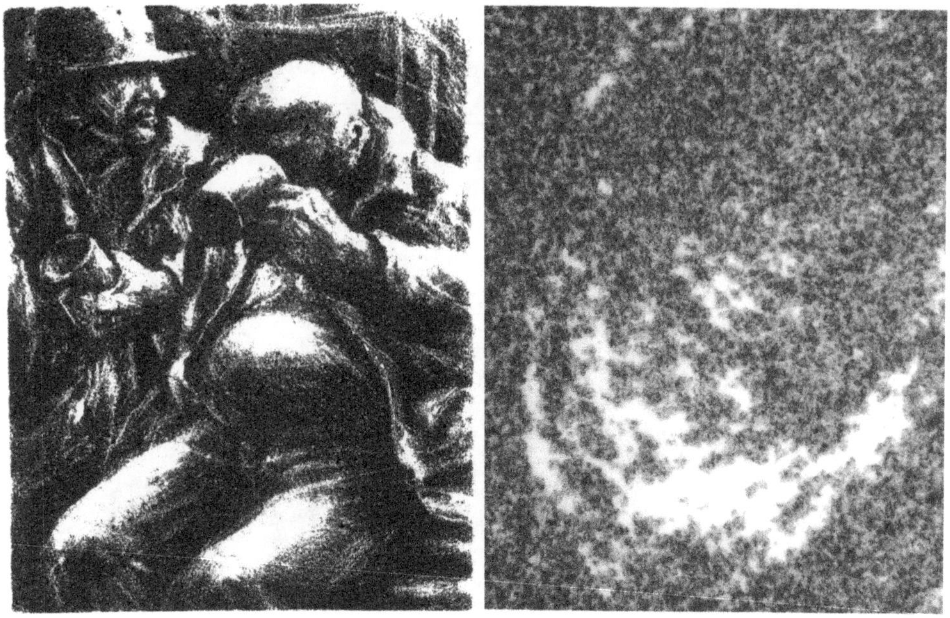

Figure 5.107 Phil Bowden, *Untitled* (1999), stone lithograph (courtesy: Phil Bowden).

Screenprinting

Contemporary flat-bed screenprinting employs a fine mesh stretched across a screen, which is coated with a photosensitive layer of emulsion and dried. The screen is exposed through an opaque positive image to UV light and the screen is washed to remove the unhardened areas of the stencil to reveal the exposed mesh and stencil areas hardened by light. The screen is dried. Although traditionally performed by hand, most is now automated. Semi-viscous pigmented ink is squeezed through the mesh using a rubber squeegee on to the paper below. A separate screen is used for each colour and a row of fixed screens move up and down to allow the substrate to move underneath at a controlled speed. In the down position colours are squeegeed on to the paper underneath.

Through the process of over-layering bright, saturated translucent or opaque colours, combined with halftone, hard edge or painterly marks, screenprint was quickly recognised to contain something unique. Screenprint inks can be mixed to specification and pigmented inks can be matched to oil paints or gauche. Lastly, the amount of control the artist had over the density, saturation and translucency of colour enabled the designer to engage with colour in a wholly novel way.

Screenprinting is still widely used across many industries and especially in designs for posters and display, and the creative arts for fine art printing. It is suitable for low-volume, specialised and functional work. Due to its versatility, clean edges, accuracy and inexpensive setup costs, it can be used to print on to many substrates, including textiles, glass and ceramics, paper and plastics. Compared to UV wideformat printing, once the screens are set up, the printing process is fast and enables the user to be in control of the process. There are a range of inks specially developed to obtain surface effects for sign and display (glosses and varnishes), decorative effects for packaging (powder additives such as glitters, metallics), textured surfaces for clothing (expanding Puff Ink is an additive to Plastisol inks, which raises the print off the garment to create a 3D feel), thick deposits for raised or embossed surfaces for Braille (UV curable inks) and adhesives for foiling and flocking.

The images in Figures 5.108 to 5.110 demonstrate the versatility of screenprinting, which has the ability to print sharp clean colours, photographic images, text, as well as texture. The last figure also

Figure 5.108 Sarah Bodman and Tom Sowden, *Cape Cod* (2012), screenprint (courtesy: Sarah Bodman and Tom Sowden).

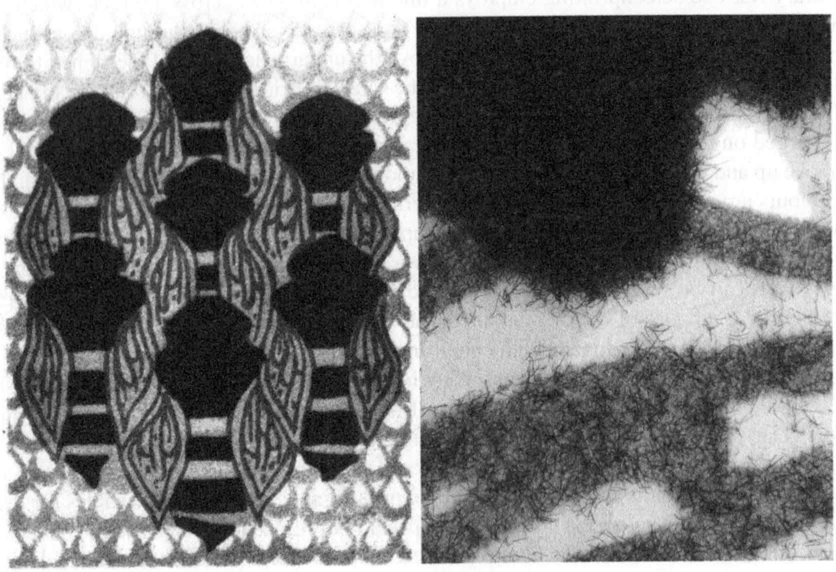

Figure 5.109 Katie Wallis, *Cell Block* (2012), screenprint with black flocking (courtesy: Katie Wallis) (https://katiewallisprint.com).

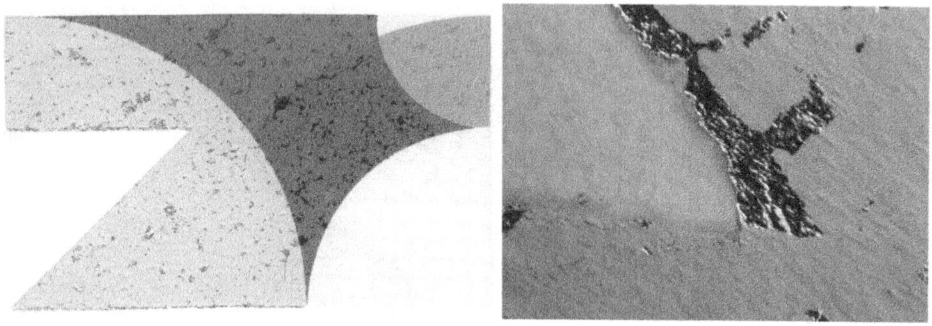

Figure 5.110 Stephen Hoskins, *Untitled* (1996), screenprint (courtesy: Stephen Hoskins).

demonstrates the importance of white ink and also the ability to lay down lighter colours over darker colours, which was known to screenprinters several decades before the arrival of inkjet technology.

Rotary screenprinting, or a continuous web cylindrical mesh screen, employs cylindrical blank pre-sensitised screens that are coated with a layer of wax resist, which is sprayed on as the pattern. The wax is removed leaving the bare mesh. In rotary screenprinting, the squeegee is fixed inside the roller and ink is pumped into the inside of the cylinder. The paper and cylinder move synchronously and the squeegee blade is pulled across the mesh, squeezing the paste or ink through the mesh on to the paper. As well as continuous printing, it can also be used to print on to curved products such as bottles.

Case Study 14: Relief Woodblock Printing

The aim of this case study is to explore methods for creating relief multicolour blocks using 2.5D printing. Relief printing is the oldest and most enduring method of image making and reproduction.

Wood engraving, letterpress, flexography and block printing are typical relief printing methods. The non-image areas of the block are cut away, and ink is then applied to the surface of the block or plate. Paper is placed on top of the plate and either rubbed by a hand-tool to the back of the paper or is passed through a press to transfer the ink from block to paper.

We can also see examples of block printing in Case Study 2: Wallpaper Design (Figures 5.5 and 5.6), which requires a complex system of pins on the side of the blocks to register the blocks along the lengths of paper.

There is an increasing interest in producing blocks that can be used for letterpress and images. A new letterpress font *A23D* is unmistakably technologically conceived (see Figure 5.111). Commissioned by Richard Ardagh of New North Press, the font was designed as a twenty-first century type, produced by type designers Scott Williams and Henrik Kubel of A2-Type, and printed by architectural model makers Chalk Studios (http://new-north-press.co.uk/project/a23d). The files were UV printed to create a wire-frame type block (reflecting the method by which they were generated (https://www.creativereview.co.uk/a2-new-north-press-3d-printed-letterpress-font).

The two important aspects of printing were, firstly, the introduction of the printing press and the ability to register moveable type in complete precision, as demonstrated by Johannes Guttenberg in the mid-fifteenth century and, secondly, in the sixteenth and seventeenth centuries, the transcription of images into tonal engravings (from paintings into prints, as scientific illustration, as caricature and satire in newspapers). This mechanisation of printing revolutionised access to reproducible images and the spread of information as words and images across the world.

An early example of relief printing is the Chinese printing from stone, which originated as a means of copying and disseminating inscriptions on stone. The Chinese method, like all relief printing, used

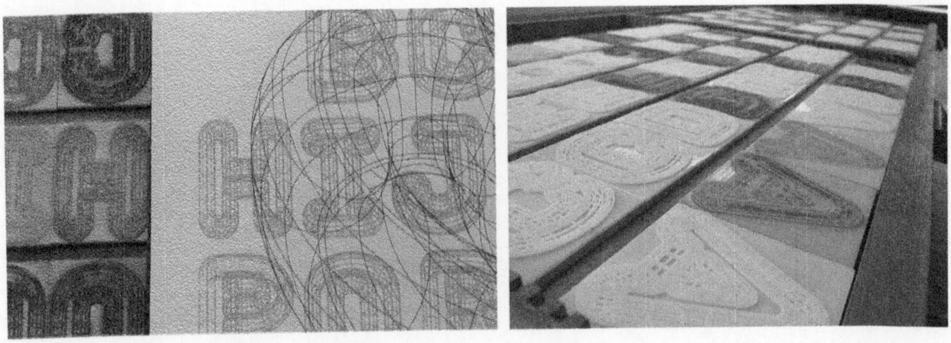

Figure 5.111 A23D: a 3D-printed letterpress font commissioned by Richard Ardagh of New North Press, designed by A2-Type and fabricated by Chalk Studios.

the same principles, where all the raised lines that projected above the indented background were printed black and the carved or indented lines remained white.

Artists working in the sixteenth century, such as Albrecht Dürer (1471–1528), Lucas Cranach the Elder (1472–1553) and Hans Burgkmair (1473–1531), were particularly interested in the combination of colour and line to create multicolour tonal woodblock prints (Gnann, 2012). The coloured tonal block aided textural differentiation, and greater contrast and highlights. In some works by Hans Burgkmair a further tone block was added, and Ugo da Carpi (1480–1523), working in Italy about 20 years later, used four tonal blocks. As exampled in his work entitled *Diogenes* (c.1527), Ugo da Carpi experimented using different colour variations to create a highly dynamic range through colour and line shading.

Chiaroscuro printing from the sixteenth century was certainly more problematic as it required the registration of two or more different tonal blocks that had to be cut and printed in perfect registration. The process represented a challenge for the artist, for example, drafting the image, cutting the blocks to accurately fit, and register more than one block. Traditionally, relief prints are created through the combination of a raised printing surface and the surface qualities of the printing woodblock (end-grain hardwood, chipboard, linoleum, acrylic), which is then inked and printed under great downward pressure. The surface of the wood is manipulated by engraving, cutting and abrasion (traditional tools include sharp V-shape or U-shaped chisels). In order to obtain a high-quality printed image, the artist needs to cut well-defined lines and when inking the block to achieve solid areas of uniform colour.

When working with line alone, as in the example of chiaroscuro woodblock prints, shading was used to indicate distance, shape and texture, through the use of the direction of hatching lines to indicate the orientation and texture of an object. This is beautifully and simply described in a woodblock print by Lucas Cranach the Elder (1509). In Figure 5.112, St Christopher is depicted stepping out of the river on to a bank. The flow of the river is interrupted by his walking-staff and indicated by a c-shaped movement of lines, and the muscles of his leg and his hands strain as he steps on to the steep, undulating river bank that is littered with small flat round pebbles and tufts of grass. His stick is knotted and rough, and his cloak falls in heavy folds around him. The flat vertical rocks behind him stand tall, with sharp angular edges and in between grow small dense prickly trees. We can gain all this information about the materials and surface textures of St Christopher's environment through Cranach's use of a myriad of equally spaced lines to describe smooth surfaces, and thickening or diminishing of lines to suggest surface modulations.

Inspired by the traditional Chiaroscuro multicolour woodblock printing techniques of the sixteenth century, and in discussion between researchers at CFPR and Océ Print technologies (the members of this production team were: (1) Melissa Olen, Carinna Parraman and Peter McCallion at the Centre for Fine Print Research, University of the West of England, Bristol, and (2) Teun Baar and Maria V. Ortiz

Figure 5.112 Lucas Cranach the Elder (1472–1553), *St Christopher* (1509), woodblock, Rosenwald Collection, National Gallery of Art, Washington, USA (courtesy: National Gallery of Art).

Segovia at Océ Print Logic Technologies, Paris), the idea came to explore methods for creating multi-colour print-quality woodblocks using additive printing technologies. For this project we decided use a range of computer numerically controlled (CNC) additive and subtractive methods to create the blocks.

This idea builds on investigations into photomechanical helio relief printing at CFPR in 1999 (Figure 5.113), which explored multi-woodblock colour printing using halftone separations and helio relief printing processes (Anderton, 2003). Helio relief printing uses a photosensitive film emulsion, which was adapted from an industrial method for making signs, and developed as a fine art medium by Eric Vontillius and Deli Sacilotto at Graphicstudio, University of South Florida (1987). The method for obtaining an image involves exposing photosensitive film (Image Pro Super) through a detailed photographic negative, which is then washed (developed) with water and applied to a substrate (wood, stone, glass, etc.). The unmasked area is sandblasted to achieve a relief image. The protective film is resistant to abrasives, allowing the substrate to be blasted relatively deeply, creating a detailed relief. This surface can be inked and printed or the relief can be used sculpturally.

A series of preliminary tests explored ways to characterise the different machine and tool parameters. The project sampled a range of methods to generate blocks using additive (fabrication) multi layering methods and subtractive methods (laser or CNC engraving) to understand whether it was possible to

Figure 5.113 Frank Tinsley, *Dark Boat*, helio relief (1999) (courtesy: Frank Tinsley, CFPR print archive).

create a range of 'hand-made' marks using traditional cutting tools, to understand machine limitations and to find out whether machine-made blocks still *look like* machine-made blocks. Following the traditional printmaking process, the blocks were inked and printed in the same way on an Albion relief press. The results of the target tests and the Chiaroscuro two-colour print demonstrated a range of qualities that were promising, but required refinement. We found the rasterisation process of the bitmapped greyscale image to vector format unrefined, with jagged edges and loss of detail. We also needed to rule out the difference between poor file interpolation and the limits of the materials, milling tools or print heads.

The objective for the next step was to improve image quality, our hypothesis being that a vector-based file may improve the jagged edges. The types of points, lines and curves that are used in vector imagery to create paths in 3D and 2D software can be transcribed into the same paths for laser engraving or machining. The benefits of this design process included the ability to define points and contours that can be scaled up or down without loss in the sharpness of the line, and did not require the shape to be reinterpreted on a pixel-by-pixel basis. In order to test the idea that a vector-based file could produce a more accurate block, a new image was generated. Two layers were designed separately, which, when superimposed, created a composite image (Figure 5.114).

In our preliminary testing, many of the detailed engraved lines in the work were of a similar width. In order to test the limits of the machine, different parameters were incorporated, such as a range of line thicknesses, complex patterns and solid and open areas. The central image was a Japanese-style fish that is surrounded by a highly decorative wave motif. The waves were designed using organic and halftone patterns and a variety of curling, wavy lines and decorative shapes. The fish makes full use of three colours: the paper colour, a background colour that contains some texture and contrast, and a second colour as the outline and some pattern.

In this experiment we used four different methods to create our printing blocks: 2.5D printing, CNC milling, laser cutting and hand-cut lino. One block was created using additive manufacturing

Figure 5.114 Test image, showing the two coloured layers as a complete image (artwork by Melissa Olen, CFPR).

methods – a relief block was printed using an Océ 2.5D digital printer prototype (also referred to as textural printing or elevated printing technology; http://www.projecteiger.com, Project Eiger, is a development of Océ Canon). Three blocks were created using subtractive methods by machining into existing material: laser cutting into plywood, milling into polyurethane modelling board and hand cutting into lino. The latter traditional approach was considered important as a benchmark to understand whether the traditional and looked-for qualities that were made by hand could be replicated in the machine-made blocks, which represented a subjective but important 'ground truth'. This hand-made process quickly posed a challenge in following the test image. As the image was specifically designed to pose a challenge to the machining and fabrication methods, some areas containing minute detail such as a halftone dot was time consuming and difficult to copy. Figure 5.115 shows the individual layers with the top layer on the left and the bottom layer on the right, with highlighted sections, which were used for comparison.

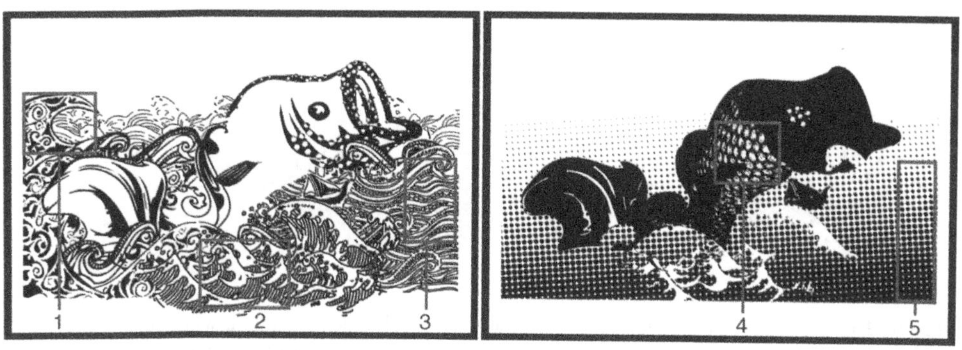

Figure 5.115 Test image, showing the separate layers: (left) top layer and (right) bottom layer, along with numbered sections for comparison between prints (courtesy: Melissa Olen, CFPR).

For each process, the file was output as a vector file (e.g. AI, PS, EPS, SVG). As performed in the preliminary tests, the blocks were machined to the same dimensions and a depth of 2 mm was used (the maximum distance between low and high parts of the block). Although a 2 mm depth or height was easily achieved by the CNC milling and 2.5D printing, the laser and hand cutting processes were less precise. With the need to adjust the power of the laser to achieve varying depths, a trade-off was identified. In order to maintain detail in the regions with fine lines, less power was necessary, which then resulted in a lower surface relief. Details of the inked relief blocks, as shown in sections 5 and 2 of Figure 5.115, were photographed and compared in Figure 5.116.

The photograph of the laser cut halftone section (top, Figure 5.116) highlights one of the biggest challenges encountered in this process. While a vector path is beneficial to creating a smooth edge, this only cuts away the immediate area around the shape. In order to remove the larger areas of negative space the laser must go back and forth over the woodblock, in a line-by-line order, and burn away the remaining space. In the areas of very fine detail, and where the first and the second pass of the laser slightly overlap, it causes more of the wood to be etched away, resulting in a loss of fine detail. This is demonstrated in both the halftone pattern and also in sections of the waves. The scorching has resulted in some of the halftone dots being erased from the block.

Similar to the laser cut woodblock, the CNC milled block also encountered problems in the halftone area. Although the small dots were not erased during the milling process, the smallest points stood 2 mm in height and just under a half of a millimetre wide. Therefore, the structure was considerably weakened, which resulted in a few of the points breaking away during the printing process. While it appears that some integrity to the structures were lost in the finer detail in the subtractive machined processes, this was not a problem with the 2.5D printed (Océ) block. One of the benefits to the additive process was that it could be built with a wider base than the 2 mm tall halftone dots, allowing it to taper to the correct size at the printing surface. This allowed the structures to maintain greater integrity and strength. While this may also be possible with certain CNC milling devices, our device would have required manual calculation and reworking of the image back in a raster format (as the variable distance in the cut is directly related to changes in grey levels).

The time required to create the individual blocks varied from approximately 45 minutes using the laser cutter, between 60 and 67 hours using the CNC milling, to several days to produce the lino cut.

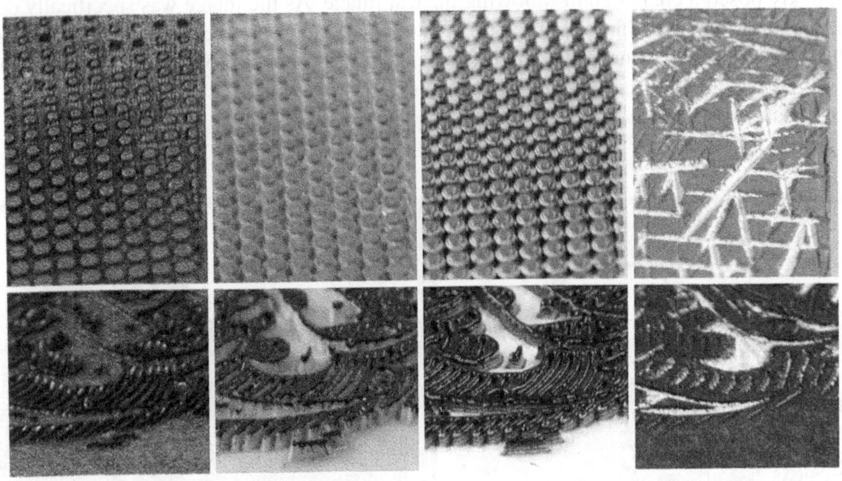

Figure 5.116 Details of the fabricated printing blocks. The blocks are inked, which highlight the differences. Processes from left to right: laser cut, CNC milled, 2.5D print and hand-cut lino. The halftone pattern in section 5 (top row) and the waves are located in section 2 (bottom row) (courtesy: Melissa Olen, CFPR)

The 2.5D printing took just over 2 hours to produce both of the blocks (which can be printed simultaneously).

All the blocks were inked and printed on to a heavyweight 400 g/m^2 cotton paper. A mid teal-blue was mixed for the background colour and a dark blue was mixed as the second colour. The blocks were printed on an Albion relief press. The prints demonstrated a rich variety of marks, each yielding their own particular qualities and exemplifying the individual limitations. Figure 5.117 compares the identified sections as illustrated in Figure 5.115.

The prints from the CNC milled blocks presented the best reproduction. These prints maintained good definition of line, including the smallest elements in the spirals, lines and bubbles in the waves of sections 1, 2 and 3. In section 4 the CNC also maintained the appearance of texture in the fish scales with more accuracy than any of the other processes. The limitation of this process, however, was the reproduction of the smallest dots in the halftone pattern, as some of the elements were missing in the print from section 5.

The Océ printing blocks were able to produce a print with well-defined shape and line at greater thicknesses, but the detail in the fine lines were lost in section 1 and the parallel lines of the waves in section 2 were broken and varied in thickness. The reproducible detail of the bubbles in section 3 and the texture of the fish scales in section 4 were also lost. Furthermore, the contour of the scales appeared more rounded in the print. These limitations can be attributed to the rounded surface of the printing block itself, as can be seen in Figure 5.116. However, this surface quality leant itself to the reproduction of the halftone pattern in section 5, where all the small dots were accurately reproduced in print.

Figure 5.117 Comparing the printed results and machining/fabricating processes (courtesy: Melissa Olen, CFPR).

Due to the limitation of the laser cutting process previously discussed in this paper, the fine detail of section 1, and all of the printed elements of section 3, failed to produce a suitable print. The areas with larger shapes and more spacing between lines and dots, as seen in sections 1, 2, 4 and 5, produced better results and maintained detail, yet the definition of the printed lines and shapes were not as clear as in the design. The halftone pattern in section 5 also presented the trade-off between accurate shape definition of the printed elements and interference from the ink transfer as a result of the low relief.

The prints produced by hand-cut lino demonstrate a high level of edge definition in both the larger shapes and the wavy lines that can be achieved through this analogue process. However, the compromise with this process is the balance of the lino material, amount of time, skill and (accuracy of) tools required for producing the smallest of the design elements. Examples include the small bubbles in section 3 and the precise placement of variable dots in the halftone pattern of section 5. It was decided to create a textural background, rather than try to re-create a halftone dot.

The on-going interest in digitally mediated processes to create prints is certainly not new, and as machined, 3D printed or fabricated blocks are used to produce print matrices for traditional printing, has demonstrated new approaches and insights towards generating a print matrix. CNC allows for printing blocks to be made on larger scales, with more detail and greater precision. However, akin to any traditional printmaking process, it is crucial to choose the most appropriate material and technique. More importantly, the setup of the image, specification requirements, file formats, time and costs in machining and materials may pose a compromise for the final print quality and barrier to production.

We identified a number of trade-offs, and each of the different processes exemplified strengths in different sections of the print. Whilst the CNC and hand cut lino demonstrated a number of strengths, these processes also required the longest time to produce.

With the ability to further modulate the relief structure and surface qualities, the Océ 2.5D digital printer prototype presents potential for refinement. An area for further investigation would be into producing a flatter print surface. Additionally, the potential exists to incorporate a microtextured surface to mimic the qualities of the other materials used in this testing. This work was supported by the Marie Curie Initial Training Networks (ITN) CP7.0 N-290154 funding.

References

Abrahams, C. (2013) *Wallpaper: A Collection of Modern Prints*, Quadrille Publishing, London.

Anderton, R. (2003) The 16th century journeyman printmaker and the invention of multi-colour printing in Europe, in *3rd Impact Multidisciplinary Printmaking Conference,* University of Cape Town, South Africa.

Attwood, P. and Powell, F. (2009) *Medals of Dishonour*, British Museum Press, London.

Baxter, W., Wendt, J., and Lin, M.C. (2004) *IMPaSTo: a realistic, interactive model for paint, in Proceedings of the 3rd international Symposium on Non-photorealistic Animation and Rendering*, ACM, Annecy, France.

Behr, C. (2010) The symbolic nature of gold in magical and religious contexts, in *Staffordshire Hoard Symposium*, British Museum.

Blatner, A., Ferwerda, J.A., and Darling, B. (2011) TangiPaint: A tangible digital painting system,in *19th Color and Imaging Conference*, Society for Imaging Science and Technology and Society for Information Display, San José, California.

Blunt, W. (1994) *The Art of Botanical Illustration*, Antique Collectors Club, Woodbridge, Suffolk.

Brusati, C. (1995) *Artifice and Illusion: Art and Writing of Samuel Van Hoogstratenby*, University of Chicago Press, Chicago, Illinois..

Cennini, C.D.A. (1954) *The Craftsman's Handbook. 'Il Libro dell'Arte'. c.1390*, Dover Publications, New York.

Chen, Z., Kim, B., Ito, D., and Wang, H. (2015) Wetbrush: GPU-based 3D painting simulation at the bristle level. *ACM Transactions on Graphics*, 34, 1–11.

Cheok, A.D., Lim, Z.S., and Tan, R.T.K. (2007) Humanistic oriental art created using automated computer processing and non-photorealistic rendering. *Computers and Graphics*, 31, 280–291.

Chu, N.S.H. and TAI, C.-L. (2005) Moxi: Real-time ink dispersion in absorbent paper. *ACM Transactions on Graphics*, 24, 504.

Clem, P.G. and Bell, N.S. (2013) *Inkjet Printing and Other Direct Writing Methods. Chemical Solution Deposition of Functional Oxide Thin Films*, Springer, Vienna.

Cohen, H. (1986) Off the shelf. *The Visual Computer*, 2, 191–194.

Colton, S. (2010) Stroke matching for paint dances, in *Sixth International Conference on Computational Aesthetics in Graphics, Visualization and Imaging* (eds P. Jepp and O. Deussen), 2010,, *Eurographics Association*, Aire-la-Ville, Switzerland, pp. 67–74.

Colton, S., Valstar, M.F., and Pantic, M. (2008) *Emotionally aware automated portrait painting, in Proceedings of the 3rd International Conference on Digital Interactive Media in Entertainment and Arts*, ACM, Athens, Greece.

Cook, J.G. (1984a) *Handbook of Textile Fibres: Man-Made Fibres*, Woodhead Publishing.

Cook, J.G. (1984b) *Handbook of Textile Fibres: Natural Fibres*, Woodhead Publishing.

Crawford, W. (1979) *The Keepers of Light*, Morgan and Morgan, New York.

Curtis, C.J., Anderson, S.E., Seims, J.E., *et al.* (1997) Computer-generated watercolor, in *Proceedings of the 24th Annual Conference on Computer Graphics and Interactive Techniques*, ACM Press/ Addison-Wesley Publishing Co.

Dawson, T.P., Curran, P.J., and Plummer, S.E. (1998) LIBERTY – Modelling the effects of leaf biochemical concentration on reflectance spectra. *Remote Sensing of Environment*, 65, 50–60.

Deussen, O., Lindemeier, T., Pirk, S.R. ander Tautzenberg, M. (2012) *Feedback-Guided Stroke Placement for a Painting Machine*, Computational Aesthetics in Graphics, Visualization, and Imaging.

Economist (2017) Adidas's high-tech factory brings production back to Germany. *The Economist*.

Eriksson, Y. (1998) *Tactile Pictures: Pictorial Representations for the Blind, 1784–1940*, Acta Universitatis Gothoburgensis.

Evans, S. (2011) *The Manor Reborn: The Transformation of Avebury Manor*, The National Trust, London.

Fenlon, W. (2014) Nvidia's dynamic super resolution is downsampling made easy. [Online]: pcgamer.com. Available at: http://www.pcgamer.com/2014/09/19/nvidias-dynamic-super-resolution-is-downsampling-made-easy [Accessed 19 November 2015].

Ferwerda, J.A. (2015) *ImpastoR: A Realistic Surface Display System*, Vision Research.

Fores, A., Ferwerda, J., and Gu, J. (2012) Toward a perceptually based metric for BRDF modelling, in *20th Color and Imaging Conference, Los Angeles, USA, November* 2012, pp. 142–148.

Fox Talbot, H.W. (1844) *The Pencil of Nature*, Longman, Brown, Green and Longmans, London.

Furferi, R., Governi, L., Vanni, N., and Volpe, Y. (2014a) Tactile 3D bas-relief from single-point perspective paintings: A computer based method. *Journal of Information and Computational Science*, 11, 5667–5680.

Furferi, R., Governi, L., Volpe, Y., *et al.* (2014b) From 2D to 2.5D, i.e. from painting to tactile model. *Graphical Models*, 76 (6), 706–723.

Gamble, W. (1897) *The Photographic Journal*, 21 (6), 6 February 1897, 126–127.

Ganapol, B.D., Johnson, L.F., Hammer, P.D., *et al.* (1998) LEAFMOD: A new within-leaf radiative transfer model. *Remote Sensing of Environment*, 63, 182–193.

Gatys, L.A., Ecker, A.S., and B, M. (2015) A neural algorithm of artistic style. arXiv [Online]: semanticscholar.org. Available at: https://pdfs.semanticscholar.org/77c5/12cbb832436e1a35ad434e6bb3d763799763.pdf.

Gnann, A. (2012) Chiaroscuro, in Arts, R.A.O.T. (ed.), Royal Academy of the Arts, London.

Goethe, J.W. (1970) *Theory of Colours (1808) 'Zur Farbenlehre'*, Massachusetts Institute of Technology, Cambridge.

Gooch, A.A. and G, B. (1999) *Using non-photorealistic rendering to communicate shape*. Available at: https://www.researchgate.net/publication/238765046_8_Using_Non-Photorealistic_Rendering_to_Communicate_Shape.

Harrod, T. (2015) *The Real Thing*, Hyphen Press, London.

Hedge, S., Gatzidid, C., and Tian, F. (2012) Painterly rendering techniques: a state-of-the-art review of current approaches. *Computer Animation and Virtual Worlds*, 24, 11.

Hertzmann, A. (1998) Painterly rendering with curved brush strokes of multiple sizes, in *Proceedings of the 25th Annual Conference on Computer Graphics and Interactive Techniques*, ACM.

Hertzmann, A. (2002) *Stroke-Based Rendering. Recent Advances in NPR for Art and Visualization*, SIGGRAPH, p. 31.

Hertzmann, A. (2003) A survey of stroke-based rendering. *IEEE Computer Graphics and Applications*, 70–81.

Hertzmann, A., Interrante, V., and Lum, E. (2002) Stroke-based rendering, recent advances In NPR for art and visualization, course notes for SIGGRAPH 2002. [Online]: http://vis.cs.ucdavis.edu/papers/ranpr2002.p.

Hoskins, L. (2005) *The Papered Wall: History, Pattern, Techniqu*, Thames & Hudson, London.

Huson, D., Hoskins, S., and Thirkell, P. (2005) Photo ceramic relief imaging, in *Proceedings of IS&T NIP & Digital Fabrication Conference*, 2005, IS&T, pp. 39–42.

Jacquemoud, S. and Baret, F. (1990) PROSPECT: A model of leaf optical properties spectra. *Remote Sensing of Environment,* 34, 75–91.

Jacquemoud, S. and Ustin, S.L. (2001) Leaf optical properties: A state of the art, in *CNES 8th International Symposium of Physical Measurements and Signatures in Remote Sensing*, pp. 223–332.

Julesz, B. (1991) Textons, the elements of texture perception and their interactions. *Nature*, 91–97.

Kaplan, M., Gooch, B., and C, E. (2000) Interactive artistic rendering, in *Proceedings of the 1st International Symposium on Non-photorealistic Animation and Rendering*, 2000, ACM Digital Library, Annecy, France, pp. 67–74.

Karagozler, E., Poupyrev, I., Fedder, G., and Suzuki, Y. (2013) Paper generators: harvesting energy from touching, rubbing and sliding, in *26th Annual ACM Symposium on User Interface Software and Technology*, St Andrews, Scotland, ACM.

Kasao, A. and Miyata, K. (2006) Algorithmic painter: a NPR method to generate various styles of painting. *The Visual Computer*, 22, 13.

Kasao, A. and Nakajima, M. (1998) A resolution independent nonrealistic imaging system for artistic use, in *Proceedings of the IEEE International Conference on Multimedia Computing and Systems*, 1998, Austin, Texas, IEEE, pp.358–367.

Klein, S., Simske, S., Adams, G., *et al.* (2012) 3D printing of transparent glass, in *NIP and Digital Fabrication Conference,* 2012, pp. 336–337.

Klein, S., Avery, M.P., Richardson, R.M., *et al.* (2015) The colour of glass, in *NIP and Digital Fabrication Conference*, 2015, pp. 89–93.

Lambert, N. (2003) *A Critical Examination of 'Computer Art'*, DPhil, University of Oxford.

Lee, J. (2001) Diffusion rendering of black ink paintings using new paper and ink models. *Computers and Graphics*, 25, 295–308.

Leung, T. and Malik, J. (2001) Representing and recognising the visual appearance of materials using three-dimensional textons. *International Journal of Computer Vision*, 43, 29–44.

Lewis, J.A. (2006) Direct ink writing of 3D functional materials. *Advanced Functional Materials*, 16, 2193–2204.

Lindemeier, T., Pirk, S., and Deussen, O. (2013) Image stylization with a painting machine using semantic hints. *Computers and Graphics*, 37, 293–301.

Loundon, G. (2015) *Object Lessons*, Ridinghouse, London.

Lovejoy, M. (2004) *Digital Currents: Art in the Electronic Age*, Routledge, London.

Lu, J., Barnes, C., Diverdi, S., and Fonkelstein, A. (2013) RealBrush: Painting with examples of physical media. *SIGGRAPH 2013 (TOG)*, 32 (4, no.117).

Maaten, L.J.P., Eric, O., and Postmac, E.O. (2010) Texton-based analysis of paintings, in *Applications of Digital Image Processing* (ed. A.G. Tescher), vol. XXXIII, SPIE, p. 14.

MacGregor, N. (2012) *A History of the World in 100 Objects*, Penguin, London.

McCallion, P. (2017) *The Development of Methods for the Reproduction in Continuous Tone of Digitally Printed Colour Artworks*, PhD thesis, University of the West of England, Bristol.

McCullough, M. (2010) Abstracting craft: The practiced digital hand, in *The Craft Reader* (ed. G. Adamson), Berg.

McMillan, W.G. and Pages, - (1890) *A Treatise on Electro-metallurgy: Embracing the Application of Electrolysis to the Plating, Depositing, Smelting, and Refining of Various Metals, and to the Reproduction of Printing Surfaces and Art-work, C.* Griffin and Company, London.

Nappi, M. (2013) Drawing w/Digits_Painting w/ Pixels: Selected artworks of the gesture over 50 Years. *Leonardo*, 46, 163–169.

Ning, X., Laga, H., Saito, S., and Nakajima, M. (2011) Contour-driven Sumi-e rendering of real photos Computers and Graphics, 35, 122–134.

O'Dowd, P., Hoskins, S., Walters, P., and Geisow, A. (2015) Modulated extrusion for textured 3D printing, in *NIP31 International Conference on Digital Printing Technologies*, Portland, Oregon, Society for Imaging Science and Technology.

Okudera, S., Sagawa, K., Nakajima, Y., *et al.* (2015) Psychological hue circle of blind people and development of a tactile color tag for clothes, in *AIC2015 TOKYO Midterm Meeting of the International Colour Association (AIC)* (eds Y. Hirohisa, K.Okajima, T. Ishida *et al.*), Ochanomizu Sola City Conference Center, Tokyo, Japan, International Colour Association (AIC).

Phong, B.T. (1975) Illumination for computer generated pictures. *Graphics and Image Processing*, 18, 311–317.

Ross, M. (2004) Interpreting the new museology. *Museum and Society*, 2, 84–103.

Rudolf, D., Mould, D., and Neufeld, E. (2003) Simulating wax crayons, in *11th Pacific Conference onComputer Graphics and Applications, 2003. Proceedings*, 8–10 October 2003, pp. 163–172.

Sayers, M.W. and Karamihas, S.M. (1998) *The Little Book of Profiling. Basic Information about Measuring and Interpreting Road Profiles*, University of Michigan.

Sayim, B. and Cavanagh, P. (2011) The art of transparency. *i-Perception*, 2, 679.

Seung-Chan, K., Israr, A., and Poupyrev, I. (2013) Tactile rendering of 3D features on touch surfaces, in *26th Annual ACM Symposium on User Interface Software and Technology*, St Andrews, Scotland, ACM.

Sheets, H.M. (2012) Looms with a view . *Artnews*, 19 September 2012. [Online]: http://www.artnews.com/2012/09/19/looms-with-a-view.

Stout, K.J. and Blunt, L. (2000) *Three Dimensional Surface Topography*, Penton Press, Kogan Page, London.

Stulik, D.C. and Kaplan, A. (2013) *The Atlas of Analytical Signatures of Photographic Processes: Cyanotype*, The Getty Conservation Institute.

Thirkell, P. (2004) *Altered Images*, Making Space Publishers, Isle of Wight.

Tresset, P. and Fol Leymarie, F. (2013) Portrait drawing by Paul the robot. *Computers and Graphics*, 37, 348363.

Ungar, S., Blades, M., and Spencer, C. (1997) Teaching visually impaired children to make distance judgements from a tactile map. *Cartographic Perspectives*, 28.

Weaver, M. (1992) *Henry Fox Talbot; Selected Texts and Bibliography*, Clio Press Ltd, Oxford.

Winkenbach, G. and Salesin, D.H. (1994) Computer-generated pen-and-ink illustration, in *Proceedings of the 21st Annual Conference on Computer Graphics and Interactive Techniques*, ACM.

Wright, S. and Stratton, J. (2007) *On the Way to Literacy*, Printing House for the Blind, Louisville.

Wu, Y.J. (2008) *The Effect of Substrate Properties on Print Attributes for Gravure Printing – From Proof to Press*, PhD Dissertation, Western Michigan University.

Xu, S., Tang, M., Lau, F.C.M., and Pana, Y. (2004) Virtual hairy brush for painterly rendering. *Graphical Models*, 66, 263–302.

Yamamoto, S., Mao, X., and Imamiya, A. (2004) Colored pencil filter with custom colors, in *12th Pacific Conference on Computer Graphics and Applications*, *Proceedings*, 6–8 October 2004, pp. 329–338.

Yang, C.-K. and Yang, H.-L. (2008) Realization of Seurat's pointillism via non-photorealistic rendering. *The Visual Computer*, 24, 303–322.

Zhang, K. and Yu, J. (2013) Generating abstract paintings in Kandinsky style, in *SIGGRAPH Asia 2013 Art Gallery*, Hong Kong, ACM, Hong Kong.

Zhang, X., Chan, K.L., and Constable, M. (2013) Exemplar-based Portrait Photograph Enhancement as informed by Portrait Paintings.

Bibliography and Further Reading

Ansty (2015) A guide to wallpaper printing, *Ansty*. [Online]: http://www.anstey.uk.com/files/documents/PDFs/Manufacturing/anstey_print_guide.pdf.

Avery, M.P., Klein, S., Richardson, R., *et al.* (2015) Through a glass clearly: the challenge of glass 3D-printing, in *Proceedings of the IS&T NIP31 International Conference on Digital Printing Technologies and Digital Fabrication,* pp. 280–284.

Baar, T., Olen, M., Parraman, C., and Ortiz Segovia, M.V. (2015) Woodblock printing as a means for 2.5D and 3D surface evaluation, in *AIC Colour 2015,* Tokyo, Japan, International Colour Association (AIC), *May* 2015, pp. 861–866.

Baar, T. and Ortiz Segovia, M.V. (2013) Colour and texture appearance modelling of 2.5D prints, in *12th Congress of the International Colour Association (AIC)* (eds L. Macdonald, S. Westland and S. Wuerger), Newcastle upon Tyne, UK.

Baar, T., Samadzadegan, S., Brettel, H., *et al.* (2014) Printing gloss effects in a 2.5D system, in *IS&T/SPIE Electronic Imaging, Proceedings of SPIE, Measuring, Modeling, and Reproducing Material Appearance,* vol. 9018, p. 90180M-8

Baar, T., Samadzadegan, S., Urban, P., and Ortiz Segovia, M.V. (2016) Interrelation between gloss and texture perception of 2.5D-printed surfaces in *IS&T/SPIE Electronic Imaging, Proceedings of SPIE, Measuring, Modeling, and Reproducing Material Appearance*, pp. 1–6.

Boyé, S., Barla, P., and Guennebaud, G. (2012) A vectorial solver for free-form vector gradients. *ACM Transactions on Graphics*, 31, 9.

Campisi, P., Neri, A., and Scarano, G. (2000) Reduced complexity modelling and reproduction of colored textures. *IEEE Transactions on Image Processing*, 9, 510–518.

Chu, N.S.H. and Tai, C.L. (2002) An efficient brush model for physically-based 3D painting, in *Proceedings of the 10th Pacific Conference on Computer Graphics and Applications*, pp. 413–421.

Clouse, D. and Voulangas, A. (2009) *The Handy Book of Artistic Printing: A Collection of Letterpress Examples, with Specimens of Type, Ornament, Corner Fills, Borders, Twisters, Wrinklers, and other Freaks of Fancy*, Princeton Architectural Press.

Collomosse, J.P. and Hall, P.M. (2003) Cubist style rendering from photographs, in *IEEE Transactions on Visualization and Computer Graphics*, 9, 443–453.

Cook, R.L. and Torrance, K.E. (1982) A reflectance model for computer graphics. *ACM Transactions on Graphics*, 1, 7–24.

Cooper, F. (2011) A gift fit for a Pope, 1500 years after its first creation, *Birmingham, Jewellery Industry Innovation Centre Report.* [Online]: http://www.bcu.ac.uk/cmsproxyimage?path=/_media/docs/frank-cooper-report.pdf.

Design Boom (2014) A23D 3D-printed letterpress font shown at London Design Festival, *Design Boom.* [Online]: http://www.designboom.com/technology/a23d-3d-printed-letterpress-font-09-17-2014/#.VBrh0ufniH8.twitter.

Diverdi, S., Krishnaswamy, A., Mech, R., and Ito, D. (2013) Painting with polygons: a procedural watercolor engine. *IEEE Transactions on Visualization and Computer Graphics,* 19, 723–735.

Dong, Y., Tong, X., Pellacini, F., and Guo, B. (2012) Printing spatially-varying reflectance for reproducing HDR images. *ACM Transactions on Graphics,* 31.

Dorsey, J. and Rushmeier, H. (2005) Digital modeling of the appearance of materials, ACM SIGGRAPH 2005 Courses.

Dorsey, J., Rushmeier, H., and Sillion, F. (2010) *Digital Modeling of Material Appearance*, Morgan Kaufmann, San Francisco.

Dyer, A.F. (2004) Henry Bradbury's First Nature-Prints of Ferns. *Fern Gazette*, 17, 18.

Elisha, D. (2009) *Printmaking + Mixed Media*, F+W Media, Inc.

Green, L. (2008) *Houses of the National Trust*, National Trust Books, London.

Haeberli, P. (1990) Paint by numbers: abstract image representations. *SIGGRAPH Computers and Graphics*, 24, 207–214.

Haller, M. (2004) Photorealism or/and non-photorealism in augmented reality, in *Proceedings of the 2004 ACM SIGGRAPH International Conference on Virtual Reality Continuum and Its Applications in Industry*, Singapore, ACM.

Hitchmough, W. (2000) *The Arts and Crafts Home*, Pavilion Press, London.

Hodgson, A. (2005) The use of gloss effects from inkjet printing for brand identification, personalisation and security, in *Proceedings of the IST International Conference on Digital Production Printing and Industrial Applications*, pp. 108–109.

Hoskins, S. (2007) *Water-Based Screenprinting*, A&C Black, London.

Hughes, A.A. and Vernon-Morris, H. (2008) *The Printmaking Bible: The Complete Guide to Materials and Techniques*, Chronicle Books.

Inglis, T.C., Inglis, S., and Kaplan, C.S. (2012) Special Section on CANS: Op Art rendering with lines and curves. *Computers and Graphics,* 36 (6), 607–621.

Kennedy, J.M. (1993) *Drawing and the Blind, Pictures to Touch*, Yale University Press, New Haven, Connecticut.

Levkowitz, H. (1997) *Color Theory and Modeling for Computer Graphics, Visualization, and Multimedia Applications*, Kluwer Academic, Boston, Massachusetts.

Li, J., Yao, L., Hendriks, E., and Wang, J.Z. (2012) Rhythmic brushstrokes distinguish Van Gogh from his contemporaries: findings via automated brushstroke extraction. *IEEE Trans. Pattern Anal. Mach. Intell.*, 34, 1159–1176.

Liu, Z. and Zhang, Z. (2011) *Face Geometry and Appearance Modeling: Concepts and Applications*, Cambridge University Press.

Martos, A. and Ruiz, B. (2013) Realistic virtual reproductions. Image-based modelling of geometry and appearance. *IEEE,* 978-1-4799-3170-5/13.

Moore, T. (1855) *The Ferns of Great Britain and Ireland by Thomas Moore and Nature Printed by Henry Bradbury*, Bradbury and Evans Whitefriars, London.

Moorlag, C.P., Zwartz, E.G., Sisler, G., *et al.* (2013) *Method of Controlling Gloss, Patent no. US8615188 B2, published* 24 December 2013.

Murray, J.K. and Cahill, S.E. (1987) Recent advances in understanding the mystery of ancient Chinese 'Magic Mirrors'. *Chinese Science*, 8, 1–8.

Ng, Y.S., Kuo, C.H., Wang, C.J., *et al.* (2003) Standardization of perceptual based gloss and gloss uniformity for printing systems (INCITS W1.1), in *Proceedings of SPIE, Image Quality and System Performance,* vol. 5294, pp. 88–93.

Ning, X., Laga, H., Saito, S., and Nakajima, M. (2011) Contour-driven Sumi-e rendering of real photos. *Computers and Graphics*, 35, 122–134.

Obein, G., Knoblauch, K., and Viénot, F. (2004) Difference scaling of gloss: nonlinearity, binocularity, and constancy. *Journal of Vision*, 4, 711–720.

Olen, M., Parraman, C., Ortiz Segovia, M.V., *et al.* (2015) Woodblock, an exploration of alternative computer aided methods for multi-colour woodblock printing, in *Impact 9 Multidisciplinary Printmaking Conference,* Hangzhou, China, China Arts Academy.

Parraman, C. (2013) The development of vector based 2.5D print methods for a painting machine, in *SPIE and IS&T Color Imaging XVIII: Displaying, Processing, Hardcopy, and Applications* (eds R. Eschbach, G.G. Marcu and A. Rizzi).

Parraman, C. (2014) The material image: Artists approaches to reproducing texture in art, in *The Image Knowledge Community, 5th International Conference on the Image*, Berlin, Germany.

Parraman, C., McCallion, P., Cohen, C., and Hoskins, S. (2013) Reproduction of texture in digitally printed artworks, in *AIC 12th Congress of the International Colour Association* (eds L. Macdonald, S. Westland and S. Wuerger), Newcastle upon Tyne, UK, pp. 577–580.

Parraman, C., O'Dowd, P., and Harding, M. (2015) Painting by numbers: Transforming fields and edges to vectors, in *AIC Colour 2015, Tokyo Midterm Meeting of the International Colour Association*, Tokyo, Japan.

Parraman, C., O'Dowd, P., and Harding, M. (2015) Painting by numbers: The development of texture in digitally printed artworks, in *Impact 9 Multidisciplinary Printmaking Conference,* Hangzhou, China Arts Academy, China.

Pavlidis, G., Koutsoudis, A., Arnaoutoglou, F., *et al.* (2007) Methods for 3D digitization of cultural heritage. *Journal of Cultural Heritage*, 8.

Pellacini, F., Ferwerda, J.A., and Greenberg, D.P. (2000) Toward a psychophysically-based light reflection model for image synthesis, in *Proceedings of the 27th Annual Conference on Computer Graphics and Interactive Techniques*, 55–64.

Pieraccini, M., Guidi, G., and Atzeni, C. (2001) 3D digitizing of cultural heritage. *Journal of Cultural Heritage*, 2.

Pierce Lhotka, B. (2011) *Digital Alchemy: Printmaking Techniques for Fine Art, Photography, and Mixed Media*, New Riders.

Rosin, P.L. and Lai, Y.K. (2013) Artistic minimal rendering with lines and blocks. *Graphical Models*, 75, 208–229.

Sansoni, G., Trebeschi, M., and Docchio, F. (2009) State-of-the-art and applications of 3D imaging sensors in industry, cultural heritage, medicine, and criminal investigation. *Sensors*, 9.

Sattler, M., Sarlette, R., and Klein, R. (2003) Efficient and realistic visualization of cloth, in *Eurographics Symposium on Rendering.*

Shoup, R. (2001) SuperPaint: An early frame buffer graphics system. *IEEE Annals of the History of Computing,* 1058–6180, 5.

Takano, R., Baba, K., Inoue, S., *et al.* (2012) Reproduction of gloss unevenness on printed paper by reflection model with consideration of mesoscopic facet, in *IS&T Proceedings of 20th Color and Imaging Conference*, pp. 206–210 (5).

Thompson, F., Wicks, G., and Crawford, G. (2015) Porous-wall hollow glass microspheres for security printing applications, in *Proceedings of IS&T NIP31 International Conference on Digital Printing Technologies and Digital Fabrication*, pp. 391–394.

Thornton, P. (1984) *Authentic Decor: The Domestic Interior 1620–1920*, Seven Dials Press, London.

Walters, P. (2008) Richard Hamilton – art, technology and a medal designed for dishonour, *CFPR*. [Online]: http://www.uwe.ac.uk/sca/research/cfpr/research/3D/Artists/hamilton/Hamilton_v4.pdf.

Walters, P., Huson, D., Parraman, C., and Stanic, M. (2009) 3D printing in colour: Technical evaluation and creative applications, in *6th Impact International Multidisciplinary Printmaking Conference*, Bristol.

Walters, P. and Thirkell, P. (2007) New technologies for 3D realization in art and design practice. *Artifact*, 1, 232–245.

Wang, J., Allebach, J. P., and Ortiz Segovia, M.V. (2014) Semantically-based 2.5D texture printing,in *IEEE International Conference on Image Processing (ICIP),* p. 4.

Warr, M. (2003) *Créer Textures et Effets de Matière*, Eyrolles.

Weyrich, T., Peers, P., Matusik, W., and Rusinkiewicz, S. (2009) Fabricating microgeometry for custom surface reflectance. *ACM Transactions on Graphics, SIGGRAPH*, 28 (32), 1–6.

White, M.A. and LeBlanc, M. (1999) Thermochromism in commercial products. *Journal of Chemistry Education*, 76, 1201–1205.

Wijntjes, M.W.A. and Pont, S.C. (2010) Illusory gloss on Lambertian surfaces. *Journal of Vision*, 10, 13.

Wu, H., Dorsey, J., and Rushmeier, H. (2011) *Physically-based interactive bi-scale material design. ACM Transactions on Graphics*, 30.

Wu, H., Dorsey, J., and Rushmeier, H. (2013) Inverse bi-scale material design, *ACM Transactions on Graphics, SIGGRAPH Asia*, 32.

Yakovlev, A.V., Milichko, V.A., Vinogradov, V.V., and Vinogradov, A.V. (2016) Inkjet color printing by interference nanostructures. *ACS Nano*, 10, 3078–3086.

Zhao, S., Hasan, M., Ramamoorthi, R., and Bala, K. (2013) Modular flux transfer: efficient rendering of high-resolution volumes with repeated structures. *ACM Transactions on Graphics (TOG)*, 32 (4, no.131).

Zhao, S., Jakob, W., Marschner, S., and Bala, K. (2012) Structure-aware synthesis for predictive woven fabric appearance. *ACM Transactions on Graphics (TOG)*, 31 (4, no. 75).

Index

aerogel, 77
Amara's Law, 71
anamorphic, 49, 50, 169
ancona, 92, 180
animation, 2, 63, 66, 88, 205, 211, 230
appearance-terms for describing, 1–5, 7, 9, 11, 15, 22, 24, 25, 28, 31, 32, 35, 36, 38, 40, 44, 52, 54, 58–62, 64, 65, 67, 72, 73, 78, 85, 86, 89, 92, 93, 99, 100, 105–107, 109–111, 127, 130, 133, 136, 137, 139, 141, 142, 161, 162, 170, 172, 173, 175, 176, 178, 179, 181, 190, 191, 197, 200–206, 208, 210, 213, 229
  bidirectional reflectance distribution function (BRDF), 78, 86–88, 109–112, 133
  characterisation, 86, 112, 211
  contrasts, 10, 15
  convex–concave, 41, 44, 46, 47, 197
  crushed relief (relievo sticciato or schiacciato), 41
  debossed, 93
  depth, 13, 16, 23, 42, 44, 45, 47, 55, 57, 63, 68, 79, 81, 82, 112, 122, 148, 150, 151, 158, 167, 169, 181, 197, 215, 217, 220, 228
  dynamic range, 11, 13, 58, 60, 96, 177, 179, 181, 224
  edge, 15, 25, 32, 39, 41, 63, 100, 111, 133, 146, 151, 211, 217, 221, 228, 230
  flattened relief (rilievo schiacciato), 41, 42, 188
gloss, 2, 3, 24, 25, 31, 63, 72, 79, 93, 98, 99, 106, 109–111, 113, 117, 141, 161, 172, 179, 183, 194, 200–202, 205
grain, 6, 59, 73, 74, 94, 133, 137, 138, 154, 172, 176, 180, 220, 224
high relief (alto-rilievo), 21
illumination, 5, 7, 13, 25, 32, 35, 36, 38, 39, 43, 44, 57, 58, 66, 80, 82, 86, 87, 106, 110, 157, 167–169, 178, 181, 195
illusion, 3, 10, 29, 31, 36, 41, 44, 46, 47, 49, 50, 67, 100, 155, 160, 167–169, 177, 180, 181, 183, 184, 230
international roughness index (IRI), 3
low-relief, 3, 15, 139, 152
macrotexture, 195, 203
microstructure, 77, 85, 195, 200
pattern, 5, 6, 14, 29, 31, 40, 52, 60, 61, 79, 82, 86, 98, 110, 112, 113, 134, 135, 137, 139, 140, 146, 147, 156, 161, 165, 166, 172, 178, 183, 190, 191, 196, 197, 210, 216, 223, 226, 228–230
properties of materials, 1–3, 8, 23–26, 29, 58–66, 74, 77, 78, 82, 85–87, 89, 96, 101, 104, 106, 116, 131, 133, 153, 161, 162, 172, 196, 201
psychological effect, 180
raking light, 21, 39, 40, 45, 58, 84, 158
reflection, 7, 25, 26, 29, 38, 57, 58, 65, 78, 79, 85, 86, 107, 112, 156, 179, 181, 183, 196, 203
refraction, 196

2.5D Printing: Bridging the Gap Between 2D and 3D Applications, First Edition. Carinna Parraman and Maria V. Ortiz Segovia.

Companion website: www.wiley.com/go/bridging2d3d

appearance-terms for describing (*Contd.*)
relief, 3, 7, 12, 15, 16, 23, 25, 26, 29, 35, 39–45, 47–49, 52, 53, 55, 57, 58, 67, 68, 79, 81–83, 86, 92, 93, 96, 98–100, 105, 109, 130, 134–136, 139, 140, 142, 145, 148, 150–156, 161, 166, 167, 177, 179–181, 184, 185, 187, 188, 190–195, 199–201, 203, 210, 214, 217–220, 223–230
relievo (cavo-relievo), 41
roughness, 2, 3, 63, 78, 85, 109, 153, 200–203
spatial relationship, 41
subsurface, 25, 86, 172, 204
surface, 1, 3, 5, 7, 8, 10, 14–16, 20, 21, 23–28, 35, 38–42, 44, 52, 53, 57, 60–63, 65–67, 72, 75, 77–86, 88–90, 93–95, 98–101, 105–111, 113, 117, 121, 124, 128–130, 133, 136, 137, 139, 140, 142, 143, 145, 150–152, 155–158, 160, 161, 166, 167, 172, 175, 178–181, 184, 190, 191, 195–197, 200–204, 206–208, 210, 214–217, 220, 221, 223–225, 228–230
tactile texture, 1, 2
texture, 1–5, 7, 8, 12, 15, 16, 20, 21, 23–26, 28–32, 35, 36, 38, 40, 49, 51, 57–60, 67, 72, 78–82, 84–86, 88, 89, 92–94, 96, 98, 99, 102, 105–107, 109–113, 133, 141–145, 150, 151, 154, 156–158, 160, 161, 175, 177, 191, 195, 197, 199, 204–206, 208, 210–213, 221, 224, 226, 229
topology, 191
visual texture, 2, 31, 86, 88, 161
archaeology, 22, 51, 57, 105, 107, 108, 112
Ars Electronica Festival, 165
artists and designers, 2–4, 11, 13–20, 22–25, 28, 31, 35, 36, 38–42, 45, 47, 49, 52, 58, 61, 63, 65–67, 70, 71, 74, 75, 77, 79, 83, 90–92, 109, 112, 119, 121–123, 127, 130, 138, 140, 142, 145–148, 150–153, 155, 156, 158, 162, 165, 167, 169, 175–177, 179, 181, 183, 188, 191–193, 203–207, 209–211, 213, 214, 221, 223, 224
Alberti, Leon Battista, 13
Alkalay, Shay, 74
Anderton, Richard, 220, 225, 230
Atkins, Anna, 128
Aure, Xavier, 79, 84, 157, 159, 160
Baar, Teun, 85, 99, 105, 109, 224
Bell, Charles, 37, 38, 86, 105, 167, 212
Bergdahl, John, 153
Bodman, Sarah, 222
Bowden, Phil, 221
Brennan, John, 217
Breton, Adela, 55–57, 66
Burgkmair, Hans, 224
Canaletto, (Giovanni Antonio Canal), 36
Carl, Eric, 144
Carpi, Uga da, 224
Cartlidge, George, 192, 193
Close, Chuck, 165
Cohen, Harold, 206, 207
Collier, John, 47
Constable, John, 12
Cotán, Juan Sanchez, 63
Cranach, Lucas (the Elder), 224, 225
Crivelli, Carlo, 180, 181
da Vinci, Leonardo, 13, 19, 20, 30, 33, 36, 66, 68, 71, 105, 106, 108
da Volpedo, Giuseppe Pelizza, 11
Day, Paul, 45, 48, 49
de Lairesse, Gerard, 176, 177
Dent, Matthew, 153, 154
de Pereda, Antonio, 63
Docherty, Thomas, 155
Dürer, Albrecht, 12, 224
Eley, Robin, 65, 66, 167
El Greco, Doménikos Theotokópoulos, 36
Estes, Richard, 38, 39
Evenhuis, Jiri, 162
Fenuta, Angelene, 162
Franz, Kai, 208
Fungai, Bernardino, 180–182
Gaudí, Antoni, 197, 198
Ghiberti, Lorenzo, 42–44, 66
Goes, Hugo, 134
Goffin, Roxanne, 215
Goltzius, Hendrick, 62, 63, 214
Goodman, Ben, 218
Greenhalgh, Shaun, 19, 29
Grien, Baldung, 36
Grosser, Benjamin, 208
Guenot, Lucy, 218
Hamen y Leon, Juanvan der, 63, 64
Hamilton, Richard, 156, 157
Heda, Willem Claeszoon, 65, 181, 182
Hockney, David, 13, 14, 18, 33

Holbein, Hans (the Younger), 49, 59, 60, 66, 67, 167
Hoskins, Stephen, 31, 44, 67, 104, 137, 192, 194, 223
Hunter, Ian, 3, 90
Huson, David, 32, 44, 67, 104–106, 110, 111, 192, 193
Ingram, Cath, 218
Ingres, Jean-Auguste Dominique, 13
Kalmyck, Fedor Ivanovich, 43, 44
Keating, Tom, 18
King, Louise, 214
Kistler, Samuel, 77
Kubel, Henrik, 223
Kusama, Yayoi, 49
Kyttanen, Janne, 162
Lorrain, Claude, 36, 37
Lucas, David, 12, 224, 225
Manchee, Doug, 11, 30, 111
Marey, Etienne-Jules, 11
Mayatt, John, 18, 19
McCallion, Peter, 14, 30, 31, 44, 67, 191, 194, 195, 224
McElheny, Josiah, 47
Mer, Yael, 74
Michelangelo, Buonarroti, 36, 43
Mor, Anthonis, 16, 62
Muybridge, Eadweard James, 11
Newell, Martin, 204
Noble, Emma, 155
O'Dowd, Paul, 165, 166, 211–213
Palladio, Andrea, 167
Parraman, Carinna, 219, 224
Parraman, Grace, 2
Peleg, Danit, 162
Perry, Grayson, 156, 165
Poupyrev, Ivan, 165
Previati, Gaetano, 11
Quinn, Marc, 165
Rembrandt (van Rijn), 15, 20, 21, 30
Sacilotto, Deli, 225
Scamozzi, Vincenzo, 167
Sharpe, Jim, 49, 50
Sowden, Tom, 222
Strazza, Giovanni, 3, 4
Sully, David, 216
Sutherland, Ivan, 210
Takizawa, Akiko, 215
Tiepolo, Giovanni Battista, 168
Tinsley, Frank, 226
Tintoretto, Jacopo, 36
Titian Titian (Tiziano Vecellio), 61
Tura, Cosimo, 183
Turner, Brad, 90
Turner, J.M.W. 20, 36
Urquiola, Patricia, 191, 193
van Eyck, Jan, 10, 15, 60, 61, 67, 177, 181
van Herpen, Iris, 22, 162
van Hoogstraten, Samuel, 65, 168–170
van Meegeren, Han, 18
Vasari, Giorgio, 20, 42, 61, 67, 167
Velázquez, Diego, 10, 11, 20, 63–65, 67
Vermeer, Johannes, 12, 13, 18, 33
Verrocchio, Andrea del, 20
Vontillius, Eric, 225
Wallis, Katie, 222
Wanders, Marcel, 75, 76
Warhol, Andy, 13, 205
Watson, Musgrave, 48
Wenner, Kurt, 169, 171
Werenskiold, Dagfin, 53
Williams, Scott, 223
Williamson, Matthew, 107, 133, 140
Wipprecht, Anouk, 162
Wisniewski, Matthias, 75, 76
Woodbury, Walter, 29, 130
Woodington, William F. 48
arts, 3, 9, 31, 54, 67, 69–71, 73, 90, 106, 107, 109, 110, 112, 113, 116, 123, 124, 129, 135, 143, 148, 157, 205, 207, 210, 221
Arts and Humanities Research Council, 90, 116, 210
Assyria, Assyrian, 10, 40, 58, 151
aura, 4, 11, 17, 18, 22, 55, 168, 179, 184
authentication, 19, 21, 97, 122, 123
azulejos, 186, 187

biodegradable, 27
biomaterials, 78, 116
biomimicry, 4, 118, 123
bioprinting, 27, 102
block-print, 134
book, 3, 9–11, 13, 19, 29, 30, 35, 36, 49, 54, 67, 69–72, 74, 89, 90, 93, 96, 112, 121, 125, 128–131, 133, 142–149, 176, 179, 183, 204
brush marks, 10, 12, 14–17, 19, 20, 21, 24, 59, 60–66, 79, 81, 101, 139, 175, 204, 206–213
Brutalist, 73
bureaus, 26, 101, 119, 120

calligraphy, 24
cameotypes, 44
camera lucida, 14
camouflage, 4
ceramic, 19, 44, 49, 51, 67, 77, 90, 93, 94, 96, 101, 103–106, 110, 111, 116, 121, 133, 138, 178, 185, 189–194, 201, 204, 221
chiaroscuro, 12, 44, 61, 181, 217, 224, 226
China, Chinese, 15, 52, 72, 179, 196,
circular economy, 28, 117–119, 124, 125
classification, 39, 88, 90, 113
CNC milling, 26, 156, 194, 196, 201, 210, 226, 228
cobalt chrome, 101, 201
coins and medals, 14, 41, 44, 95, 151, 192
colour, 1–8, 11–16, 20, 21, 23, 25–33, 36, 44, 49, 51, 54, 55, 57–60, 65–68, 71–73, 77–82, 84–86, 89–91, 93, 96–98, 100, 106, 108, 109, 112, 113, 121, 122, 127, 129–131, 133, 135–137, 139, 140, 144, 155, 156, 158–161, 165, 168, 174, 175, 179, 181, 183–185, 190, 191, 194, 197, 199–201, 203–206, 208, 210–212, 216, 217, 221, 224–226, 229
colour management, 26, 96, 106, 183, 203, 210
companies, 26, 28, 52, 89, 102, 115, 117, 119, 120, 122, 162, 191
- A2-Type, 223, 224
- Adidas, 162
- Adobe, 13, 73, 187, 188, 205, 211
- Amazon, 27, 67, 121
- Anaglypta, 136, 138, 139
- Anstey Wallpaper Company, 135
- Canon Océ, 20, 26, 96, 99, 227
- Cole and Sons, 136
- Craven Dunnill Jackfield, 191
- Cruse, 79
- Dimense, 100, 142
- D Systems, 26, 27, 102, 117, 156
- Dupont, 26, 102
- EFI VUTEk, 100
- Entic Designs, 190, 192
- Epson, 100, 184
- Estudios (Studios) Durero, 23, 28, 94, 150, 151
- Eyemat, 178
- Fab Labs, 28
- Filaflex, 162
- Flickr, 9, 86
- Formica, 178
- Foster and Partners, 90, 103
- Freedom of Creation, 162
- Fresco Shockline, Italy, 90, 94
- Fujifilm, 97, 99, 100
- Gartner, 116, 117
- Gerber, 162
- Google, 11, 165
- Hewlett-Packard, 26, 81, 107
- How Toons, 118
- Humphries Weaving, 164
- Ikea, 88, 89
- Indigo, 100, 121, 184
- iSense, 79
- Johnson Tiles, 191
- Knitic, 162, 163
- Kodak, 184
- Kurz, 183
- Landor Phototex, 142
- LEGO, 101
- Levis, 165
- Lincrusta, 136–139
- Lintor, 99
- Lithoz, 103, 104
- Luximprint, 199
- Mad Evil Scientist, 208
- Magnolia Editions, 164, 165
- Makerbot, 26, 79
- Make Things Do Stuff, 118
- MakeXYZ, 119
- Mambo Unlimited, 188
- Merck and Company, 100, 184
- Mimaki, 97–100, 145, 184
- Mutina, 191
- Nike, 162
- Nomex, 162
- Open Knit, 162, 163
- Organovo, 102, 103
- Osborne and Little, 133, 140
- Pixar, 66, 204, 205
- Quixel, 88
- Raw Edges, 74
- Recreus, 162
- Ricoh, 27
- Roland, 26, 100, 156, 184, 201
- Rostock, 165, 166
- Royal Mint, 152–155
- Scodix, 100
- Shape Shifter, 162
- Shapeways, 26, 28, 101, 119, 143
- Skanska, 103
- Smile Plastics, 74

Stratasys, 22, 26, 117, 201
Sun Chemical, 102
SwissQPrint, 99
Tagsmart, 19, 123
Tooteko, 149, 150
Uniqlo, 121
Voxel, 102
Walt Disney Pictures, 204
Wedgewood, 45
Wohlers, 117, 123
Xaar, 179
Youtube, 19, 66, 176
ZCorp, 26
computational colour science, 68, 108, 113
computer-aided design (CAD), 205, 210
constancy, 7, 32
convincing, 1–5, 15, 17, 19, 36, 42, 45, 53, 58, 63, 66, 90, 127, 154, 168, 177, 184, 204
Corning glass, 116
Council for Higher Education in Art and Design, 116, 123
cradle to cradle, 70, 124
craft, 3, 24, 30, 55, 67, 71–73, 75, 106, 107, 112, 115, 121, 135, 176, 184, 205
Crafts Council, 71, 72, 106, 107
craftsperson, 28, 69, 77, 180
cross-platform, 15, 85
customisable printing, 121

Dafen Oil Painting Village, China, 15–17
dental printing, 77, 78, 102, 112, 201
design, 3, 4, 22, 25, 26, 32, 67, 70–72, 76, 86, 88–90, 96, 98–100, 102, 108, 110–112, 115–119, 123, 124, 127, 133, 134, 139, 142, 153–155, 162, 164, 176, 188, 190, 194, 197, 199, 205, 210, 214, 223, 226, 230
Design Council, 210
designer *see* Artists and Designers
Didú project, 94, 150, 151 *see also* Estudios Durero
digital imaging, 5, 10, 13, 17, 19, 20, 24–26, 29, 31–33, 44, 55, 81, 92–96, 100, 107–112, 117, 120, 121, 125, 133, 140, 142, 148, 158, 161, 167, 169, 178, 184, 190–193, 201, 203, 205, 210, 212, 214, 227, 230
drawing, 1–3, 13, 14, 17, 19, 20, 39, 40, 43, 44, 54, 58, 67, 93, 115, 121, 136, 158, 203–209, 211, 212, 220
DRUPA, 95–97, 99, 100, 117, 120, 122, 140, 142

Egypt, Egyptians, 81, 105, 197
Egyptian faience, 104
electronic circuit, 94, 96, 109
encyclopaedia or encyclopédie, 32, 69, 70, 107, 110
e-paper, 4, 14
Estudios Durero, 23, 28, 94, 150, 151
exhibitions, 18, 71, 90, 115, 156
Altered Images, 193
Drawing from Turner, 20
Medals of Dishonour, 155, 157, 230
The Power of Making, 72
Surface Design, 72
Touching the Prado, 150
What is Luxury, 72

fibres and yarns, 60, 77, 136, 161, 164, 165
food, 1, 27, 73, 102
France, French, 10, 13, 52, 79, 82, 92, 136, 169, 170, 171, 225
fresco, 49, 54, 133, 168, 169, 170
functional ceramics, 92, 104

gallery *see* Museums and Galleries
G-code, 166, 193, 210
gelatine, 14, 44, 129, 130, 191, 194, 216
genetic algorithm, 208
gilding, 137, 179–181, 183
glass, 2, 13, 14, 18, 33, 38, 44, 47, 49, 58, 60, 77, 78, 92, 108, 109, 113, 116, 124, 129, 141, 170, 180, 185, 195–201, 204, 220, 221, 225
gloss management, 105, 203
gold, 2, 11, 60, 65, 81, 117, 136, 140, 142, 156, 179–181, 183–186, 200, 230
graphene, 77, 78, 111, 116, 125
The Great Recovery Project, 118, 119
grissailles, 167, 177

hair, 4, 5, 10, 24, 66, 146, 154
high dynamic range (HDR), 13, 94, 168
The Higher Education Funding Council, 115
Hubble Telescope, 10
human visual system, 8, 22
hybrid, 93, 99
hydroxyapatite, 78
hype cycle, 71, 116
hyper-real, 65

Illustrator *see* companies, Adobe
images, 1, 3, 9–15, 17, 19, 23, 24, 28–32, 35, 36, 38, 39, 42–44, 49, 60, 63, 66, 68, 73, 78–81, 84–86, 88–90, 92, 93, 95, 96, 98, 100, 106, 107, 109, 110, 112, 113, 121, 129, 130, 133, 136, 140, 142, 144–151, 156, 158, 160, 161, 165, 170, 178, 179, 184, 190, 192–195, 204, 205, 207–211, 216, 217, 220, 221, 223
implants, 201
infrared, 53, 122, 131
inks, 14, 15, 24, 26, 70, 92–102, 111, 116, 121, 122, 124, 133, 142, 199–201, 203, 210, 214, 221
- clear, 97–99
- CMYK, 16, 100, 142, 183, 184
- dye sublimation, 100
- eco solvent, 99, 100, 111, 140, 142
- fluorescent, 98, 100
- gloss, 25
- Latex Inks, 100, 140, 142
- Marabu Inks, 95
- NanoInks, 100
- pigment, 4, 14, 16, 21, 23, 53, 54, 60, 64, 67, 70, 77, 96, 97, 100, 101, 128, 130, 131, 138, 175, 177, 179–181, 183, 184, 191, 204, 216
- puff, 94, 100, 221
- security, 129, 200
- silver, 102, 122
- translucent, 26
- UV curable, 97, 184
- varnish, 15, 25, 79, 98, 99, 109, 142, 175, 179, 184, 203
- white, 98–100, 121, 202, 223

Internet of Things, 92, 122, 176
Italy, Italian, 3, 36, 51, 53, 60, 80, 122, 134, 168, 172–173, 178, 188, 191, 224

Japan Braille Library, 146
Japan, Japanese, 5, 25, 26, 137, 144, 146–147, 190, 196–197, 200

The Khan Academy, 118
kinetic, 39
kinkarakawakami, 137
kuan cai, 53

lace, 11, 128, 136, 166, 191
laser cut, 28, 75, 96, 101, 117, 122, 226–228, 230
lithophane, 44, 192
Living Paintings charity, 144, 146, 148, 149

macromelt, 27, 103
magazine, 9, 18, 30, 67, 90, 107, 111, 115, 120, 121, 123, 216
magic mirrors, *Makkyo*, 196, 197
marble, 3, 4, 11, 49, 51, 53, 60, 133, 168, 172–179, 191
Materials Council, UK, 90
materials in waiting, 70, 77
material resources, 1–3, 5, 7, 9, 11, 15, 16, 21–30, 35, 36, 39, 42, 44, 49, 53, 55, 58, 60–62, 64–75, 77, 78, 81, 87–94, 96, 98–102, 104–113, 115–117, 119, 120, 122–125, 127, 130, 133, 136, 138, 145, 161, 162, 164–166, 168, 169, 177–181, 183, 188, 195, 197, 200, 201, 203–208, 210, 214, 220, 224, 226, 230
- Architonic, London, 90
- Design Boom, 90
- The Institute of Making, University College, London, 90
- Materfad, Barcelona, 92
- Materialbiblioteket, Stockholm, 92
- The Material Lab, London, 91
- Materials Council, UK, 90
- Materio, Paris, Belgium, Prague, Shanghai, 92
- Matrec, Ancona, 92
- Matweb, Material property database 92
- Modulator, Berlin, 92
- ReMaterialise Library, London, 90, 150

materials science and engineering (MSE), 77
mauveine, 77, 116
measurement methods for capturing, measuring and modelling, 10, 17, 19, 20, 58, 59, 66, 69, 78, 79, 81, 82, 84, 87, 88, 92, 130, 156, 178
- bidirectional reflectance distribution function (BRDF), 78, 87, 133
- coordinate measuring machine (CMM), 152
- Fourier transform infrared spectroscopy (FTIR), 53
- goniophotometry, 79
- goniospectrophotometry, 81
- infrared imaging, 10, 21, 53, 105, 158, 160
- interferometry/structure light scanning, 78, 79

Kubelka-Munk, 132
laser scanning, 78–82, 105–109, 156, 158
multifocal microscopy, 79, 81
photogrammetry, 67, 80, 105–108, 157
polynomial texture mapping (PTM), 67, 81, 82, 106, 107, 111
profilometry, 79, 81, 153
radiative transfer model (RTM), 131, 132
ray tracing, 132
reflectance transformation imaging (RTI), 80–82, 106–108, 112
scanner, 22, 25, 28, 78–83, 105, 109, 112, 158
stereophotography, 79
X-ray, 10, 19, 21, 158–160
X-ray fluorescence (XRF), 53
metal, 11, 14, 27, 38, 39, 51, 58, 62, 64, 77, 93, 101, 112, 113, 129, 151, 152, 155, 181, 184, 190, 195, 197, 201, 204, 214, 220
metamaterials, 121
microfactories, 119
microtexture, 214
mirror, 49, 121, 168, 196, 197, 220
Moore's Law, 119, 124, 129
multicolour, 44, 223–225, 230
multisensory, 1, 30
museums and galleries, 3, 9, 11, 16–22, 28–30, 33, 35–37, 39, 40, 43, 47, 51, 52, 55, 56, 58–61, 63, 64, 66–68, 71, 79, 80, 82–84, 105, 106, 112, 121, 123, 129, 134, 135, 142, 143, 148, 150, 153, 155–158, 164, 165, 168, 172, 173, 176, 177, 180–183, 187, 192, 197, 207, 225, 230
British Museum, London, 18, 40, 43, 172, 173
The Capodimonte National Museum, Naples, 61
The Computer Museum, California, 207
Gallerie dell'Accademia, Venice, 168
National Gallery, London, 59–61, 63, 64, 66–68, 80, 82, 84, 157, 158, 168, 180–183
National Gallery of Scotland, Edinburgh, 64
The National Gallery of Art, Washington, 61, 64, 68, 225
The National Museum of Tiles, Lisboa, 187
National Science and Media Museum, Bradford, 88, 129
The National Trust, 136, 137, 170, 175
Palazzo Nuovo, Rome, 172
The Prado, Madrid, 20, 150
Rembrandthuis, Amsterdam, 21
Rijksmuseum, Amsterdam, 16, 17, 20, 37, 62, 63, 176, 177, 182
Tate Britain, London, 20, 49
The Thyssen-Bornemisza National Museum, Madrid, 177
Victoria and Albert Museum, London, 71, 134, 135

nanofoams, 77
nanomaterials, 77, 116
The National Theatre, London, 176
non-photographic rendering, 204, 206

Old Master, 13, 15, 18, 19, 23, 65, 67, 149, 150, 205, 210
open source ecology (OSE), 118
optics, 7, 9, 31, 105, 109, 111, 112, 195, 199

packaging, 15, 25, 92, 94–100, 111, 117, 121, 122, 144, 183, 221
paint, 2, 3, 5, 6, 11, 15–17, 20, 36, 53, 54, 60, 65, 66, 75, 86, 89, 90, 108, 133, 135, 139, 158, 161, 175, 176, 179, 188, 197, 205–210, 214, 215, 230
paint marketing, 89
pastiglia, 179–181
Pencil of Nature, 128
perception, 8–11, 15, 16, 18, 21–24, 29–32, 38, 42, 49, 57, 58, 66–68, 82, 85, 86, 106, 107, 109, 167, 180
perspective, 3, 10, 13, 18, 19, 23, 28, 40, 42, 43, 45, 57, 58, 70, 77, 78, 149–151, 158, 168, 169, 181
photoceramic relief, 44
photography, 3, 10, 11, 13, 14, 22, 29, 64, 66, 67, 80, 130, 153
Photoshop *see* Companies, Adobe
piezo, 33, 93, 97
pigment, 4, 14, 21–23, 53–54, 60–64, 70–77, 96–101, 128–131, 138, 174–177, 179–184, 191, 204, 216–221
pixel, 3, 80, 81, 85, 105, 158, 160, 195, 203, 208–211
plastography, 44, 194
polychromy, 49, 53–55, 172, 180, 183
Pompeii, 54, 172

practitioners, scientists and writers, 26, 70, 72, 73, 207
Anderson, Chris, 75, 105
Arouet, Franois Marie, 9, 10, 69
Auer, Alois, 129
Babbage, Charles, 14
Bacon, Francis, 69
Balin, Paul, 136
Beale, Gareth, 53, 54, 67
Bennett, Alan, 18, 109
Berger, John, 10, 29
Bradbury, Henry Riley, 129–131
Brewster, David, 44, 46, 67
Cennini, Cennino, 175, 179–181, 230
Chamberlain, Neville, 119
Chambers, Ephram, 69, 108
Charney, Daniel, 31, 72, 106, 107
Chevreul, Michel Eugene, 206
Chrétien, Gilles-Louis, 13
Constable, Martin, 59, 66, 205, 234
copper Frank, 184, 235
Dawes, J.C., 119
Dehn, Jakki, 90
Diderot, Denis, 69, 70, 107
Didi-Huberman, Georges, 10, 29
Donald, Farnsworth, 164
Elkins, James, 10, 13, 18, 29, 31
Eriksson, Yvonne, 110, 145
Falco, Charles M., 13
Faraday, Michael, 129
Farthing, Stephen, 20
Fourdrinier, Henry and Sealy, 139
Fox Talbot, William Henry, 129
Franquelo, Manuel, 82, 83, 158
Fuferi, 149–150
Gershenfeld, Neil, 28, 31, 120, 124
Goethe, Johann Wolfgang von, 193
Gombrich, E.H., 3, 29, 31, 36, 67, 68
Gregory, Richard, 31, 36, 44, 67, 68
Herschel, Sir John Frederick William, 128
Hull, Charles, 26, 30
Ives, Frederick E., 14
Jacquard, Joseph Marie, 59, 164, 165
Jakubowski, Marcin, 118
Kaufman, Josh, 19, 30
Kemp, Martin, 13, 30, 66
Kingsley, Charles, 127
Klein, Susanne, 31, 109, 125, 199, 200
Knott, Stephen, 72, 107
Kurzweil, Ray, 207
le Rond d'Alembert, Jean-Baptiste, 69
Lipson, Hod, 31, 32, 102
Lovejoy, Margot, 205
Lowe, Adam, 111, 165
McLuhan, Marshal, 71
Miodownik, Mark, 72, 73, 78, 107, 112, 125
Mitchell, W.J.T. 9, 30, 115, 124
Moore, Gordon, 119
Moore, Thomas, 129–131
Munz, John, 25, 26, 30
Omarini, Sergio, 54, 67
Perkin, William Henry, 77, 116
Poitevin, Alphonse, 14
Polanyi, Michael, 71, 108, 115, 124
Rood, Ogden Nicholas, 206
Rousseau, Jean-Jacques, 69
Shneiderman, Ben, 71, 108, 205
Smith, Henry, 129
Snow, C.P., 71, 108
Thirkell, Paul, 14, 30, 31, 44, 67, 192, 193
Thomas, Sophie, 119
Tulley, Gever, 118
Voltaire (*see* Practitioner, Arouet, François Marie)
Walton, Frederick, 137, 139
Webster, Ken, 117, 118, 124
Wedgewood, Josiah, 14, 44
Weschler, Lawrence, 13, 33
printing-methods for printing, 1, 3, 8, 12, 14–16, 21, 22, 24–33, 35, 38, 44, 53, 63, 69–72, 75, 77, 78, 84–86, 90, 92–103, 105, 106, 109–113, 115–123, 125, 127–130, 132–136, 139–143, 145, 150, 155, 156, 158, 161, 162, 165–167, 169, 178, 180, 183, 184, 190, 191, 194, 199–204, 206, 210–217, 220, 221, 223–230
analogue, 12, 24, 25, 31, 63, 72, 81, 93, 94, 96, 118, 121, 133, 140, 167, 204, 209, 210, 214, 230
aquatint, 129, 214, 215
Benday Dots, 14
block, 12, 93, 133–136, 139, 142, 223, 227
collotype, 14
cyanotype, 128
decorative printing, 15, 199, 221, 226
direct metal laser sintering (DMLS), 27, 201
direct write technologies (DW), 101, 212
engraving machine, CNC, 154, 225, 226
etching, 12, 39, 101, 214–217

flexo or flexographic printing, 133, 136, 142, 184, 216, 217
flocking, 136, 221, 222
fused deposition modelling (FDM)27
halftone, 11, 14–16, 93, 96, 129, 195, 201, 209, 210, 216, 217, 221, 225–230
helio, 220, 225, 226
inkjet, 3, 14, 15, 17, 19, 25, 27, 30, 31, 33, 78, 92, 96, 97, 100–102, 109, 111, 116, 124, 141, 142, 178, 184, 190, 193, 202, 210, 223
intaglio, 25, 39, 41, 95, 140, 214, 216, 217
Letratone, 6, 14
letterpress, 25, 121, 217, 218, 223, 224
lithography, 14, 15, 92, 97, 220
photo etching, 215, 216
photogravure, 14, 215, 216
photomechanical, 3, 13, 31, 44, 67, 96, 129, 130, 191, 214, 216, 225
piezoelectric inkjet, 178
raster image processing (RIP), 210
rotogravure, 25, 133, 139, 214
screenprinting, 25, 94–97, 100–102, 139, 142, 190, 214, 221–223
security printing, 98, 102, 117, 122, 125, 200
selective laser sintering, 27, 201
stereolithography, 26, 27, 30
stochastic, 14, 86, 132, 195, 216, 217
tactile prints, 94, 143
woodblock, 12, 25, 53, 93, 136, 223–225, 228
Woodburytype, 14, 44, 130, 191, 194
woodcut, 14
Print Isn't Dead magazine, 121
printmaking, 226, 230
print-on-demand, 16
Project Eiger, 28, 227
Project Gutenberg, 128
Project Silkworm, 166
pteridomania, 127

QR codes, 122
QuickTime, 73

radio-frequency identification (RFID), 122
representation, 3, 9, 12, 26, 31, 32, 35, 40, 57, 58, 60, 64, 68, 92, 106, 155, 202, 204, 208, 210
reproduction, 1, 4, 7–9, 11–13, 15, 16, 19, 20, 22–25, 28–30, 32, 38, 55, 58, 67, 78, 82, 84–86, 92, 93, 96, 110, 111, 120, 121, 127, 130, 133, 152, 156, 158, 178–180, 200, 223, 229
research groups and universities, 20, 22, 27, 30, 32, 33, 52–54, 65–68, 72, 78, 81–84, 86–88, 90, 96, 102–105, 108–110, 112, 116, 118, 144, 147, 148, 156, 157, 162, 163, 166, 192, 199, 200, 210, 213, 224, 230
Birmingham Jewellery Quarter, 156, 184
Centre for Bits and Atoms, MIT, 120
Centre for Fine Print Research (CFPR), 84, 104, 105, 156, 157, 159, 160, 165, 166, 192–195, 210–215, 224–229
Columbia University, 87, 102
Computer Graphics and Geometry Laboratory, 196
Cornell University, 86
Creative Machines Lab, 102
CultLab3D, Fraunhofer IGD, 82
CUReT, 86, 87
Delft University of Technology, 21, 176
Deloitte, 122–124
Design Academy in Eindhoven, 96
Disney Research Pittsburgh, 162, 163
École Polytechnique Fédérale de Lausanne (EPFL), 195
Ellen MacArthur Foundation, 118, 125
Euramet, 78
Factum Arte Foundation, 80, 82, 83, 158, 165
Fraunhofer Institute, 82
Getty Research Institute, 68, 82, 83, 106, 110
Graphicstudio, University of South Florida, 225
Harvard John A. Paulson School of Engineering and Applied Sciences (SEAS), 121, 123
HP Labs, 28, 199
Institute for Research in Computer Science and Automation, 82
The Institute for Visualisation of History, 58
Japan Women's University, 144
The Jewellery Industry Innovation Centre, Birmingham, 184
Kingston University, 90
Loughborough University, 27, 103
Massachusetts Institute of Technology Media Lab (MIT), 22, 86, 102

research groups and universities, (*Contd.*)
MERL Lab (Mitsubishi Electric Research Laboratories), 86, 87
Northwestern University, Illinois, 77
Opensurfaces, Cornell University, 86, 89, 105
Rembrandt Research Project, 30
Research Institute of National Printing Bureau in Japan, 200
Rochester Institute of Technology, 82, 112, 205
Rolex Learning Center, 195
Sapienza University, 54
Singapore University of Technology and Design (SUTD), 102, 112
Smithers Pira, 161
South Dakota School of Mines and Technology, 200
Swansea University, 102
University of Colorado Boulder, 147, 148
University of Oxford, 86
Utrecht University, 86
VisTex, Massachusetts Institute of Technology Media Lab, 86
re-shoring and in-shoring, 119
resilience, 118
Roman, 44, 54, 151, 172, 197
Rosetta Stone, 18, 32

scagliola (scaglia), 173, 174
scalable vector graphics (SVG), 85, 101, 204, 208–212, 226, 228
science, technology, engineering, mathematics (STEM), 123
sculpture, 4, 35, 39, 41, 53, 67, 68, 172
sfumato, 36, 177
sgraffito, 181, 188
shadow, 13, 15, 43, 44, 168, 181
shape, 4, 30, 58, 75, 85, 86, 95, 101, 102, 106, 107, 112, 113, 123, 125, 131, 144, 146, 151, 154, 155, 157, 160, 162, 172, 178, 196, 206, 224, 226, 228–230
silver, 11, 14, 47, 100, 102, 122, 128, 140, 141, 152, 156, 184, 195
software
Autocad, 90, 211
Inkscape, 211
MacPaint, 205
Microsoft Fresh Paint, 206
Real Brush, 206
Renderman, 66, 205
Rhino, 90
Sketchfab, 58, 84
SuperPaint, 205
Tangipaint, 205, 230
TurboCAD, 90
Zbrush, 3D, 90
Spain, Spanish, 36, 134, 137, 150, 162, 185, 186, 189, 190, 191
spatial deception, 167
stained glass, 49, 180, 197, 198
standards, 9, 15, 26, 84, 85, 93, 96, 161, 195, 204
American Society for Testing and Materials (ASTM), 26
International Color Consortium (ICC), 26
International Organization for Standardization (ISO), 84
The Utah Teapot, 204
still-life, 11, 167
stone, 3, 6, 18, 32, 39, 40, 51, 53, 60, 67, 73, 77, 170, 172, 176, 177, 179, 220, 221, 223, 225
stroke-based rendering, 206, 211
super materials, 77
sustainability, 27, 73, 117, 118
The Swedish Library of Talking Books, 145

tacit knowledge, 3, 70, 71, 93, 205
tactile picture books or Touch to, 36, 93, 110, 142, 147
taxonomy of texture, 2, 5
angular, 224
bumpy, 23, 197
burnished, 180, 181, 217
coarse, 23
cold, 23, 38, 183
concrete, 92, 101, 103, 174
cool, 38, 75, 101
corrugated, 23
crisp, 63
crystalline, 23, 172
crumpled, 66
damp, 38
dense, 23, 54, 156, 165
diaphanous, 4, 196
distressed, 191
dull, 7, 181
elastic, 23
embossed, 93, 94, 98, 121, 127, 130, 133, 136–139, 143, 180, 184, 197, 221

engraved, 25, 42, 62, 95, 137, 139, 140, 152, 155, 216, 226
etched, 41, 129, 190, 195–197, 215, 216, 220, 228
fibrous, 23, 136
fine, 16, 25, 28, 29, 81, 140, 199, 214, 228, 230
finished, 27, 93, 105, 128, 174, 180, 201, 208
flaky, 23
fresh, 2, 131, 133
glassy, 180
glazed, 64, 105, 188, 192, 194
glossy, 66, 99, 110, 201, 203
grainy, 23
greasy, 220
hard, 1, 3, 39, 54, 95, 145, 220
incised, 39, 40, 52, 53, 217
irregular, 6, 118
jagged, 52, 205, 226
knitted, 161, 162, 178
layered, 52, 60, 156
lined, 10, 59
lustrous, 161
malleable, 39
matte, 23, 62, 99, 201–203
matted, 66
metallic inks, 100, 111, 133
moiré, 217
mosaic, 49, 51, 178, 197
opaque ink, 95
padded, 135
patterned, 4, 5, 125, 134, 143, 172, 184, 185, 190, 200
pixelated, 191
plastic, 2, 27, 66, 73, 94, 129, 149, 217, 220
pleated, 62
pliable, 197
polished, 28, 36, 64, 73, 172, 174, 177, 178, 196, 197
porous, 23, 113, 172, 220
pressed, 25, 95, 127, 129, 130, 135, 137, 155, 190, 214, 217
prickly, 2, 217, 224
reflective, 23, 38, 79, 81, 86, 177, 183, 196, 204
resinous, 129
ribbed, 185
ridged, 133
rough, 1, 2, 23, 28, 32, 57, 145, 174, 196, 202, 224
sandy, 196
scaled, 154, 226
scratched, 214
shiny, 2, 23, 38, 111
silky, 23, 62
skin, 3, 4, 16, 21, 27, 29, 64, 66, 86
slimy, 2
slippery, 2, 23, 57
smooth, 1, 2, 23, 41, 62, 63, 95, 100, 140, 141, 145, 147, 161, 172, 180, 184, 191, 195, 197, 204, 205, 211, 217, 224, 228
smudged, 206
soft, 1, 23, 94, 95, 100, 102, 111, 129, 137, 144–146, 155, 161, 178
spiky, 23
sponge, 204
stamped, 180
stiff, 1, 23
textural, 1, 5, 25, 28, 29, 38, 44, 86, 154, 208, 224, 227, 230
thick, 15, 53, 60, 64, 78, 94, 95, 133, 156, 197, 204, 216, 221
thin, 15, 54, 99, 101, 128, 165, 179, 197, 204, 214, 216, 220
tough, 54, 129
transparent, 23, 79, 97, 149, 217
varnished, 10, 73, 79
velvety, 62
wavy, 146, 197, 226, 230
wooden, 51, 53, 64, 73, 133, 134, 136, 137, 160, 168, 174, 186, 187
wool, 92, 135, 136, 161, 179
woollen, 174
woven, 21, 136, 141, 156, 161, 162, 164, 165, 180
wrinkled, 1
TED lecture, 118
texel, 85, 208, 210
textile, 38, 59, 61, 65, 75, 78, 96, 100, 102, 118, 120, 121, 133, 134, 161, 162, 164, 166, 167, 190, 221
texton, 85, 106, 107, 195, 210
The Tinkering School, 118
The Tolley Marbles, 172, 173
topography, 30, 81, 86, 99, 105, 108, 109, 153, 158, 202
toxic, 27
translucency, 4, 23–25, 44, 58, 63, 72, 82, 197, 209, 221
trick of the eye, 67, 168

trompe l'oeil, 22, 23, 38, 65, 67, 133, 167–169, 176, 177
truth to materials, 67, 73

Ukiyo-e, 25
Universal Fabricator, 75
university *see* Research groups and universities
Utah Teapot, 204

velvet, 2, 5, 23, 24, 59–62, 136, 146, 161, 179, 180, 183, 217
virtual reality, 26, 58, 105, 108, 112
virtual reality modeling language (VRML), 26

wallpaper, 15, 100, 127, 133–137, 139–142, 189, 190, 214, 216, 223, 230
watercolour, 55, 56, 58, 156, 206, 208, 213, 214
William Morris and Arts and Crafts, 73, 127, 134, 135
wonderstuff, 90, 106, 124
wood, 2, 3, 5, 12, 39, 51–53, 58–60, 64, 72–74, 77, 92–94, 98, 99, 107, 108, 134, 137–139, 141, 154, 177, 178, 182, 185, 191, 214, 218, 220, 223–225, 228
workflow, 3, 8, 25, 26, 32, 70, 72, 78, 82, 84, 85, 94, 106, 117, 133, 142, 153, 158–160, 180, 183, 203, 210
World-Wide-Web, 9
wrapping buildings, 170

yarns *see* fibres and yarns

zirconia, 201

Printed and bound by CPI Group (UK) Ltd, Croydon, CR0 4YY
16/09/2026
14963479-0003